Senjuti Sinharoy • Yun Kang •
Vagner Benedito
Editors

The *Medicago truncatula* Genome

Springer

Editors
Senjuti Sinharoy
National Institute of Plant
Genome Research
New Delhi, India

Yun Kang
Oklahoma State University
Stillwater, OK, USA

Vagner Benedito
Davis College of Agriculture
Natural Resources and Design
West Virginia University
Morgantown, WV, USA

ISSN 2199-4781 ISSN 2199-479X (electronic)
Compendium of Plant Genomes
ISBN 978-3-030-90756-3 ISBN 978-3-030-90757-0 (eBook)
https://doi.org/10.1007/978-3-030-90757-0

This Springer imprint is published by the registered company Springer Nature Switzerland AG
The registered company address is: Gewerbestrasse 11, 6330 Cham, Switzerland

Compendium of Plant Genomes

Series Editor

Chittaranjan Kole, President, International Climate Resilient Crop Genomics Consortium (ICRCGC) and International Phytomedomics & Nutriomics Consortium (IPNC), Genome India International (GII), Nayabad, Kolkata, West Bengal, India

Whole-genome sequencing is at the cutting edge of life sciences in the new millennium. Since the first genome sequencing of the model plant *Arabidopsis thaliana* in 2000, whole genomes of about 100 plant species have been sequenced and genome sequences of several other plants are in the pipeline. Research publications on these genome initiatives are scattered on dedicated web sites and in journals with all too brief descriptions. The individual volumes elucidate the background history of the national and international genome initiatives; public and private partners involved; strategies and genomic resources and tools utilized; enumeration on the sequences and their assembly; repetitive sequences; gene annotation and genome duplication. In addition, synteny with other sequences, comparison of gene families and most importantly potential of the genome sequence information for gene pool characterization and genetic improvement of crop plants are described.

Interested in editing a volume on a crop or model plant?
Please contact Prof. C. Kole, Series Editor, at ckoleorg@gmail.com

More information about this series at http://link.springer.com/bookseries/11805

This book series is dedicated to my wife Phullara and our children Sourav and Devleena

Chittaranjan Kole

Preface to the Series

Genome sequencing has emerged as the leading discipline in the plant sciences coinciding with the start of the new century. For much of the twentieth century, plant geneticists were only successful in delineating putative chromosomal location, function, and changes in genes indirectly through the use of a number of "markers" physically linked to them. These included visible or morphological, cytological, protein, and molecular or DNA markers. Among them, the first DNA marker, the RFLPs, introduced a revolutionary change in plant genetics and breeding in the mid-1980s, mainly because of their infinite number and thus potential to cover maximum chromosomal regions, phenotypic neutrality, absence of epistasis, and codominant nature. An array of other hybridization-based markers, PCR-based markers, and markers based on both facilitated construction of genetic linkage maps, mapping of genes controlling simply inherited traits, and even gene clusters (QTLs) controlling polygenic traits in a large number of model and crop plants. During this period, a number of new mapping populations beyond F2 were utilized and a number of computer programs were developed for map construction, mapping of genes, and for mapping of polygenic clusters or QTLs. Molecular markers were also used in the studies of evolution and phylogenetic relationship, genetic diversity, DNA fingerprinting, and map-based cloning. Markers tightly linked to the genes were used in crop improvement employing the so-called marker-assisted selection. These strategies of molecular genetic mapping and molecular breeding made a spectacular impact during the last one and a half decades of the twentieth century. But still they remained "indirect" approaches for elucidation and utilization of plant genomes since much of the chromosomes remained unknown and the complete chemical depiction of them was yet to be unraveled.

Physical mapping of genomes was the obvious consequence that facilitated the development of the "genomic resources" including BAC and YAC libraries to develop physical maps in some plant genomes. Subsequently, integrated genetic–physical maps were also developed in many plants. This led to the concept of structural genomics. Later on, emphasis was laid on EST and transcriptome analysis to decipher the function of the active gene sequences leading to another concept defined as functional genomics. The advent of techniques of bacteriophage gene and DNA sequencing in the 1970s was extended to facilitate sequencing of these genomic resources in the last decade of the twentieth century.

As expected, sequencing of chromosomal regions would have led to too much data to store, characterize, and utilize with the-then available computer software could handle. But the development of information technology made the life of biologists easier by leading to a swift and sweet marriage of biology and informatics, and a new subject was born—bioinformatics.

Thus, the evolution of the concepts, strategies, and tools of sequencing and bioinformatics reinforced the subject of genomics—structural and functional. Today, genome sequencing has traveled much beyond biology and involves biophysics, biochemistry, and bioinformatics!

Thanks to the efforts of both public and private agencies, genome sequencing strategies are evolving very fast, leading to cheaper, quicker, and automated techniques right from clone-by-clone and whole-genome shotgun approaches to a succession of second-generation sequencing methods. The development of software of different generations facilitated this genome sequencing. At the same time, newer concepts and strategies were emerging to handle sequencing of the complex genomes, particularly the polyploids.

It became a reality to chemically—and so directly—define plant genomes, popularly called whole-genome sequencing or simply genome sequencing.

The history of plant genome sequencing will always cite the sequencing of the genome of the model plant Arabidopsis thaliana in 2000 that was followed by sequencing the genome of the crop and model plant rice in 2002. Since then, the number of sequenced genomes of higher plants has been increasing exponentially, mainly due to the development of cheaper and quicker genomic techniques and, most importantly, the development of collaborative platforms such as national and international consortia involving partners from public and/or private agencies.

As I write this preface for the first volume of the new series "Compendium of Plant Genomes," a net search tells me that complete or nearly complete whole-genome sequencing of 45 crop plants, eight crop and model plants, eight model plants, 15 crop progenitors and relatives, and three basal plants is accomplished, the majority of which are in the public domain. This means that we nowadays know many of our model and crop plants chemically, i.e., directly, and we may depict them and utilize them precisely better than ever. Genome sequencing has covered all groups of crop plants. Hence, information on the precise depiction of plant genomes and the scope of their utilization are growing rapidly every day. However, the information is scattered in research articles and review papers in journals and dedicated Web pages of the consortia and databases. There is no compilation of plant genomes and the opportunity of using the information in sequence-assisted breeding or further genomic studies. This is the underlying rationale for starting this book series, with each volume dedicated to a particular plant.

Plant genome science has emerged as an important subject in academia, and the present compendium of plant genomes will be highly useful to both students and teaching faculties. Most importantly, research scientists involved in genomics research will have access to systematic deliberations on the plant genomes of their interest. Elucidation of plant genomes is of interest not only for the geneticists and breeders, but also for practitioners of an array of plant science disciplines, such as taxonomy, evolution, cytology, physiology, pathology, entomology, nematology, crop production, biochemistry, and obviously bioinformatics. It must be mentioned that information regarding each plant genome is ever-growing. The contents of the volumes of this compendium are, therefore, focusing on the basic aspects of the genomes and their utility. They include information on the academic and/or economic importance of the plants, description of their genomes from a molecular genetic and cytogenetic point of view, and the genomic resources developed. Detailed deliberations focus on the background history of the national and international genome initiatives, public and private partners involved, strategies and genomic resources and tools utilized, enumeration on the sequences and their assembly, repetitive sequences, gene annotation, and genome duplication. In addition, synteny with other sequences, comparison of gene families, and, most importantly, the potential of the genome sequence information for gene pool characterization through genotyping by sequencing (GBS) and genetic improvement of crop plants have been described. As expected, there is a lot of variation of these topics in the volumes based on the information available on the crop, model, or reference plants.

I must confess that as the series editor, it has been a daunting task for me to work on such a huge and broad knowledge base that spans so many diverse plant species. However, pioneering scientists with lifetime experience and expertise on the particular crops did excellent jobs editing the respective volumes. I myself have been a small science worker on plant genomes since the mid-1980s and that provided me the opportunity to personally know several stalwarts of plant genomics from all over the globe. Most, if not all, of the volume editors are my longtime friends and colleagues. It has been highly comfortable and enriching for me to work with them on this book series. To be honest, while working on this series I have been and will remain a student first, a science worker second, and a series editor last. And I must express my gratitude to the volume editors and the chapter authors for providing me the opportunity to work with them on this compendium.

I also wish to mention here my thanks and gratitude to the Springer staff, particularly Dr. Christina Eckey and Dr. Jutta Lindenborn for the earlier set of volumes and presently Ing. Zuzana Bernhart for all their timely help and support.

I always had to set aside additional hours to edit books beside my professional and personal commitments—hours I could and should have given to my wife, Phullara, and our kids, Sourav and Devleena. I must mention that they not only allowed me the freedom to take away those hours from them but also offered their support in the editing job itself. I am really not sure whether my dedication of this compendium to them will suffice to do justice to their sacrifices for the interest of science and the science community.

New Delhi, India Chittaranjan Kole

Preface

It is indisputable that *Medicago truncatula* (barrel medic) has been an invaluable model for legume species as well as to understand key plant features that are not present in the simpler model, *Arabidopsis thaliana*. Indeed, *M. truncatula* has been an essential tool to reveal the genetic, biochemical, and physiological processes of symbiotic nitrogen fixation, seed development, specialized metabolism, leaf development, and disease resistance in plants, to name a few.

This book explores the state of the art of *M. truncatula* as a model for unveiling mechanisms not only of legumes but plants in general. We start navigating the dynamics of gene expression through transcriptomics resources in Chap. 1. This chapter examines the resources available and provides a vision on how to utilize transcriptomic datasets in light of new tools, such as artificial intelligence, to discover hidden patterns to explain biological processes.

Chapter 2 explores the use of genome-wide association analyses (GWAS) to find associations between genome sequence variation and quantitative and qualitative traits. While such analyses only provide association, this approach has been uniquely useful to create strong hypotheses on gene functions to establish genetic causation of important traits.

In *M. truncatula*, the tobacco retrotransposable element *Tnt1* has been very useful in applying forward as well as reverse genetics by providing the research community with mutants from a population that virtually fully covers its genome. The *Tnt1* mutant population was extensively characterized phenotypically as well as exhaustively sequenced and deposited on a searchable database. Chapter 3 provides an update on this resource.

In the same token, Chap. 4 explores the mutant collection created via Fast Neutron Bombardment (FNB), which produces mostly deletions of varied sizes. This resource has been invaluable to understand the development of leaf complexity in legumes and the identification of key genes involved in this process.

Chapter 5 discusses the usefulness of *M. truncatula* as a reservoir for the identification of novel resistance loci for studying disease resistance in legumes, especially powdery mildew, an important fungal disease in beans that can cause significant yield losses in hot and humid climates.

As a legume model, *M. truncatula* has been uniquely positioned as a tool to study symbiotic nitrogen fixation. Chapter 6 explores the transcriptional networks during the development of the root nodule. The de novo development of this organ upon a biotic stimulus (rhizobial infection) under a specific environmental condition (low nitrogen in the soil) and the permissive development of an intracellular symbiosis through the creation of a semi-organelle (the symbiosome) in the cortical cells of the forming nodule can be studied to understand legume biology, and they can also be further explored to understand endosymbioses and organogenesis in general.

Furthermore, Chap. 7 analyzes the dynamics of gene expression from large transcriptomic datasets in order to integrate information of nodule development through a systems biology perspective. As our knowledge of the process evolved and new tools are created to integrate different datasets, such as genetic variation, transcriptional dynamics, metabolomics, and cellular/anatomic imaging, we will be able to consolidate mechanisms of nodule development and symbiotic nitrogen fixation that will be useful to plant biologists as well as of practical use for legume breeders.

Chapters 8–10 bring textbook examples of how *M. truncatula* genetic resources, such as transcriptional databases and a mutant collection, can be used to identify novel genes involved in biological processes. Chapter 8 reveals a novel gene that is key for symbiotic nitrogen fixation in legumes. Chapter 9 shows how this model has been used to study the development of complex leaves, especially leaf blade growth patterns. Chapter 10 explores the roles that *WOX* genes play in plant development, including leaf pattern formation.

Chapters 11 and 12 demonstrate the usefulness of *M. truncatula* as a tool to study seed development. Chapter 11 explores the early onset of seed development from a genomic perspective, whereas Chap. 12 focuses on seed maturation and desiccation, which is fundamental for seed viability and longevity.

At last, Chap. 13 examines the genetic control of organ size development using *M. truncatula* as a model.

In toto, this book celebrates the humble barrel medic as a pillar of legume genomics as well as a tool that goes beyond to extend our knowledge of plant biology. The tools and resources available to the *M. truncatula* research community, allied to exquisite and talented scientists distributed worldwide that work together as an inclusive community, are expanding the knowledge and creating new challenges in this continued endeavor. We hope you enjoy reading these chapters, learn something new, and feel enthused and empowered to continue or start this journey of discoveries with us.

New Delhi, India — Senjuti Sinharoy
Stillwater, USA — Yun Kang
Morgantown, USA — Vagner Benedito

Contents

Contributors

Kaustav Bandyopadhyay Amity Institute of Biotechnology, Amity University Haryana, Gurgaon, India

Vagner Augusto Benedito Division of Plant and Soil Sciences, Davis College of Agriculture, Natural Resources and Design, West Virginia University, Morgantown, WV, USA

Akanksha Bhardwaj National Institute of Plant Genome Research, New Delhi, India

Jingya Cai Department of Biological Sciences and BioDiscovery Institute, University of North Texas, Denton, TX, USA

Divya Chandran Laboratory of Plant–Microbe Interactions, Regional Centre for Biotechnology, NCR Biotech Science Cluster, Faridabad, Haryana, India

Garima Chauhan National Institute of Plant Genome Research, New Delhi, India

Jianghua Chen CAS Key Laboratory of Topical Plant Resources and Sustainable Use, CAS Center for Excellence in Molecular Plant Sciences, Xishuangbanna Tropical Botanical Garden, Chinese Academy of Sciences, Kunming, Yunnan, China

Rujin Chen MOE Key Laboratory of Cell Activities and Stress Adaptations, Lanzhou University, Lanzhou, China;
School of Life Sciences, Lanzhou University, Lanzhou, China

Yuhui Chen School of Life Sciences, Lanzhou University, Lanzhou, China

Rebecca Dickstein Department of Biological Sciences and BioDiscovery Institute, University of North Texas, Denton, TX, USA

Lucas Gontijo Silva Maia Division of Plant and Soil Sciences, Davis College of Agriculture, Natural Resources and Design, West Virginia University, Morgantown, WV, USA

Arunima Gupta Laboratory of Plant–Microbe Interactions, Regional Centre for Biotechnology, NCR Biotech Science Cluster, Faridabad, Haryana, India

Megha Gupta Laboratory of Plant–Microbe Interactions, Regional Centre for Biotechnology, NCR Biotech Science Cluster, Faridabad, Haryana, India; Kalinga Institute of Industrial Technology, Bhubaneswar, Orissa, India

Yun Kang Oklahoma State University, Stillwater, OK, USA

Hao Lin Biotechnology Research Institute, Chinese Academy of Agricultural Sciences, Beijing, China

Jaiana Malabarba Institut Agro, INRAE, IRHS, Université d'Angers, Beaucouzé, France;
Univ Angers, Institut Agro, INRAE, IRHS, SFR QUASAV, Angers, France

Yingying Meng Biotechnology Research Institute, Chinese Academy of Agricultural Sciences, Beijing, China

Kirankumar S. Mysore Institute for Agricultural Biosciences, Oklahoma State University, Ardmore, OK, USA

Raja Sekhar Nandety Institute for Agricultural Biosciences, Oklahoma State University, 3210 Sam Noble Parkway, Ardmore, OK, USA;
Cereal Crops Research Unit, United States Department ofAgriculture-Agricultural Research Service, Edward T. Schafer Agricultural Research Center, Fargo, ND, USA

Lifang Niu Biotechnology Research Institute, Chinese Academy of Agricultural Sciences, Beijing, China

Muhammet Şakiroğlu Department of Bioengineering, Adana Alparslan Türkeş Science and Technology University, Adana, Turkey

Senjuti Sinharoy National Institute of Plant Genome Research, New Delhi, India

Million Tadege Department of Plant and Soil Sciences, Institute for Agricultural Biosciences, Oklahoma State University, Ardmore, OK, USA

Taylor Troiani Department of Biological Sciences and BioDiscovery Institute, University of North Texas, Denton, TX, USA

Vijaykumar Veerappan Department of Biological Sciences and BioDiscovery Institute, University of North Texas, Denton, TX, USA;
Department of Biology, Eastern Connecticut State University, Willimantic, CT, USA

Jerome Verdier Institut Agro, INRAE, IRHS, Université d'Angers, Beaucouzé, France;
Univ Angers, Institut Agro, INRAE, IRHS, SFR QUASAV, Angers, France

Hui Wang College of Grassland Science and Technology, China Agricultural University, Beijing, China

Jiangqi Wen Institute for Agricultural Biosciences, Oklahoma State University, Ardmore, OK, USA

Christina Laureen Wyman Division of Plant and Soil Sciences, Davis College of Agriculture, Natural Resources and Design, West Virginia University, Morgantown, WV, USA

Prashant Yadav Amity University Haryana, Gurgaon, India

Lina Yang Division of Plant and Soil Sciences, Davis College of Agriculture, Natural Resources and Design, West Virginia University, Morgantown, WV, USA

Zeyong Zhang MOE Key Laboratory of Cell Activities and Stress Adaptations, Lanzhou University, Lanzhou, China;
School of Life Sciences, Lanzhou University, Lanzhou, China

Abbreviations

δ13C	Carbon-13 isotope discrimination
ABA	Abscisic acid
ABI	Abscisic acid insensitive
ABRE	ABA response element
Ac/Ds	Activator–dissociator
aCGH	Array-based comparative genomic hybridization
AD	Arbitrary degenerate
AGLF	AGAMOUS-like flower
AI	Artificial intelligence
AM	Arbuscular mycorrhiza
AON	Autoregulation of nodulation
ARF	Auxin response factor
ARP	Actin-related Protein
AS	Asymmetric leaves
ASL18	Asymmetric leaves 2-like
BC	Back crossed
BOP	Blade on petiole
BR	Brassinosteroid
BRI	Brassinosteroid receptor
BS	Big seed
BZR	Brassinazole resistant
C_2H_2	Cystine-2/ histidine-2
CAPS	Cleaved amplified polymorphic sequences
CBS	Cystathionine-β-Synthase-like
CC	Coiled-coil
CCaMK	Calcium calmodulin-dependent protein kinase
CE	Cytokinin response elements
CEP	C-terminally encoded peptide
ChIP	Chromatin immunoprecipitation
CHS	Chalcone synthase
CK	Cytokinin
CKX	Cytokinin oxidase/dehydrogenase enzymes
CLE	CLAVATA3/embryo surrounding region related
CLV	CLAVATA
CMLM	Compressed MLM
CNVs	Copy number variations
CO	Cortex-specific transcript

CoGe	Comparative genomics
CRA	Compact root architecture
CRE	Cytokinin response
CSSP	Common symbiosis signaling pathway
CYC	CYCLIN
CYC-box	CYC-RE contains a palindromic sequence
CYC-RE	Cyclops responsive cis element
DAP	Days after pollination
DASH	DOF acting in seed embryogenesis and hormone accumulation
DIB	Dwarf and increased branching
DMI	Does not make infections
DNF	Defective in nitrogen fixation
$dNod^{+/-}$, Fix^{+}	Delayed nodule formation and effective nitrogen fixation
DS	Desiccation sensitive
DT	Desiccation tolerance
EDS	Enhanced disease susceptibility
EFD	Ethylene response factor required for nodule differentiation
EMS	Ethyl methane sulfonate
ENOD	Early nodulin
EOD	Enhancer of DA1
ERN	Ethylene response factor required for nodulation
EST	Expressed sequence tags
ET	Ethylene
ETI	Effector-triggered immunity
EXO	Exocyst
EXP	Expansin
FAO	Food and Agriculture Organization
FCL	Fused compound leaf
Fix-	N-fixation defective
Fix−	Non-functional
FLOT	Flotillin
FNB	Fast neutron bombardment
FST	Flanking sequencing tags
FUS3	FUSCA 3
GA	Gibberellin
GAN	Generative adversarial network
GID	Gibberellin insensitive dwarf
GIF	GRF-interacting factor
GIF/AN	GRFs-ANGUSTIFOLIA
GL	Glabra
GLM	General linear model
GPAR	Genetic profile–activity relationship
GRF	Growth-regulating factor
gRNA	Guide RNA
GWAS	Genome-wide association studies
HAN	HANABA TARANU

HDL	Headless
HDT	Histone deacetylases
HD-ZIP	Homeobox leucine zipper
HPI	Hours post-inoculation
HR	Hypersensitive response
IKU	HAIKU
IMGAG	Medicago Genome Annotation Group
InDels	Insertion/deletions
INRA	Institut National de la Recherche Agronomique, France
IPD3	Interacting protein of DMI3
IRLC	Inverted repeat lacking clade
IT	Infection thread
JA	Jasmonate
JCVI	J. Craig Venter Institute
KAN	Kanadi
KIX8/9	Kinase-inducible interacting 8/9
KLU	Kluh
KNOX	Knotted1-like homeobox
L1L	Leafy cotyledon1-like
LATH	Lathyroides
LBD	Lateral organ boundaries domain
LCM	Laser-capture microdissection
LD	Linkage disequilibrium
LEA	Late embryogenesis abundant
LEC	Leafy cotyledon
LeGOO	Legume graph-oriented organizer knowledge base
LIN	Lumpy infection
LIPME	Laboratory of Plants Microbes and Environment Interactions
LIS	Legume Information System
LL	Lateral leaflets
LL	little leaf
lncRNA	Long-non-coding RNA
LOG	Lonely guy
LRR	Leucine-rich repeat
LTR	Long-terminal repeat
LYK	LysM domain-containing receptor kinase
MAP	Mitogen-activated protein
MeJA	Methyl jasmonate
mini	Mini seed
MIO	Mini organ
MLM	Mixed linear model
MLO	Mildew resistance locus O
MNP	Mini plant
MP	Monopteros
MtGEA	M. truncatula gene expression atlas
Mu	Mutator
Mya	Million years ago

NAD	Nodules with activated defense
NBCL	NOOT-BOP-COCH-LIKE
NCGR	Center for Genome Resources
NCM	Nodule central meristem
NCR	Nodule-specific cysteine rich
NF	Nodulation factors
NFC	Nitrogen fixation clade
NFP	Nod factor perception
NF-Y	Nuclear factor Y
NGA	NGATHA
NGS	Next generation sequencing
NHEJ	Non-homologous end joining
Nif	Nitrogen fixation
NIN	Nodule inception
NLP	NIN-like protein
NLR	Nucleotide-binding/leucine-rich repeat
Nod^-	Non-nodule
$Nod^{+/-}/Fix^-$	Defective nodule formation and nitrogen fixation
Nod^+/Fix^-	Ineffective nodule
$Nod^+/Fix^{+/-}$	Partially ineffective nodule
$Nod^{++}/Fix^{+/-}$	Supernodulation
NPD	Nodule-specific PLAT domain
NPL	Nodulation pectate lyase gene
NRAMP	Natural resistance-associated macrophage protein transporter family
NRE	Nodulation responsive element
NRI	Noble Research Institute
NS	Narrow sheath
NSF	National Science Foundation
NSP	Nodulation signaling pathway
NVM	Nodule vascular meristem
OSU	Oklahoma State University
OTUB	Ovarian tumor domain-containing ubiquitin aldehyde-binding protein
PA	Proanthrocyanidins
PALM	Palmate-like pentafoliata
PAM	Protospacer adjacent motif
PAMPs	Pathogen-associated molecular patterns
PAR	Medicago truncatula proanthocyanidin regulator
PBM	Peribacteroid membrane
PBS	Peribacteroid space
PE	Paired end
PEBV	Pea early browning virus
PEG	Polyethylene glycol
PHB	Phabulosa
PINNA	Pinnate-like pentafoliata
PKL	Pickle
PLT	Plethora

PPD	Peapod
PR	Pathogenesis related
PRRs	Pattern recognition receptors
PTI	Pattern-triggered immunity
PZ	Periphery zone
QRL	Quantitative resistance loci
qRT-PCR	Quantitative reverse-transcription PCR
QTL	Quantitative trait loci
QTN	Quantitative trait nucleotides
RDN	Root-determined nodulation
REM	Remorin
REP	Resistance to E. pisi race
RFO	Raffinose oligosaccharide family
RIL	Recombinant inbred line
RK	Receptor kinase
RNAi	RNA interference
RNS	Root nodule symbiosis
ROS	Reactive oxygen species
ROT	Rotundifolia
RPG	Rhizobium polar growth
RPT	Root Phototropism
RPW	Resistance to powdery mildew
RSD	Regulator of symbiosome differentiation
SA	Salicylic acid
SAM	Shoot apical meristem
SAP	Sterile apetala
SCF	SKP1/cullin/F-box
SCR	Scarecrow
scRNA-Seq	Single-cell transcriptomics
Seqcap	Sequence capture
SERF	Somatic embryo-related factor
SGL	Single leaflet
SHB	Short hypocotyl under blue
SHR	Short root
SLB	Small leaf and bushy
SLM	Smooth leaf margin
SM	Symbiosome membrane
SNF	Sucrose non-fermenting
SNF	Symbiotic nitrogen fixation
SNPs	Single nucleotide polymorphisms
SRA	Sequence read archive
SSP	Seed storage protein
STACH	Stachyose
STF	Stenofolia
STP	Sugar transport protein
STY	Short internodes/stylish
SUC	Sucrose
SUC	Sucrose transporter

SUNN	SUper numeric nodule
SWI3	Switch3
Sym	Symbiosis
TAIL-PCR	Thermal asymmetric interlaced PCR
TBD	Terminal bacteroid differentiation
TChAP-seq	Tandem chromatin affinity purification sequencing
TCP	Teosinte branched1/cycloidea/proliferating cell factor
TCS	Two-component system
T-DNA	Transfer DNA
TF	Transcription factors
TGW	Thousand-grain weight
TIGR	The Institute for Genomic Research
TIR	Toll-interleukin receptor
TL	Terminal leaflet
T_m	Melting temperature
TNL	TIR-NLR
Tnt1	Tobacco retrotransposon
TRAP	Translating ribosome affinity purification
VAMP	Vesicle-associated membrane proteins
VIGS	Virus-induced gene silencing
VIT	Vacuolar iron transporter
VPY	VAPYRIN
VTL	Vacuolar iron transporter like
WC	Water content
WGS	Whole-genome sequencing
WOX	WUSCHEL-related homeobox
WRI	Wrinkled
WUS	WUSCHEL
YAB	Yabby
YSL	Yellow stripe like
ZIP	Zrt-, Irt-like protein

1 The Model Legume, *Medicago truncatula* in the Genomic Era: Speeding Up Discoveries in Legume Biology

Christina Laureen Wyman, Lucas Gontijo Silva Maia, Lina Yang, and Vagner Augusto Benedito

Abstract

Knowing when and where genes are expressed is key to understanding the development and physiology of an organism. The fine characterization of the transcriptional activity of a model organism provides insights on how tissues differentiate, organs are built and function, and how these organs exchange molecules with the rest of the organism. The *Medicago truncatula* research community has benefited from high-quality, extensive transcriptomic datasets, and analyses tools early in the establishment of this species as a legume model. This chapter visits the importance and uniqueness of *M. truncatula* as a model to understand legumes and the transcriptomic resources available to do so. We also provide a vision on how to employ large transcriptional datasets in the context of development and physiology by employing artificial intelligence methods to model a framework for biological processes based on gene expression dynamics.

1.1 Legume Evolution and Features

Legumes (Fabaceae, formerly Leguminosae) encompass one of the most evolutionarily successful botanical families, after only the Asteraceae and the Orchidaceae (SOTWP 2017). This family emerged in the late Cretaceous and diversified in the early Paleocene, circa 65 million years ago (Lyson et al. 2019; Herrera et al. 2019; Koenen et al. 2019, 2020). Legumes currently comprise 20,856 species grouped into six subfamilies, of which the largest clades are the Caesalpinioideae (that now includes the mimosoid clade) and the Papilionoideae (Azani et al. 2017). The success of this family of cosmopolitan distribution can be attributed mainly to the conquest of nitrogen-poor environments due to a mutualistic root symbiosis with bacteria that allows the plant to indirectly make use of the abundant, but typically unavailable atmospheric nitrogen. Furthermore, many species are also

Christina Laureen Wyman, Lucas Gontijo Silva Maia and Lina Yang—Authors with equal contribution.

C. L. Wyman · L. Gontijo Silva Maia · L. Yang · V. A. Benedito (✉)
Division of Plant and Soil Sciences, Davis College of Agriculture, Natural Resources and Design, West Virginia University, Morgantown, WV 26506–6108, USA
e-mail: vagner.benedito@mail.wvu.edu

S. Sinharoy et al. (eds.), *The Medicago truncatula Genome*, Compendium of Plant Genomes,
https://doi.org/10.1007/978-3-030-90757-0_1

highly adapted to additional stresses, such as dry environments and acidic, low-P soils.

Legumes also comprise the second most important crop family, second only to cereals (Poaceae). Species of agricultural and industrial relevance produce a diversity of products, such as food and ingredients (e.g., soybean, common beans, chickpea, lentils, peas, peanuts, lupin, carob, tamarind, kudzu, rooibos, guar, licorice, acacia, *Astralagus*); pasture, feed, and fodder (e.g., soybean, alfalfa, barrel medic, clover, vetch, stylo, lupin, mesquite); dyes (e.g., indigo, logwood, sappanwood, and of historical importance during the colonization of the Americas, brazilwood, which was exploited almost to extinction and is now a protected species in Brazil); lumber (e.g., acacias, rosewoods); and biodiesel (*Millettia pinnata*, formerly *Pongamia pinnata*: Biswas et al. 2011), not to mention those used for medicinal and ornamental purposes as well as invasive and noxious species (e.g., kudzu).

As plant models, legumes can provide fundamental insights to important plant features that cannot be assessed in the plant model *par excellence*, *Arabidopsis thaliana*. Such distinctive traits include symbiotic nitrogen fixation in root nodules (in many legumes and a few examples outside this family but within the Rosids II clade), mycorrhizal fungi-root symbiotic association, a distinctive specialized metabolism (e.g., isoflavonoids and a multitude of alkaloids, polyphenols, and saponins; Wink 2013), development of complex leaf and flower patterns, presence of glandular trichomes, and nutrient accumulation (protein, oil, starch) in seeds. Thus, the study of model legumes is essential to advance our understanding of plant biology as a whole and to provide insights for breeding programs and sustainable cropping practices.

1.2 Concurrent Model Legume Species and the Uniqueness of *Medicago truncatula* (Barrel Medic)

All three established model legumes belong to the Papilionoideae subfamily: soybean (*Glycine max*), *Lotus japonicus*, and *Medicago truncatula* (Table 1.1). They have similar generation times (about three months, seed-to-seed) and are prolific and autogamous. They have well-annotated genome sequences (Sato et al. 2007; Schmutz et al. 2010; Young et al. 2011; Kamal et al. 2020), in addition to publicly available transcriptome datasets (Lamblin et al. 2003; Benedito et al. 2008; Severin et al. 2010; Verdier et al. 2013; Carrère et al. 2021), and a plethora of genetic resources, such as mutant collections (Tadege et al. 2008; Pislariu et al. 2012; Tsuda et al. 2015; Małolepszy et al. 2016).

M. truncatula (Figs. 1.1 and 1.2) is unique among the legume models in that it develops indeterminate nodules with persistent meristems, a more primitive trait (Sprent and James 2007; Ren 2018). The inverted repeat-lacking clade (IRLC) of Hologalegina (galegoids), to which Medicago and most cultivated temperate legume crops belong (e.g., chickpea, fava bean, lentil), is characterized by the loss of a 25 kb inverted repeat region in the plastidial genome. Moreover, and most importantly, this clade (along with the Dalbergioid tribe) controls bacteroid differentiation by a massive expression of nodule-specific cysteine-rich (NCR) peptides (Czernic et al. 2015; Wang et al. 2017; Yang et al. 2017). Therefore, understanding nodulation and symbiotic nitrogen fixation in temperate legume crops can be best achieved by studying *M. truncatula* within the model options available. *M. truncatula* is also closely related to the autotetraploid, perennial crop, alfalfa (*M. sativa*), which is the

Table 1.1 Comparison of the three model legume systems

	Medicago truncatula	*Lotus japonicus*	*Glycine max*
Faboideae subfamily clade	Hologalegina (IRLC)	Hologalegina (robinioid)	Phaseolid (millettioid)
Nodule development	Indeterminate	Determinate	Determinate
Bacteroid division after infection	Endoreduplication	No endoreduplication	No endoreduplication
Form of N exported	Amides	Amides	Ureides
Genome size	430 Mb	500 Mb	1.15 Gb
Genome ploidy	Diploid (N = 8)	Diploid (N = 8)	Allotetraploid (N = 20)

Medicago truncatula is our model of choice organism due to its small, diploid genome size, ample availability of genetic resources, and development of indeterminate nodules

most cultivated forage legume worldwide. The alfalfa genome is highly syntenic to that of *M. truncatula* (Li et al. 2014; von Wettberg et al. 2019), and these species are estimated to have diverged circa 5.3 million years ago (Chen et al. 2020). However, the ploidy (2n = 4X = 32) and allogamous fertilization system preclude alfalfa from being suitable as a model genetic system.

1.3 Genomic and Transcriptomic Resources for *Medicago truncatula*

This organism was established as a model species in the 1990s to understand symbiotic nitrogen fixation (Barker et al. 1990; Journet et al. 2002). The advent of expressed-sequence tags (EST) sequencing databases hosted by The Institute for Genomic Research (TIGR; now J. Craig Venter Institute, JCVI) propelled this species into molecular genetics and solidified it as a biological model by offering the research community with the (now decommissioned) *Medicago* EST database (Fedorova et al. 2002). This database guided the early transcriptomic analyses of the species and the annotation of the *M. truncatula* genome by the International Medicago Genome Annotation Group (IMGAG), as well as the design of the 50 k Medicago GeneChip commercial microarray produced by Affymetrix. This EST-based microarray allowed the construction of the *M. truncatula* Gene Expression Atlas (MtGEA: Benedito et al. 2008; He et al. 2009) by the Samuel Roberts Noble Foundation (now, Noble Research Institute, NRI). Along with the genome sequencing efforts, the MtGEA was a seminal tool to launch this

Fig. 1.1 *Medicago truncatula* cv. "A17." **a** Trifoliate leaf. Trichomes are visible on the leaf and stem. An anthocyanin-rich core next to the midvein base of leaflets develops in some accessions. **b** Twenty-one-day-old plants growing in turface/vermiculite substrate under optimal growth conditions in a greenhouse

Fig. 1.2 *Medicago truncatula* cv. "A17" indeterminate nodules (28-days post-inoculation with *Sinorhizobium meliloti* strain *Sm1021-LacZ*). The larger nodule is elongated and pink due to accumulation of leghemoglobin and indicative of active nitrogen fixation. The smaller, developing nodule still lacks leghemoglobin. Scale bar: 500 μm

species to its current status as a *bona fide* plant model. The initial MtGEA dataset counted with three experiments (a set with all organs of the mature plant, a nodule development time course, and a seed development series), which allowed exploring the molecular genetics of the organism. Since the commercial microarray design was based on the TIGR's Medicago EST database, which each EST was based on contigs (created from overlapping cDNA sequences), once the Medicago genome sequence and annotation were available, efforts were made to map the Affymetrix probesets to the annotated Medicago transcriptome. As expected, the 50,902 Medicago probesets on the chip did not exactly match 1-to-1 to the Medicago transcriptome. Due to lack of maintenance, funding, and priority changes at the Noble Research Institute, the original MtGEA is being decommissioned. An interesting platform that hosts high-quality Medicago gene expression data during nodule development is the Indian-based *Arachis hypogaea* Nodule Developmental Gene Expression (http://14.139.61.8/AhNGE/index.php), even though the platform is focused on peanut. A small portion of the Medicago Affymetrix data is also available at the Medicago eFP Brower (http://bar.utoronto.ca/efpmedicago/cgi-bin/efpWeb.cgi), which is maintained by the University of Toronto. This tool creates pictographs of plant tissues colored with expression levels of the gene of interest. The Affymetrix GeneChip technology has been surpassed by RNA-Seq due to its flexibility for allowing data revisions through remapping of reads and reanalysis of transcription quantification considering new annotations of the genome, as well as the cost–benefit of an open sequencing platform (instead of being gridlocked by probesets designed in the mid-2000s).

The new generation of sequencing approaches provided a leap in biology by allowing a comprehensive assessment of the whole transcriptome in each sample at a fast pace and affordable cost. The NCBI's Sequence Read Archive (SRA) currently holds more than two thousand publicly accessible *M. truncatula* RNA-Seq samples, including a thousand samples from Illumina pair-ended samples. This information can be used to explore gene expression patterns across a multitude of conditions and an in-depth analysis of gene network associations.

Filling the void left by the MtGEA termination by the Noble Research Institute, a new tool for Medicago transcriptomics, has recently emerged: the MtExpress (Carrère et al. 2021). This platform not only hosts the legacy MtGEA database (https://lipm-browsers.toulouse.inra.fr/pub/expressionAtlas/app/mtgeav3) but also brings high-quality RNA-Seq data along with analysis tools provided by Laboratory of Plants Microbes and Environment Interactions (LIPME) at INRA-CNRS in Toulouse, France (https://lipm-browsers.toulouse.inra.fr/pub/expressionAtlas/app/rnaseqv1) (Carrère et al. 2021).

1.4 Additional Resources Available for Functional Genetic Analyses of *Medicago truncatula*

The genome sequencing and gene annotation efforts of *M. truncatula* cultivar "A17" are at the heart of research aimed to understand the molecular genetics of development and physiology of the species. The fifth and latest version, MtrunA17r5.0-ANR (Mt5.0), was released in 2018 by the *Institut National de la Recherche Agronomique* (INRA, France). It encompasses 430 Mb of genome sequence split into eight chromosomes that hold 44,623 annotated protein-coding and 4081 long-non-coding RNA (lncRNA) genes (https://medicago.toulouse.inra.fr/MtrunA17r5.0-ANR/; Pecrix et al. 2018). Expanding prior efforts to identify symbiotic genes in *M. truncatula* (Benedito et al. 2008; Young et al. 2011; Roux et al. 2014), Pecrix et al. (2018) investigated the symbiotic panorama imbued in this genome. They identified 270 symbiosis-related gene islands and 1960 genes expressed exclusively or highly induced during nodule development and nitrogen fixation, which may be key to understanding the evolution and mechanisms of this trait in plants.

The Legume Federation (legumefederation.org) is a consortium funded by the UnitedStates National Science Foundation (NSF) focused on comparative legume genomic data and offers links to current species-specific genomic information and tools. Legume Graph-Oriented Organizer knowledge base (LeGOO: www.legoo.org; Carrère et al. 2020) is a European hub focused on *M. truncatula* genomics data and tools, in addition to interspecific comparative analysis tools. In turn, the Legume Information System (LIS: https://legumeinfo.org/) is an integrative platform focused on linking crop traits to genetic and genomic information to advance legume breeding. The plant genomics platform Phytozome v.13 (phytozome-next.jgi.doe.gov) currently displays the *M. truncatula* genomic features (version Mt4.0v1: phytozome-next.jgi.doe.gov/info/Mtruncatula_Mt4_0v1) along with 12 other legume species, which allows not only the exploration of the genome, but also the performing preliminary comparative genetic analysis across the plant kingdom.

We cannot overlook tools and pan-plant platforms that can also serve the legume genomics research community. Particularly, Cyverse (formerly called iPlant Collaborative) is the United States-based cyber environment for data storage, sharing, and analysis tools. It provides software and computational power to run analyses by scientists with varying levels of bioinformatics experiences (Merchant et al. 2016). This platform has been seminal to legume biology by allowing researchers to perform analyses and storing genomic data for the Legume Federation and other platforms. Comparative Genomics (CoGe: https://genomevolution.org/coge) is a Cyverse-powered pan-species platform for researchers to execute comparative genomic investigation.

The Noble Research Institute (NRI) was home to the most extensive collection of DNA-insertion mutants of all legumes: the *M. truncatula* mutant collection (http://medicago-mutant.noble.org/mutant/: Tadege et al. 2008; Pislariu et al. 2012). The tobacco retrotransposon *Tnt1* was used to create and flag insertion events in the *Medicago truncatula* genome of ecotype “R108,” in which insertion events preferentially occur in gene-rich regions (Tadege et al. 2008). The mutant database features 21,741 *Tnt1* lines containing 470,000 random insertions in the genome, which theoretically covers ~90% of all genes in the *M. truncatula* genome (Sun et al. 2018, 2019). The insertion can also work as a tag to identify where insertions are located in the mutant genome. In addition, seeds of a *M. truncatula* cv. “A17” mutant collection containing 7624 lines based on fast neutron bombardment (FNB), which mostly creates deletions, are also available to researchers at the NRI’s *M. truncatula* mutant collection (Rogers et al. 2009; Chen and Chen 2018). This collection has been recently transferred to Oklahoma State University (https://medicago-mutant.dasnr.okstate.edu/mutant/index.php).

The Medicago HapMap Project (http://www.medicagohapmap.org/; https://medicagohapmap2.org/), led by the University of Minnesota and hosted by the National Center for Genome Resources (NCGR), offers genome sequence variation data and analysis of 384 inbred lines not only of *M. truncatula*, but also several other species within the genus, thus delivering a substantial coverage of the genetic diversity within this group. This project also provides seeds of 338 accessions for researchers interested in genome-wide association studies (GWAS: Kang et al. 2015, 2019).

In toto, the aspiration of a previous generation of biologists assessing the genome of organisms comprehensively has recently come true. We now have access to virtually complete genome sequences and transcriptome profiles of cells, tissues, and organs. The legume geneticist can now analyze and interpret these datasets in light of unique features of the plant's biochemistry and physiology. Indeed, gene networks can be promptly built from large transcriptomic datasets using statistical tools, and hypotheses on gene functions and gene interactions can be rapidly evaluated with genetic variants and insertional mutants available in germplasm collections.

1.5 The Next Stage of Legume Genomics

The future of Medicago genomics will bring us dynamic models of gene expression for a better understanding of how key legume traits develop (e.g., nodule organogenesis) and how symbiotic nitrogen fixation is controlled and sustained. In order to achieve this goal, more detailed single-cell transcription analysis is needed. Studies like the single-cell transcriptomics (scRNA-Seq) conducted on shoot apical meristem of maize (Satterlee et al. 2020) could be used to provide valuable information to nodule development modeling and training datasets.

Although very useful for biological insights until now, the construction of gene networks should surpass conventional statistical analyses (e.g., linear correlations as in the WGCNA package: Langfelder and Horvath 2008) and start adopting artificial intelligence (AI, a.k.a. deep learning) approaches, such as conditioning generative adversarial network (GAN) (Méndez-Lucio et al. 2020), genetic profile-activity relationship (GPAR) (Gao et al. 2021), and others (Wang et al. 2020). Indeed, AI can integrate different sets of *omics* data to reveal novel relationships and patterns in biological organisms.

Therefore, AI can potentially be used to model the evolution of genome features (e.g., dynamics of gene family radiation, genomic transpositions during speciation, and their relationship with the appearance of novel traits) and integrate genome landscapes, transcriptional dynamics, as well as metabolomics and cell imaging data to predict biochemical and developmental features of an organism. Along with transcriptional datasets derived from genetic and environmental disturbances (e.g., mutations, gene editing and variant analyses, development under diverse environments), deep learning algorithms can be very powerful to reveal relationships between gene functions and how genetics and biochemistry are integrated in the cell, organs, and the organism as a whole.

The developmental program of an organism can be understood as a complex, uninterruptedly operative algorithm that coordinates molecules and molecular interactions within a cell that incorporates environmental cues and intercellular communication as inputs and spews out coordinated cellular phenotypes that lead to the development of a whole, multicellular organism. Since the ultimate objective of systems biology is to understand the script of life (Schrödinger 1944), we will have to integrate all types of *omics* data in the cell and to contemplate the subcellular localizations of molecules and their intermolecular interactions.

The bridge between systems biology and artificial intelligence is still in its infancy. Nonetheless, it has the potential to become the next conceptual jump in biology in this quest to unveil the cellular and organismal algorithms. Due to its relative simplicity, the development of an organ such as the legume nodule from cortical root cells upon the stimulus of bacterial infection can serve as an ideal systems biology model to use artificial intelligence to unveil how cells differentiate, how organs are formed, and how an organism integrates its physiology.

Acknowledgements This work was supported by Hatch funds from the United States Department of Agriculture (USDA, Hatch Project WVA00686).

References

Azani N, Babineau M, Bailey CD, Banks H, Barbosa AR, Pinto RB, Boatwright JS, Borges LM, Brown GK, Bruneau A, Candido E (2017) A new subfamily classification of the Leguminosae based on a taxonomically comprehensive phylogeny: the Legume Phylogeny Working Group (LPWG). Taxon 66 (1):44–77

Barker DG, Bianchi S, Blondon F, Dattée Y, Duc G, Essad S, Flament P, Gallusci P, Génier G, Guy P, Muel X (1990) *Medicago truncatula*, a model plant for studying the molecular genetics of the *Rhizobium*-legume symbiosis. Plant Mol Biol Report 8(1):40–49

Benedito VA, Torres-Jerez I, Murray JD, Andriankaja A, Allen S, Kakar K, Wandrey M, Verdier J, Zuber H, Ott T, Moreau S, Niebel A, Frickey T, Weiller G, He J, Dai X, Zhao PX, Tang Y, Udvardi MK (2008) A gene expression atlas of the model legume *Medicago truncatula*. Plant J 55(3):504–513

Biswas B, Scott PT, Gresshoff PM (2011) Tree legumes as feedstock for sustainable biofuel production: opportunities and challenges. J Plant Physiol 168 (16):1877–1884

Carrère S, Verdier J, Gamas P (2021) MtExpress, a comprehensive and curated RNAseq-based Gene Expression Atlas for the model legume *Medicago truncatula*. Plant Cell Physiol 10:pcab110

Carrère S, Verdenaud M, Gough C, Gouzy J, Gamas P (2020) LeGOO: an expertized knowledge database for the model legume *Medicago truncatula*. Plant Cell Physiol 61(1):203–211

Chen Y, Chen R (2018) Physical mutagenesis in Medicago truncatula using fast neutron bombardment (FNB) for symbiosis and developmental biology studies. In: Functional genomics in Medicago truncatula, (pp. 61–69). Humana Press, New York, NY

Chen H, Zeng Y, Yang Y et al (2020) Allele-aware chromosome-level genome assembly and efficient transgene-free genome editing for the autotetraploid cultivated alfalfa. Nat Commun 11:2494

Czernic P, Gully D, Cartieaux F, Moulin L, Guefrachi I, Patrel D, Pierre O, Fardoux J, Chaintreuil Chen Y, Chen R (2018) Physical mutagenesis in *Medicago truncatula* using fast neutron bombardment (FNB) for symbiosis and developmental biology studies. In: Functional genomics in Medicago truncatula. Humana Press, New York, NY, pp 61–69

Czernic P, Gully D, Cartieaux F, Moulin L, Guefrachi I, Patrel D, Pierre O, Fardoux J, Chaintreuil C, Nguyen P, Gressent F (2015) Convergent evolution of endosymbiont differentiation in dalbergioid and inverted repeat-lacking clade legumes mediated by nodule-specific cysteine-rich peptides. Plant Physiol 169(2):1254–1265

Dai X, Zhuang Z, Boschiero C, Dong Y, Zhao PX (2021) LegumeIP V3: from models to crops—an integrative gene discovery platform for translational genomics in legumes. Nucleic Acids Res 49(D1):D1472–D1479

Fedorova M, Van de Mortel J, Matsumoto PA, Cho J, Town CD, VandenBosch KA, Gantt JS, Vance CP (2002) Genome-wide identification of nodule-specific transcripts in the model legume *Medicago truncatula*. Plant Physiol 130(2):519–537

Gao S, Han L, Luo D, Liu G, Xiao Z, Shan G, Zhang Y, Zhou W (2021) Modeling drug mechanism of action with large scale gene-expression profiles using GPAR, an artificial intelligence platform. BMC Bioinform 22(1):1–3

He J, Benedito VA, Wang M, Murray JD, Zhao PX, Tang Y, Udvardi MK (2009) The *Medicago truncatula* gene expression atlas web server. BMC Bioinform 10(1):1–9

Herrera F, Carvalho MR, Wing SL, Jaramillo C, Herendeen PS (2019) Middle to Late Paleocene Leguminosae fruits and leaves from Colombia. Aust Syst Bot 32(6):385–408

Journet EP, Van Tuinen D, Gouzy J, Crespeau H, Carreau V, Farmer MJ, Niebel A, Schiex T, Jaillon O, Chatagnier O, Godiard L (2002) Exploring root symbiotic programs in the model legume *Medicago truncatula* using EST analysis. Nucleic Acids Res 30 (24):5579–5592

Kamal N, Mun T, Reid D, Lin JS, Akyol TY, Sandal N, Asp T, Hirakawa H, Stougaard J, Mayer KF, Sato S (2020) A chromosome-scale *Lotus japonicus* Gifu genome assembly indicates that symbiotic islands are not general features of legume genomes. BioRxiv. https://doi.org/10.1101/2020.04.17.042473

Kang Y, Torres-Jerez I, An Z, Greve V, Huhman D, Krom N, Cui Y, Udvardi M (2019) Genome-wide association analysis of salinity responsive traits in *Medicago truncatula*. Plant Cell Environ 42(5):1513–1531

Kang Y, Sakiroglu M, Krom N, Stanton-Geddes J, Wang M, Lee YC, Young ND, Udvardi M (2015) Genome-wide association of drought-related and biomass traits with HapMap SNPs in *Medicago truncatula*. Plant Cell Environ 38(10):1997–2011

Koenen EJ, Ojeda DI, Steeves R, Migliore J, Bakker FT, Wieringa JJ, Kidner C, Hardy OJ, Pennington RT, Bruneau A, Hughes CE (2019) Large-scale genomic sequence data resolve the deepest divergences in the legume phylogeny and support a near-simultaneous evolutionary origin of all six subfamilies. New Phytol 225(3):1355–1369

Koenen EJ, Ojeda DI, Bakker FT, Wieringa JJ, Kidner C, Hardy OJ, Pennington RT, Herendeen PS, Bruneau A, Hughes CE (2020) The origin of the legumes is a complex paleopolyploid phylogenomic tangle closely associated with the Cretaceous-Paleogene (K-Pg) mass extinction event. Syst Biol. syaa041

Lamblin AF, Crow JA, Johnson JE, Silverstein KA, Kunau TM, Kilian A, Benz D, Stromvik M, Endre G, VandenBosch KA, Cook DR (2003) MtDB: a database for personalized data mining of the model legume *Medicago truncatula* transcriptome. Nucleic Acids Res 31(1):196–201

Langfelder P, Horvath S (2008) WGCNA: an R package for weighted correlation network analysis. BMC Bioinform 9(1):1–3

Li J, Dai X, Liu T, Zhao PX (2012) LegumeIP: an integrative database for comparative genomics and transcriptomics of model legumes. Nucleic Acids Res 40(D1):D1221–D1229

Li X, Wei Y, Acharya A, Jiang Q, Kang J, Brummer EC (2014) A saturated genetic linkage map of autotetraploid alfalfa (*Medicago sativa* L.) developed using genotyping-by-sequencing is highly syntenous with the *Medicago truncatula* genome. G3: Genes Genomes Genet 4(10):1971–1979

Lyson TR, Miller IM, Bercovici AD, Weissenburger K, Fuentes AJ, Clyde WC, Hagadorn JW, Butrim MJ, Johnson KR, Fleming RF, Barclay RS (2019) Exceptional continental record of biotic recovery after the Cretaceous-Paleogene mass extinction. Science 366 (6468):977–983

Małolepszy A, Mun T, Sandal N, Gupta V, Dubin M, Urbański D, Shah N, Bachmann A, Fukai E, Hirakawa H, Tabata S (2016) The *LORE1* insertion mutant resource. Plant J 88(2):306–317

Méndez-Lucio O, Baillif B, Clevert DA, Rouquié D, Wichard J (2020) *De novo* generation of hit-like molecules from gene expression signatures using artificial intelligence. Nat Commun 11(1):1

Merchant N, Lyons E, Goff S, Vaughn M, Ware D, Micklos D, Antin P (2016) The iPlant collaborative: cyberinfrastructure for enabling data to discovery for the life sciences. PLoS Biol 14(1):e1002342

Pecrix Y, Staton SE, Sallet E, Lelandais-Brière C, Moreau S, Carrere S, Blein T, Jardinaud MF, Latrasse D, Zouine M, Zahm M et al (2018) Whole-genome landscape of *Medicago truncatula* symbiotic genes. Nat Plants 4(12):1017–1025

Pislariu CI, Murray JD, Wen J, Cosson V, Muni RR, Wang M, Benedito VA, Andriankaja A, Cheng X, Jerez IT, Mondy S (2012) A *Medicago truncatula* tobacco retrotransposon insertion mutant collection with defects in nodule development and symbiotic nitrogen fixation. Plant Physiol 159(4):1686–1699

Ren G (2018) The evolution of determinate and indeterminate nodules within the Papilionoideae subfamily. Doctoral dissertation, Wageningen University. https://library.wur.nl/WebQuery/wurpubs/532640

Rogers C, Wen J, Chen R, Oldroyd G (2009) Deletion-based reverse genetics in *Medicago truncatula*. Plant Physiol 151(3):1077–1086

Roux B, Rodde N, Jardinaud MF, Timmers T, Sauviac L, Cottret L, Carrère S, Sallet E, Courcelle E, Moreau S, Debellé F (2014) An integrated analysis of plant and bacterial gene expression in symbiotic root nodules using laser-capture microdissection coupled to RNA sequencing. Plant J 77(6):817–837

Sato S, Nakamura Y, Asamizu E, Isobe S, Tabata S (2007) Genome sequencing and genome resources in model legumes. Plant Physiol 144(2):588–593

Satterlee JW, Strable J, Scanlon MJ (2020) Plant stem-cell organization and differentiation at single-cell resolution. Proc Natl Acad Sci 117(52):33689–33699

Schmutz J, Cannon SB, Schlueter J, Ma J, Mitros T, Nelson W, Hyten DL, Song Q, Thelen JJ, Cheng J, Xu D (2010) Genome sequence of the palaeopolyploid soybean. Nature 463(7278):178–183

Schrödinger E (1944) What is life? The physical aspect of the living cell. Cambridge University Press, UK, 194 pp. ISBN 0-521-42708-8

Severin AJ, Woody JL, Bolon YT, Joseph B, Diers BW, Farmer AD, Muehlbauer GJ, Nelson RT, Grant D, Specht JE, Graham MA (2010) RNA-Seq Atlas of *Glycine max*: a guide to the soybean transcriptome. BMC Plant Biol 10(1):1–6

SOTWP—State of the World's Plants (2017) Naming and counting the world's plant families. http://stateoftheworldsplants.org/2017/report/SOTWP_2017_1_naming_and_counting_the_world_s_plant_families.pdf

Sprent JI, James EK (2007) Legume evolution: where do nodules and mycorrhizas fit in? Plant Physiol 144 (2):575–581

Sun L, Ge Y, Bancroft AC, Cheng X, Wen J (2018) FNBtools: a software to identify homozygous lesions in deletion mutant populations. Front Plant Sci 9:976

Sun L, Gill US, Nandety RS, Kwon S, Mehta P, Dickstein R, Udvardi MK, Mysore KS, Wen J (2019) Genome-wide analysis of flanking sequences reveals that *Tnt1* insertion is positively correlated with gene methylation in *Medicago truncatula*. Plant J 98 (6):1106–1119

Tadege M, Wen J, He J, Tu H, Kwak Y, Eschstruth A, Cayrel A, Endre G, Zhao PX, Chabaud M, Ratet P (2008) Large-scale insertional mutagenesis using the *Tnt1* retrotransposon in the model legume *Medicago truncatula*. Plant J 54(2):335–347

Tsuda M, Kaga A, Anai T, Shimizu T, Sayama T, Takagi K, Machita K, Watanabe S, Nishimura M, Yamada N, Mori S (2015) Construction of a high-density mutant library in soybean and development of a mutant retrieval method using amplicon sequencing. BMC Genom 16(1):1–8

Verdier J, Torres-Jerez I, Wang M, Andriankaja A, Allen SN, He J, Tang Y, Murray JD, Udvardi MK (2013) Establishment of the *Lotus japonicus* Gene Expression Atlas (LjGEA) and its use to explore legume seed maturation. Plant J 74(2):351–362

Wang H, Cimen E, Singh N, Buckler E (2020) Deep learning for plant genomics and crop improvement. Curr Opin Plant Biol 54:34–41

Wang Q, Yang S, Liu J, Terecskei K, Ábrahám E, Gombár A, Domonkos Á, Szűcs A, Körmöczi P, Wang T, Fodor L (2017) Host-secreted antimicrobial peptide enforces symbiotic selectivity in *Medicago truncatula*. Proc Natl Acad Sci 114(26):6854–6859

von Wettberg EJ, Ray-Mukherjee J, Moriuchi K, Porter SS (2019) *Medicago truncatula* as an ecological, evolutionary, and forage legume model: new directions forward. The Model Legume Medicago truncatula 27:31–40. https://onlinelibrary.wiley.com/doi/abs/10.1002/9781119409144.ch03

Wink M (2013) Evolution of secondary metabolites in legumes (Fabaceae). S Afr J Bot 89:164–175

Yang S, Wang Q, Fedorova E, Liu J, Qin Q, Zheng Q, Price PA, Pan H, Wang D, Griffitts JS, Bisseling T (2017) Microsymbiont discrimination mediated by a host-secreted peptide in *Medicago truncatula*. Proc Natl Acad Sci 114(26):6848–6853

Young ND, Debellé F, Oldroyd GE, Geurts R, Cannon SB, Udvardi MK, Benedito VA, Mayer KF, Gouzy J, Schoof H, Van de Peer Y et al (2011) The *Medicago* genome provides insight into the evolution of rhizobial symbioses. Nature 480(7378):520–524

Genome-Wide Association Studies in *Medicago truncatula*

Muhammet Şakiroğlu and Yun Kang

Abstract

Legumes are an integral part of a sustainable agroecosystem for their ability to form symbiosis with nitrogen-fixing rhizobia and deposit a significant amount of atmospheric nitrogen to agricultural land. In order to sustain and expand the contribution of legumes to agricultural production and ecosystem services, a leap forward advancement in legume genomics should be captured in all major legumes. One of the main advancements in understanding legume genomics has been the extensive investigation and utilization of model plants. *Medicago truncatula* was suggested and subsequently established as the model system for legumes and for understanding plant–microbe symbiotic associations. A wide array of genomics and bioinformatics tools were developed, and these tools provided an ample frame to understand different aspects of the Medicago genome and to delineate many veiled aspects of plant–microbe interactions. As part of an international consortium consisting of multiple universities and research institutes, the *Medicago* HapMap project was initiated, and over 300 genetically diverse inbred lines were re-sequenced, and genetic markers (single nucleotide polymorphisms (SNPs) and insertion/deletions (Indels)) have been publicly released. With the availability of high-density HapMap genetic markers, genome-wide association studies (GWAS) have been performed to explore the association between quantitative differences of a phenotype and genetic variation in a number of traits in *M. truncatula*. Here, we summarize the advances in the *M. truncatula* HapMap project as well as provide an up-to-date report of association studies in this model legume.

2.1 Introduction

Legumes are an integral part of a sustainable agroecosystem for their ability to form symbiosis with nitrogen-fixing rhizobia and deposit a significant amount of atmospheric nitrogen to the agricultural land. The global economic contribution of the cultivated legumes from nitrogen fixation alone is estimated to be US $10 billion based on the 40–60 million tons of N-fixed annually (Graham et al. 2004; Smil 1999). In order to sustain and expand the contribution of legumes to agricultural production and ecosystem services, a leap forward achievement needs

M. Şakiroğlu
Department of Bioengineering, Adana Alparslan Türkeş Science and Technology University, Adana 01250, Turkey

Y. Kang (✉)
Oklahoma State University, Stillwater, OK, USA
e-mail: yunkang@okstate.edu

S. Sinharoy et al. (eds.), *The Medicago truncatula Genome*, Compendium of Plant Genomes,
https://doi.org/10.1007/978-3-030-90757-0_2

to be captured in all major legumes. Such an ambitious goal is only conceivable via comprehensive understanding of the plant genome and extensive utilization of the cutting edge technologies.

Contemporary genomic tools have been essential to uncover many aspects of plant genomes previously obscured to researchers and plant breeders. One of the major advancements in the understanding of legume genomics is due to the extensive investigation of model species. As the first comprehensive plant model, the complete sequencing of the Arabidopsis genome was released in the year 2000, and a wealth of functional genomics studies was subsequently conducted (The Arabidopsis Genome Initiative 2000). While these studies addressed a wide array of queries regarding the basic plant biology, Arabidopsis has a basic shortcoming as it does not manifest symbiotic associations with mycorrhizal fungi or rhizobia, which limits its use for genomic studies unveiling symbiotic plant–microbe associations (Fedorova et al. 2002). Therefore, a need for a model system that could provide a comprehensive understanding of plant–microbe symbiotic associations using genomic approaches was advocated early in the genomics era (Barker et al. 1990; Cook et al. 1997).

Along with *Lotus japonicus*, *Medicago truncatula* was suggested and subsequently emerged as a model system of choice for legumes and for understanding plant–microbe symbiotic associations due to its small diploid genome, short generation time, and available protocol for genetic transformation mediated by *Agrobacterium tumefaciens*, and subsequent in vitro regeneration (de Bruijn 2020). Since then, *M. truncatula* has been well-established as a comprehensive genomics model system exceeding the expectation of simply being a model system for plant–microbe interactions (Carrere et al. 2020; de Bruijn 2020). In addition, wide array of genomics and bioinformatics tools were developed and used for *M. truncatula*, including a well-annotated genome, a comprehensive gene atlas, well-characterized mutant collections, and the Medicago HapMap project (Benedito et al. 2008; Carrére et al. 2020; Stanton-Geddes et al. 2013). Such tools provided an ample frame to understand different aspects of the Medicago genome and help to delineate many veiled aspects of plant–microbe interactions. The tools available for Medicago have aided to execute comparative genomics studies and, aided by comparative legume genomics, to crossover the knowledge into other legume forage and food crops.

2.1.1 Utilization of Genome-wide Association Mapping

Genome-wide association studies (GWAS) were initially developed in human genetics where controlled mapping populations are inconceivable and where introducing artificial mutations is not practically or ethically attainable (Manolio et al. 2009; Slate et al. 2009). This approach was later adopted by animal and plant breeding studies (Ogura and Busch 2015). In association mapping, the statistical significance of the association between quantitative differences of a phenotype and the genetic variation in a genome is tested using a diverse set of individuals (Sakiroglu et al. 2012; Sakiroglu and Brummer 2017).

Although initially a number of earlier marker systems were suggested and used in a number of plant systems for GWAS, their power was limited as a reflection of their genome coverage potential (Sakiroglu et al. 2012). Therefore, the true power of GWAS depends on the availability of a set of polymorphisms that are densely dispersed across the genome, and single nucleotide polymorphisms (SNP) are the markers serving this purpose. SNPs are abundant across genomes and have been discovered and validated in high numbers in *M. truncatula* (Pecrix et al. 2018; Stanton-Geddes et al. 2013). The resolution of mapping causative polymorphism is quite high in GWAS studies in comparison to classical quantitative trait locus (QTL) analyses thus providing a clear superiority in identifying true causative polymorphisms that underlie a phenotype of interest (Ogura and Busch 2015).

2.1.2 Genome Sequencing and HapMap Marker Development in Medicago truncatula

The whole genome sequencing efforts of the model legume species *M. truncatula* were conducted between 2002 and 2011, and a reference genome of accession A17 was initially released by an international consortium (Young et al. 2011). While the first *M. truncatula* genome draft released, Mt3.5, has been extensively utilized to tackle various aspects of legume genetics, the genome was later re-sequenced multiple times (Young and Zhou 2019). In 2014, an improved *M. truncatula* A17 genome version generated by very deep Illumina sequencing and a new version, Mt4.0, was released. The latest whole genome re-sequencing using PacBio technology was performed by INRA (Institut National de la Recherche Agronomique, France), and the latest version, Mt 5.0, was released in 2018 (Pecrix et al. 2018).

In 2009, the *Medicago* HapMap project was undertaken, and a total of over 300 accessions were subsequently sequenced with Illumina next generation technology to discover single nucleotide polymorphisms (SNPs) across the Medicago genome. Among these accessions, 30 (HM001–HM030) were deeply sequenced with over 20-fold coverage, and the rest were sequenced with at least fivefold coverage. The first *M. truncatula* HapMap v3.5 contained 288 accessions. In the subsequent phylogenetic analysis, some of the accessions were found actually not to be *M. truncatula* but belonging to a sister taxon, while some accessions were excluded due to their high similarity (Yoder et al. 2014). In the Mt4.0 HapMap SNP data release, only 262 accessions were included in the main data file and were recommended to be used for GWAS analysis (http://www.medicagohapmap.org/downloads/mt40). Among the 262 accessions, 35 were Tunisian lines (HM210-HM245, while HM216 is not *M. truncatula*), and the seeds need to be ordered separately from Tunisia on condition of project collaboration, which is the primary reason why these accessions were excluded in most published *M. truncatula* GWAS studies.

In 2018, after the *M. truncatula* A17 genome was re-sequenced by PacBio, the Mt4.0 SNPs were transferred to the new PacBio genome assembly and the Mt5.0 HapMap SNPs were subsequently released (Pecrix et al. 2018). The raw Mt5.0 HapMap SNP file released in June 2018 contains 39,368,576 SNPs without any SNPs in heterozygous status, which is in contrast to most HapMap SNP datasets (https://medicago.toulouse.inra.fr/MtrunA17r5.0-ANR/).

2.1.3 GWAS Targeting Biomass, Yield, Maturity, and Phenological Traits

As one of the first and most comprehensive studies to target the genetic variation associated with plant height, trichome density, flowering time, and nodulation in *M. truncatula*, GWAS were performed with >6 million SNPs in a mapping panel of 226 accessions (Stanton-Geddes et al. 2013). The phenotypic measurements from the target phenotypes were performed in a replicated greenhouse trial under controlled conditions. The GWAS for plant height, flowering time, and trichome density was performed using the efficient mixed linear model approach expedited (Kang et al. 2010; Zhang et al. 2010) that was implemented in the TASSEL 3.0 software (Bradbury et al. 2007). A previously uncharacterized set of genes was identified adjacent to the associated SNPs along with a number of genes previously characterized. These genes were reported to contain genetic variations likely responsible for naturally occurring variations in the traits of interest (Stanton-Geddes et al. 2013).

Crop yield is a major trait of interest in virtually any crops, and plant diseases pose major threats to secure yield. Indeed, identifying the mechanisms of plant resistance to a number of biological agents is a major goal in crop genetics (Poland et al. 2009). Improvement efforts in most breeding programs focus on selecting resistant cultivars that have accumulated single resistance

(R) genes manifesting Mendelian inheritance, which are easily overcome by new strains of pathogens (Jones and Dangl 2006; Lannou 2012). More recently, quantitative resistance in which multiple genetic loci are targeted has been suggested as an alternative strategy for obtaining long-term resistance (Palloix et al. 2009). The genomic control of quantitative resistance (also known as partial resistance) has been targeted using classical bi-parental QTL-based approaches as well as quantitative resistance loci (QRL) have been identified (Bonhomme et al. 2014). As the disease resistance is phenotyped as a quantitative trait in this approach, GWAS using high-density genetic markers has been successfully performed to discover the components associated with the control of disease resistance (Ingvarsson and Street 2011; Rafalski 2010).

In addition to all the other quality traits, improving nutritional seed quality remains a paramount goal and a challenge in breeding programs of grain legumes. Nonetheless, the genomic control of the seed nutritional quality has not been fully elucidated. Particularly, the differential accumulation of globulins, as the major parameter in the seed nutritional value of legumes, has not been thoroughly investigated with the cutting edge genomic tools until recently. The model legume *M. truncatula* has been employed recently to query for multiple seed size and composition traits with genome-wide association studies in two studies (Le Signor et al. 2017; Chen et al. 2021). In the study by Le Signor et al. (2017), *M. truncatula* accessions with known abundance of 7S and 11S globulins in seeds were employed in combination with 6,013,644 HapMap SNPs. The results revealed genomic regions and genes carrying polymorphisms are linked to globulin variations. In the subsequent analyses, these results were cross-compared with pea (*Pisum sativum*) and pinpointed a set of candidate genes for the regulation of globulin abundance. A number of key candidate genes involved in transcription, chromatin remodeling, post-translational modifications, transport, and targeting of proteins to storage vacuoles were also identified. The authors concluded that GWAS in *M. truncatula* successfully identified genes that improve seed nutritional value in legumes (Le Signor et al. 2017).

In a subsequent comprehensive study by Chen et al. (2021), 32 seed traits related to size/color and composition were analyzed in 162 *M. truncatula* accessions, and GWAS were performed using 4,852,061 Mt5.0 SNPs. The seed composition traits analyzed included seed protein content as well as macro and micro element concentrations. Using two GWAS methods, 79 quantitative trait nucleotides (QTNs) were identified as regulating seed size, and 41 QTNs were identified for seed composition related to nitrogen (i.e., storage protein) and sulfur (i.e., sulfur-containing amino acid) concentrations/contents. In addition, a strong positive correlation was revealed between seed size and protein content. Considering the high synteny between *M. truncatula* and other legumes crop species, the ample information provided by this study could be valuable for other legumes.

2.1.4 GWAS Targeting Genetic Control of Abiotic Stress Factors

There is a vast array of abiotic stress factors limiting plant production, while there are increasing demands for more production of primary goods to sustain the food needs of an ever-increasing world population. The clash of elevated yield demands with increasing yield-threatening, abiotic stress factors require an unprecedented amount of investments into understanding and improving plant abiotic stress resistance. To achieve such a goal, it is mandatory to understand at the genomic level how plants are adapted to various abiotic stress factors and to apply such knowledge in breeding programs. Model plants have been invaluable to understand the genomic control of stress responses in plants, and GWAS studies were undertaken in *M. truncatula* to elucidate the genomic response to abiotic stresses, such as drought and salinity.

Drought is one of the most prominent abiotic stress factors causing a dramatic reduction in crop yields and elevating global food insecurity (Zhang et al. 2018). The adverse effects of drought on plant production have been projected to increase in severity and reach in the context of global climate change (Swann 2018). Therefore, understanding the genomic mechanisms of drought resistance and improving crop drought tolerance to ensure food supplies is a major goal of crop research. The study conducted by Kang et al. (2015) targeted genetic control of the traits related to drought tolerance/adaptation in *M. truncatula* using the GWAS approach. Biomass and drought-related traits were measured in either 220 or 84 *M. truncatula* HapMap accessions. The traits evaluated were shoot biomass, maximum leaf size, specific leaf weight, stomatal density, trichome density, and shoot carbon-13 isotope discrimination (δ13C) under well-watered conditions, along with leaf chlorophyll and dry weight changes under PEG treatment. Genome-wide association analyses were performed utilizing the general linear model (GLM), the standard mixed linear model (MLM), and compressed MLM (CMLM) in the software TASSEL. A large number of candidate genes and chromosome regions containing SNP markers were found to be associated with the traits of interest under diverse analysis models. Among the associated SNPs, those that are confined to a genomic region under linkage disequilibrium (LD) were further investigated to identify potential candidate genes (Table 2.1). For instance, a 0.5 Mbp region on Chromosome 2 harboring a gene for a plasma membrane intrinsic protein, PIP2, was associated with plant biomass implying that the allelic variation could be used to increase dry matter yields via marker assisted selection. Similarly, genetic variation of a gene for a disulfide isomerase-like protein showed strong association with both shoot biomass and leaf size and could also be useful for breeding (Kang et al. 2015).

Recently, a total of 132 diverse *M. truncatula* HapMap genotypes were used to identify genes involved in salinity stress using GWAS. A study conducted in a controlled growth chamber was performed, in which plants grown in soil were subjected to a step-wise increase in NaCl concentrations, from 0 through 0.5% and 1.0% to 1.5% (257 mM). A number of traits were measured, including vigor, shoot biomass, shoot water content, leaf chlorophyll content, leaf size, and leaf and root concentrations of proline and major ions (e.g., Na^+, Cl^-, K^+, Ca^{2+}). A total of 2,528,531 SNPs that passed quality criteria were utilized to identify candidate genomic regions. The standard mixed linear model (MLM) was applied during the association studies to test for association between SNPs and phenotypic traits using the software TASSEL (Kang et al. 2019). The authors also performed linkage disequilibrium analyses among top 214 SNPs that were associated with multiple traits. The top 100 associations identified in the GWAS of each phenotypic trait and the principal components of correlated traits were further investigated. The results revealed a total of 12 genomic regions associated with salinity responsive traits. As, most candidate SNPs were localized in specific sectors of the genome, which were selected as candidate regions. The candidate SNPs located on Chromosome 2 found in this study are indicated in Fig. 2.1. The candidate genes from the regions of interest were further narrowed down based on previously published studies, in silico, and subsequent transcriptional analyses of extreme genotypes. The results of these analyses comprehensively yielded seven candidate genes encoding a vacuolar H^+-ATPase, two transcription factors, two proteins involved in vesicle trafficking, one peroxidase, and a protein of unknown function (Kang et al. 2019). Further functional analyses of these genes will potentially clarify their role in plant salinity stress response, which can then be applied in the breeding of more resilient genotypes.

2.1.5 GWAS Targeting Genetic Control of Plant-Microbe Interactions

A wide array of microbes are capable of invading living plant cells and developing a complex relationship with the host. The nature of this relationship ranges from mutualistic symbiosis to

Table 2.1 Summary of GWAS studies conducted in *M. truncatula*

References	Trait(s) of interest	Number of SNPs used	Number of lines used	Log (p value) for significance	Number of significant associations	A subset of potential candidate genes
Stanton-Geddes et al. (2013)	Flowering time, plant height, trichome density, rhizobial nodule symbiosis	6,344,526	226	6	>1000	Calmodulin CAML3, NFP, SERK2, MtnodGRP3A, chit4, MtN5, Calmodulin CAML2, MCA8, MtnodGRP1B, MtNRT1.3, MtHMGR3, MtMMPL1
Bonhomme et al. (2014)	Resistance to *Aphanomyces euteiche*	5,107,697	179	5	44	F-box protein coding GAMYB transcription factor
Kang et al. (2015)	Biomass/drought adaptation and dehydration tolerance	1,839,676	220	7	>1000	Wall-associated kinase, aquaporin PIP2, protein disulfide isomerase-like protein, telomere recombination domain protein, Exonuclease
Yoder et al. (2014)	Response to climate	1,918,637	202	5	>1000	F-box protein, 3-isopropylamate dehydrogenase, calcineurin B-like protein, TMV resistance N, ABC transporter ATP-binding protein/permease, PDS5-like sister-chromatid cohesion protein, protein kinase
Burgarella et al. (2016)	Response to climate of flowering phenology genes	5206	174	FDR 0.1	167	VERNALIZATION INSENSITIVE-like (VIL), COP9 signalosome complex subunits (CNS) CONSTANS (CO), FAR-RED IMPAIRED RESPONSE (FARe)
Le Signor et al. (2017)	Seed protein content	6,013,644	175	7	>1000	Inositol polyphosphate 5-phosphatase 11, Sec14p-like phosphatidylinositol transfer protein, RNA-dependent RNA polymerase, kinase, peptidase S8, subtilisin-related, ser-active site
Rey et al. (2017)	Susceptibility to *Phytophthora palmivora*	5,329,189	172	7	2	RAD1
Kang et al. (2019)	Salinity resistance	2,528,531	132	4	>200	Non-clathrin coat protein zeta1-COP, synaptobrevin-like protein, C2 domain protein, hypothetical protein (Medtr2g436940), mTERF protein, FAD-linked oxidoreductase

(continued)

Table 2.1 (continued)

References	Trait(s) of interest	Number of SNPs used	Number of lines used	Log (p value) for significance	Number of significant associations	A subset of potential candidate genes
Michno et al. (2020)	Nodulation	1984 (selected from Stanton-Geddes et al. 2013)	226	5	292	MEDTR4G027195, Pectinesterase/pectinesterase inhibitor,TPX2 (targeting protein for Xklp2) family protein, synaptotagmin-1-related, cysteine-rich RLK (receptor-like kinase) protein
Chen et al. (2021)	Seed size, color, and compositions	4,852,061	162	7 or 5	120	Zinc-fnger transcription factor, CAAT-binding transcription factor, WD40/BEACH domain protein, RNA-binding domain protein

pathogenesis. The formation of a host-microbe interface in these biotrophic interactions is one of the fundamental stages of plant–microbe interaction. A specialized membrane compartment during this interaction can be formed, which plays a major role in the facilitation of controlled nutrient and signal exchanges between plants and the microbe (Huisman et al. 2015). The genetic control of host-microbe interfaces has implications for a successful symbiotic relationship between host plants and arbuscular mycorrhiza (AM) fungi (peri-arbuscular membrane) and between legumes and rhizobia (peribacteroid membrane), and the extrahaustorial membrane surrounding haustoria formed by pathogenic fungi and oomycetes has implication for plant productivity and resistance to pathogens. To facilitate AM fungal symbiosis, plant genomes encode a set of genes required for microbial perception and accommodation. Nevertheless,

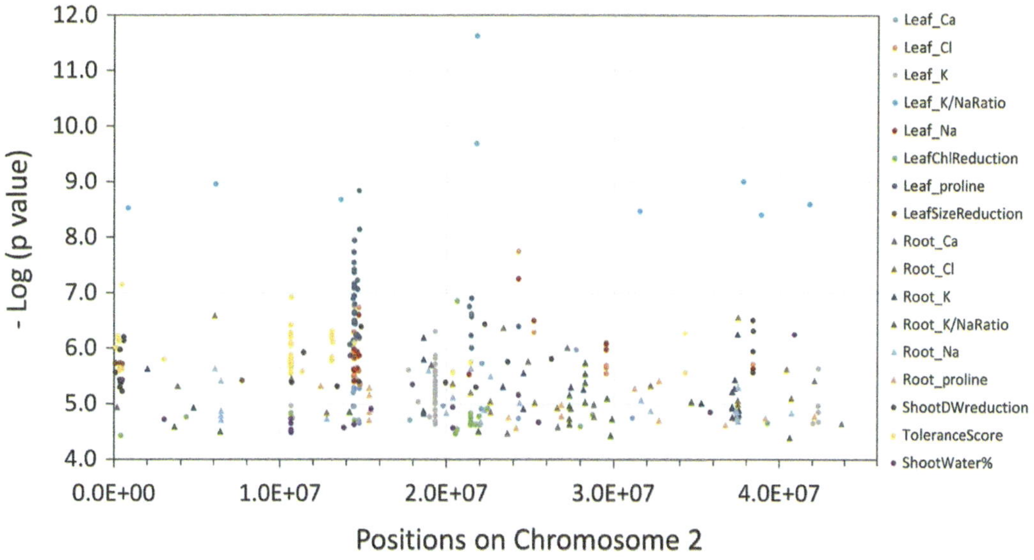

Fig. 2.1. Top potential causative single nucleotide polymorphisms (SNPs) associated with salinity stress resistance on Chromosome 2 identified by genome-wide association studies. All SNPs shown are among the 100 SNPs with the lowest *P* values for each trait. Reprint from Kang et al. (2019)

the infection by filamentous root pathogens is also attributed to some degree to the same set of genes.

A genome-wide association mapping approach has been utilized to identify genes contributing to the colonization of *M. truncatula* roots by the pathogenic oomycete *Phytophthora palmivora* (Rey et al. 2017). *Phytophthora* species are known to invade roots of a broad range of plant species as causal agents of root rot and can form haustoria in plant roots (Rey and Schornack, 2013; Wang et al. 2011). A total of 172 lines from the Medicago HapMap collection were phenotyped for *P. palmivora* resistance in replicated trials, and GWAS were carried out using 5,329,189 SNPs. SNPs that are the most significantly associated with plant colonization response were identified on Chromosome 4 at the nucleotide positions 43,024,916 and 43,026,700, which is upstream of a *RAD1* gene. RAD1 is a GRAS transcription regulator that positively controls AM symbiosis. RAD1 transcription was induced both in response to AM fungus and *P. palmivora*. The authors also adopted a reverse genetics approach, and the results indicated that a reduction in RAD1 transcript levels reduced the colonization by AM fungi as well as by *P. palmivora,* and a *rad1* mutant was impaired in the full colonization of both. Therefore, the authors concluded that the *RAD1* gene identified in the GWAS study in *M. truncatula* has roles that exceed symbiotic interactions, such as microbe-induced root development as well as interactions with various beneficial and pathogenic microbes (Rey et al. 2017).

The soil-borne root pathogen *Aphanomyces euteiches* is an oomycete that naturally infects *M. truncatula* and a number of economically important legume crops, such as pea (*Pisum sativum*) and alfalfa (*M. sativa*) (Gaulin et al. 2007; Moussart et al. 2008). Identifying genetic regions involved in the manifestation of the resistance to the root pathogen *A. euteiches* in legumes could contribute to the development of more resistant cultivars. A total of 5.1 million SNPs were utilized in a genome-wide association study performed using both in vitro and greenhouse phenotyping data gathered from 179 HapMap genotypes. GWAS results yielded a number of candidate genes, and two independent major loci on Chromosome 3 were identified repeatedly under different phenotypic methods. When evaluated in depth, the top candidate SNPs were found to reside within the promoter and coding regions of an F-box protein coding gene. The subsequent validation studies using qRT-PCR and bioinformatic analyses with 20 lines indicated that the resistance is associated with mutations directly affecting the interaction domain of the F-box protein. The results also demonstrate that GWAS in Medicago are a powerful method to identify loci involved in resistances against *A. euteiches* and potentially other pathogens (Bonhomme et al. 2014).

In another study, GWAS were utilized to identify nodulation-related genes in *M. truncatula*. A total of >6 million SNPs in a mapping panel of 226 accessions were used in this study (Stanton-Geddes et al. 2013). The phenotypic measurements from the target phenotypes were performed in a replicated greenhouse trial under control conditions. The GWAS for five nodule-related traits were performed using the efficient mixed linear model approach implemented in the TASSEL 3.0 software program (Bradbury et al. 2007). The authors reported a wealth of candidate SNPs responsible for variations in nodule traits. Due to the high number of genetic variations identified, which is practically unattainable to study, they considered the 200 candidate SNPs with smallest *P* values for each trait (Stanton-Geddes et al. 2013).

In a follow-up nodulation study, GWAS approach was combined with co-expression networks to identify genes involved in the nodule traits in *M. truncatula* (Michno et al. 2020). Here, the authors re-analyzed GWAS results generated by Stanton-Geddes et al. (2013) by combining GWAS results with published co-expression networks. In this study, co-expressed genes manifest strong edges in networks, implying interrelated biological functions (Fig. 2.2). Therefore, linking genes of unknown functions to those of known functions becomes possible based on the expression network (Michno et al. 2020). The study concluded that integrating gene co-

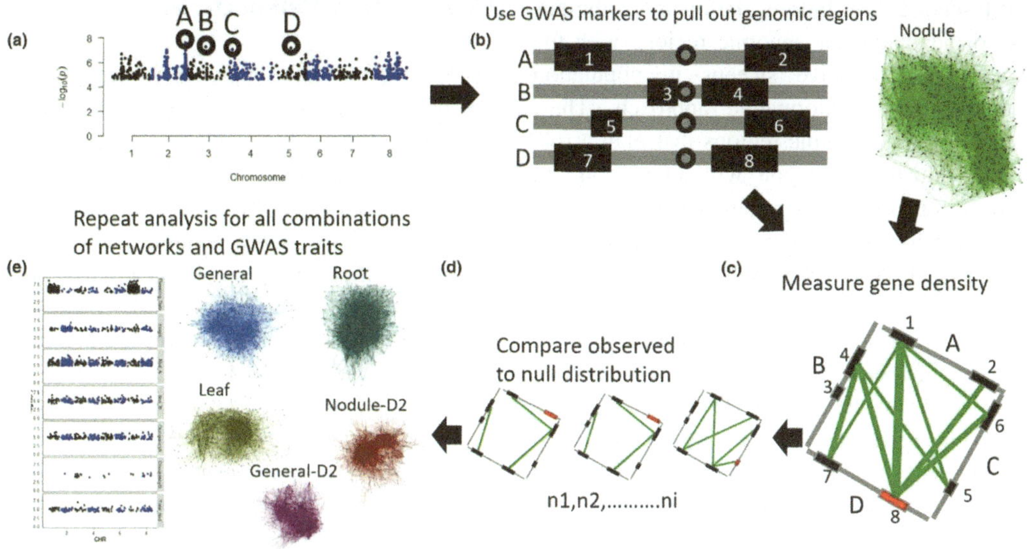

Fig. 2.2 GWAS and co-expression pipeline using Camoco. **a** Manhattan plot represents DNA markers used as input for Camoco, and bold black circles represent a subset of markers used for illustrative purposes. **b** Regions along a chromosome from previously selected markers are represented as gray bars, and genes are represented as black rectangles. Black circles represent a SNP from (**a**). **c** Genes from previously identified intervals are then selected from the co-expression network for per-gene network density measurements. Colored lines represent the strength of co-expression between two genes in a co-expression network. Wider lines represent gene pairs that are more strongly co-expressed. The red box represents the current gene being measured for density. **d** Per-gene density measurement of random subnetworks equal in size to the testing set. **e** Other GWAS traits and networks used for analysis. Reprint from Michno et al. (2020)

expression network analyses with GWAS in *M. truncatula* for targeting nodulation traits yielded a group of genes in the immediate vicinity of top significant GWAS markers that are highly co-expressed. A previously validated nodulation gene, *PEN3-like*, and several other genes with nodulation-related annotations were among these genes identified, including already characterized genes as well as various uncharacterized ones, which could be newly detected genes involved in the nodulation in *M. truncatula* (Michno et al. 2020). This result clearly demonstrates the usefulness of GWAS to create strong hypotheses that inform functional genetics.

The capacity of plants to form different symbioses through their roots has given them the ability to secure an improved acquisition of limiting macro- and micro-nutrients from the soil. In addition to the AM symbiosis, root nodule symbiosis between legumes and bacteria of the family Rhizobiaceae has attracted an elevated attention. Understanding molecular mechanisms involved in the nodulation and nitrogen fixation has been a general interest for enhancing plants' capacity to increase symbiosis and nitrogen fixation as well as to transfer the mechanism to other (potentially non-legume) cultivated plants (Rogers and Oldroyd 2014). Thus, identifying genomic regions and genes involved in the nodulation in *M. truncatula* has been of interest, and several approaches have been used to identify genes and pathways, including the screening and analysis of *Tnt1* mutant lines (Veerappan et al. 2019).

2.1.6 GWAS Targeting Genetic Control of Adaptation

The study of how natural populations adapted to various environmental conditions has implications for understanding plant evolution (Leimu and Fischer 2008). Indeed, the quest for the targets of adaptation or genomic regions with the signature of selective sweeps in population genomics is an informative approach. Thus, identifying the genomic regions that play a role in the climatic adaptation will allow both, to identify functionally important genes as well as contribute to understanding of adaptation itself (Yoder et al. 2014). A GWAS study performed with a dataset of 202 *M. truncatula* accessions targeted loci responsible for adaptation to three climatic gradients (annual mean temperature, precipitation in the wettest month, and isothermality) using 1,918,637 SNPs. The authors used mixed linear models (MLM) implemented in the software program TASSEL to identify candidate SNPs. Instead of setting a standard threshold, the authors selected top 200 candidate SNPs associated with all the three climatic gradients. The patterns of diversity detected among the SNPs located around candidate genes were consistent with the history of soft selective sweeps acting on the loci underlying adaptation to climate (Yoder et al. 2014).

Another association study targeted genetic control of environmental heterogeneity in *M. truncatula* using a set of 224 candidate flowering genes (Burgarella et al. 2016). The study aimed to evaluate the potential role of candidate genes in the adaptation to climatic variations using the Medicago core collection CC192 containing 174 accessions. Candidate gene approaches are noted to be particularly powerful for identifying strong marker-trait associations if a particular set of genes with prior knowledge or expectation is available. The study concluded that flowering time arbitrates adaptation to climatic conditions via genes located upstream in the flowering pathways. A total of 12 candidate flowering genes were identified for phenological adaptation to climate. Among the positive associations, four genes reside on Chromosome 7 where a flowering time QTL was previously localized (Burgarella et al. 2016). This is yet another example of the usefulness of GWAS to correctly identify genes related to important traits in *M. truncatula*.

2.1.7 Future Perspectives

GWAS is a powerful approach to detect causative polymorphisms in the genome leading to phenotypic variation, which have been exploited extensively in plants. One of the major advantages is that it does not require prior knowledge of QTLs, and the entire genome can be thoroughly investigated for possible causative polymorphisms. *M. truncatula* has been a very effective model system to investigate a wide spectrum of legume and plant genomes. The availability of a dense set of HapMap markers in this genetic model provides an invaluable tool to investigate the genetic control of virtually any traits in *M. truncatula*. Since there are several closely related species to *M. truncatula* that are agriculturally important species, such as alfalfa and other Faboideae, the findings can potentially be extrapolated to closely related crops. Therefore, GWAS in *M. truncatula* are a valuable tool for the basic understanding of plant biology as well as revealing novel information about the genetic control of agronomically important traits.

References

Barker DG, Bianchi S, Blondon F, Dattée Y, Duc G, Essad S, Flament P, Gallusci P, Génier G, Guy P (1990) Medicago truncatula, a model plant for studying the molecular genetics of theRhizobium-legume symbiosis. Plant Mol Biol Report 8(1):40–49

Benedito VA, Torres-Jerez I, Murray JD, Andriankaja A, Allen S, Kakar K, Wandrey M, Verdier J, Zuber H, Ott T (2008) A gene expression atlas of the model legume Medicago truncatula. Plant J 55(3):504–513

Bonhomme M, André O, Badis Y, Ronfort J, Burgarella C, Chantret N, Prosperi J-M, Briskine R, Mudge J, Debellé F (2014) High-density genome-wide association mapping implicates an F-box encoding gene in M edicago truncatula resistance to A phanomyces euteiches. New Phytol 201(4):1328–1342

Bradbury PJ, Zhang Z, Kroon DE, Casstevens TM, Ramdoss Y, Buckler ES (2007) TASSEL: software for association mapping of complex traits in diverse samples. Bioinformatics 23(19):2633–2635

Burgarella C, Chantret N, Gay L, Prosperi J-M, Bonhomme M, Tiffin P, Young ND, Ronfort J (2016) Adaptation to climate through flowering phenology: a

case study in Medicago truncatula. Mol Ecol 25 (14):3397–3415
Carrere S, Verdenaud M, Gough C, Gouzy J, Gamas P (2020) LeGOO: an expertized knowledge database for the model Legume Medicago truncatula. Plant Cell Physiol 61(1):203–211
Chen, Z., Lancon V, Le-Signor C, She Y, Kang Y, Jerome V (2021) Genome-wide association study identified candidate genes for seed size and seed composition improvement in *M. truncatula.* Sci Reports 2021(11):4224
Cook DR, VandenBosch K, de Bruijn FJ, Huguet T (1997) Model legumes get the nod. Plant Cell 9(3):275
de Bruijn FJ (2020) The model Legume Medicago truncatula, 2 Volume Set. Wiley
Fedorova M, Van de Mortel J, Matsumoto PA, Cho J, Town CD, VandenBosch KA, Gantt JS, Vance CP (2002) Genome-wide identification of nodule-specific transcripts in the model legume Medicago truncatula. Plant Physiol 130(2):519–537
Gaulin E, Jacquet C, Bottin A, Dumas B (2007) Root rot disease of legumes caused by Aphanomyces euteiches. Mol Plant Pathol 8(5):539–548
Graham MA, Silverstein KA, Cannon SB, VandenBosch KA (2004) Computational identification and characterization of novel genes from legumes. Plant Physiol 135(3):1179–1197
Huisman R, Bouwmeester K, Brattinga M, Govers F, Bisseling T, Limpens E (2015) Haustorium formation in Medicago truncatula roots infected by Phytophthora palmivora does not involve the common endosymbiotic program shared by arbuscular mycorrhizal fungi and rhizobia. Mol Plant Microbe Interact 28 (12):1271–1280
Ingvarsson PK, Street NR (2011) Association genetics of complex traits in plants. New Phytol 189(4):909–922
Jones JD, Dangl JL (2006) The plant immune system. Nature 444(7117):323–329
Kang Y, Sakiroglu M, Krom N, Stanton-Geddes J, Wang M, Lee Y-C, Young ND, Udvardi M (2015) Genome-wide association of drought-related and biomass traits with HapMap SNPs in Medicago truncatula. Plant, Cell Environ 38(10):1997–2011
Kang Y, Torres-Jerez I, An Z, Greve V, Huhman D, Krom N, Cui Y, Udvardi M (2019) Genome-wide association analysis of salinity responsive traits in Medicago truncatula. Plant, Cell Environ 42(5):1513–1531
Kang HM, Sul JH, Service SK, Zaitlen NA, Kong S, Freimer NB, Sabatti C, Eskin E (2010) Variance component model to account for sample structure in genome-wide association studies. Nat Genet 42 (4):348–354
Lannou C (2012) Variation and selection of quantitative traits in plant pathogens. Ann Rev Phytopathol 50
Le Signor C, Aimé D, Bordat A, Belghazi M, Labas V, Gouzy J, Young ND, Prosperi J-M, Leprince O, Thompson RD (2017) Genome-wide association studies with proteomics data reveal genes important for synthesis, transport and packaging of globulins in legume seeds. New Phytol 214(4):1597–1613
Leimu R, Fischer M (2008) A meta-analysis of local adaptation in plants. PloS One 3(12):e4010
Manolio TA, Collins FS, Cox NJ, Goldstein DB, Hindorff LA, Hunter DJ, McCarthy MI, Ramos EM, Cardon LR, Chakravarti A (2009) Finding the missing heritability of complex diseases. Nature 461 (7265):747–753
Michno J-M, Liu J, Jeffers JR, Stupar RM, Myers CL (2020) Identification of nodulation-related genes in Medicago truncatula using genome-wide association studies and co-expression networks. Plant Direct 4(5): e00220
Moussart A, Even M-N, Tivoli B (2008) Reaction of genotypes from several species of grain and forage legumes to infection with a French pea isolate of the oomycete Aphanomyces euteiches. Eur J Plant Pathol 122(3):321–333
Ogura T, Busch W (2015) From phenotypes to causal sequences: using genome wide association studies to dissect the sequence basis for variation of plant development. Curr Opin Plant Biol 23:98–108
Palloix A, Ayme V, Moury B (2009) Durability of plant major resistance genes to pathogens depends on the genetic background, experimental evidence and consequences for breeding strategies. New Phytol 183 (1):190–199
Pecrix Y, Staton SE, Sallet E, Lelandais-Brière C, Moreau S, Carrere S, Blein T, Jardinaud M-F, Latrasse D, Zouine M (2018) Whole-genome landscape of Medicago truncatula symbiotic genes. Nat Plants 4(12):1017–1025
Poland JA, Balint-Kurti PJ, Wisser RJ, Pratt RC, Nelson RJ (2009) Shades of gray: the world of quantitative disease resistance. Trends Plant Sci 14(1):21–29
Rafalski JA (2010) Association genetics in crop improvement. Curr Opin Plant Biol 13(2):174–180
Rey T, Schornack S (2013) Interactions of beneficial and detrimental root-colonizing filamentous microbes with plant hosts.Genome Biol 14(6):121
Rey T, Bonhomme M, Chatterjee A, Gavrin A, Toulotte J, Yang W, André O, Jacquet C, Schornack S (2017) The Medicago truncatula GRAS protein RAD1 supports arbuscular mycorrhiza symbiosis and Phytophthora palmivora susceptibility. J Exp Bot 68(21–22):5871–5881
Rogers C, Oldroyd GE (2014) Synthetic biology approaches to engineering the nitrogen symbiosis in cereals. J Exp Bot 65(8):1939–1946
Sakiroglu M, Brummer EC (2017) Identification of loci controlling forage yield and nutritive value in diploid alfalfa using GBS-GWAS. Theor Appl Genet 130 (2):261–268
Sakiroglu M, Sherman-Broyles S, Story A, Moore KJ, Doyle JJ, Brummer EC (2012) Patterns of linkage disequilibrium and association mapping in diploid alfalfa (M. sativa L.). Theor Appl Genet 125(3):577–590

Slate J, Gratten J, Beraldi D, Stapley J, Hale M, Pemberton JM (2009) Gene mapping in the wild with SNPs: guidelines and future directions. Genetica 136 (1):97–107

Smil V (1999) Nitrogen in crop production: an account of global flows. Global Biogeochem Cycles 13(2):647–662

Stanton-Geddes J, Paape T, Epstein B, Briskine R, Yoder J, Mudge J, Bharti AK, Farmer AD, Zhou P, Denny R, May GD, Erlandson S, Yakub M, Sugawara M, Sadowsky MJ, Young ND, Tiffin P (2013) Candidate genes and genetic architecture of symbiotic and agronomic traits revealed by whole-genome, sequence-based association genetics in Medicago truncatula. PLoS One 8(5):e65688.

Swann AL (2018) Plants and drought in a changing climate. Curr Climate Change Reports 4(2):192–201

The Arabidopsis Genome Initiative (2000) Analysis of the genome sequence of the flowering plant Arabidopsis thaliana. Nature 408(6814):796–815.

Veerappan V, Troiani T, Dickstein R (2019) Whole genome sequencing of symbiotic nitrogen fixation mutants from the Medicago truncatula Tnt 1 mutant population to identify relevant Tnt 1 and MERE 1 insertion sites. The Model Legume Medicago Truncatula 1019–1026

Wang Y, Meng Y, Zhang M, Tong X, Wang Q, Sun Y, Quan J, Govers F, Shan W (2011) Infection of Arabidopsis thaliana byPhytophthora parasitica and identification of variation in host specificity. Mol Plant Pathol. 12(2):187–201

Yoder JB, Stanton-Geddes J, Zhou P, Briskine R, Young ND, Tiffin P (2014) Genomic signature of adaptation to climate in Medicago truncatula. Genetics 196(4):1263–1275

Young ND, Debellé F, Oldroyd GE, Geurts R, Cannon SB, Udvardi MK, Benedito VA, Mayer KF, Gouzy J, Schoof H (2011) The Medicago genome provides insight into the evolution of rhizobial symbioses. Nature 480(7378):520–524

Young ND, Zhou P (2019) The sequenced genomes of Medicago truncatula. The Model Legume Medicago Truncatula 828–834

Zhang Z, Ersoz E, Lai C-Q, Todhunter RJ, Tiwari HK, Gore MA, Bradbury PJ, Yu J, Arnett DK, Ordovas JM (2010) Mixed linear model approach adapted for genome-wide association studies. Nat Genet 42 (4):355–360

Zhang H, Li Y, Zhu J-K (2018) Developing naturally stress-resistant crops for a sustainable agriculture. Nat Plants 4(12):989–996

Tnt1 Insertional Mutagenesis in *Medicago truncatula* for Gene Function Analysis

3

Raja Sekhar Nandety, Jiangqi Wen, and Kirankumar S. Mysore

Abstract

Legumes play a key role in the sustainable agriculture to fix atmospheric nitrogen through a mutualistic symbiotic process with nitrogen-fixing bacteria. *Medicago truncatula* is a diploid model legume and has high levels of diversity, and its two primary accessions Jemalong A17 and R108 were primarily being used for development of genetic resources for advancement of legume biology. Both the *M. truncatula* accessions, A17 and R108, have been sequenced. *Tnt1*, a retrotransposon from tobacco, was used to generate insertion mutants in *M. truncatula* R108. Approximately, ~21,000 insertion lines have been generated and publicly available. This resource was pivotal for many forward and reverse genetic studies in *M. truncatula* during the past 13 years. In this chapter, we compare and contrast the genome sequences of A17 and R108, describe the development of *Tnt1* insertion lines, describe various methods used to identify the flanking sequence tags (FSTs), and show distribution of FSTs in the *M. truncatula* genome.

R. S. Nandety · J. Wen · K. S. Mysore (✉)
Institute for Agricultural Biosciences, Oklahoma State University, 3210 Sam Noble Parkway, Ardmore, OK 73401, USA
e-mail: kmysore@okstate.edu

R. S. Nandety
Cereal Crops Research Unit, United States Department ofAgriculture-Agricultural Research Service, Edward T. Schafer Agricultural Research Center, Fargo, ND 58102, USA

3.1 Introduction

The legume family is the third largest land plant family with more than 19,000 known species (Graham and Vance 2003). Legumes play a key role in the sustainable agriculture to fix atmospheric nitrogen through a mutualistic symbiotic process with nitrogen-fixing bacteria, rhizobia (Gepts et al. 2005; Young and Udvardi 2009; Lee et al. 2018; Sun et al. 2019). Legumes are important species in not only ecosystems but also agricultural crops such as soybean, common bean, pea, and alfalfa. Legumes and legume products are important direct protein and oil sources for humans and animals. In the 1990s, *Medicago truncatula* was chosen as one of the model legumes for fundamental research on legume biology and symbiotic nitrogen fixation (SNF). In the early 2000s, scientists realized that SNF shares pathways with the arbuscular mycorrhizal (AM) symbiosis. *M. truncatula,* thus, became a model for AM symbioses as well. Furthermore, unlike the majority of angiosperms that show maternal inheritance, *M. truncatula* exhibits biparental inheritance of plastid DNA, making it an excellent model to study plastid

S. Sinharoy et al. (eds.), *The Medicago truncatula Genome*, Compendium of Plant Genomes,
https://doi.org/10.1007/978-3-030-90757-0_3

genetics in angiosperms (Matsushima et al. 2008).

Medicago truncatula is a diploid model legume closely related to the tetraploid forage crop, alfalfa (*Medicago sativa*) and is quite amenable to genetic studies due to its relatively small genome size of 450 MB when compared to other legumes. *M. truncatula* has high levels of genetic diversity and has synteny with other legumes with much larger and more complex genomes (Burks et al. 2018). Two different ecotypes, Jemalong A17 and R108, of *M. truncatula* are being used as a model. The past few years have seen a sequencing revolution era, through which genomic landscapes of model organisms have exploded (Young and Bharti 2012). Such a genomic revolution in legumes has resulted in identification of several nodulation genes in legumes with a wide variety of molecular functions (Mun et al. 2016; Roy et al. 2020). Phylogenetically, R108 is one of the most distant *M. truncatula* accessions from A17 (Blanca et al. 2011), but owing to its high transformation efficiency, R108 is more attractive for genetic studies (Tadege et al. 2005). A17 is the accession used for generating the *M. truncatula* reference genome but is highly recalcitrant to transformation and regeneration. Phylogenetically, R108 is one of the most distant *M. truncatula* accessions from A17 (Branca et al. 2011).

3.2 Genome Differences Among *M. truncatula* Accessions

Currently, the *M. truncatula* genome information is available for the two accessions, R108 and Jemalong A17, as mentioned above. The A17 genome was sequenced using the conventional BAC sequence assembly, and the gaps were filled with Illumina shotgun sequences (Young et al. 2011). Most of the *M. truncatula* genomics data from expressed sequence tags (ESTs) (Bell et al. 2001) to gene expression atlas (Benedito et al. 2008; He et al. 2009) were generated in the A17 background. The reference genome sequence of genotype A17 was completed and released in 2011 (Young et al. 2011). The assembly of the A17 genome sequence came with significant updates in 2014 (Mt4.0; Tang et al. 2014) and 2018 (Mt5.0; Pecrix et al. 2018). Further, the new A17 v5.0 was annotated with 44,623 protein-coding genes, 4081 long non-coding RNAs (lncRNAs), and 24,645 intact transposable elements (Pecrix et al. 2018). Besides the popular accessions, R108 and A17, used in research, about 226 accessions from *M. truncatula* were shotgun sequenced and compared (Stanton-Geddes et al. 2013). Genome-wide association studies (GWAS) to identify candidate genes and the genetic architecture underlying phenotypic variation in plant height, trichome density, flowering time, and nodulation resulted in the identification of greater than 6 million SNPs from across 226 accessions (Stanton-Geddes et al. 2013). Based on pollen viability and genetic crosses, Kamphuis et al. proved that A17 bears a reciprocal translocation that distinguishes it from R108 and other *M. truncatula* accessions (Kamphuis et al. 2007). Genome sequencing and assembly show that the reciprocal translocation involves the long arms of Chromosomes 4 and 8 and has a size of ~ 30 Mb (Tang et al. 2014). Furthermore, the A17 plastid DNA has a large ~45 kb inversion relative to the R108 plastid DNA. Kamphuis et al. (2007) showed that all F1 individuals derived from crosses involving A17 have 50% pollen viability or less. The translocation in A17 might result in inaccurate synteny analysis between *M. truncatula* and other legume genomes (Pecrix et al. 2018), and the inversion causes aberrant recombination in genetic crosses, including crosses between A17 and R108 (Kamphuis et al. 2007). Due to the significant differences from genomics to physiological responses, it was necessary to fully sequence the R108 genome.

The first draft assembly of *M. truncatula* R108 (R108 v1.0) was constructed using a combination of PacBio, Dovetail, and BioNano technologies as described by Moll et al. (2017). This version of R108 assembly has 909 scaffolds and 402 Mbp sequence length. It also captures 55,706 genes compared to 50,894 genes in the A17 assembly of Mt4.0, indicating that the gene space in the R108 assembly is largely complete.

In addition, this assembly of R108 genome identified more than 22.7 Mb of novel sequence that is not present in the A17 reference assembly, representing 5.7% of the R108 genome. These "novel" sequences are likely a mix of sequences that are truly novel in the R108 genome as well as sequences that represent in both genomes, but they have diverged beyond our ability to detect them or sequences that are in the A17 genome, but it did not make it into the A17 assembly (Moll et al. 2017). Out of nearly 23 Mb of novel R108 sequence, 1.6 Mb represent novel R108 coding sequence that could not be found in the A17 assembly (Moll et al. 2017). In a side-by-side comparison between A17 Mt4.0 and R108 v1.0, Moll et al. found one R108 scaffold (scf005, 16.4 Mb) spanning the upper arm of Chromosome 4 and the lower arm of Chromosome 8 in A17 and another two scaffolds (scf015, 12.0 Mb and scf002, 17.6 Mb) together spanning the upper arm of Chromosome 8 plus the lower arm of Chromosome 4. This observation confirms a chromosomal-scale translocation between the reference A17 and the widely used R108 accession observed by Kamphuis et al. (2007). A17 chromosomal mapping is far from a perfect gold standard given the evolutionary distance between A17 and R108. Joined segments of R108 scaffolds that map to different A17 chromosomes may still map to the same R108 chromosome. The current R108 v1.0 does not provide pseudo chromosomes like Mt4.0. Because of the distance between R108 and the A17 reference as well as the inability to interbreed them to create a genetic map, having a second high-quality *M. truncatula* reference in R108 has been a priority in the *M. truncatula* community. Further, complete R108 genome will be an important resource for the R108 functional community to support gene functional characterization in the *Tnt1* lines.

In an effort to explore structural variation among the 15 de novo assembled *M. truncatula* genomes, Zhou et al. (2017) further confirmed that R108 (HM340) has 5.07 million SNPs with A17 (HM101), which accounts for 2.37% of the sequenced genomes (Zhou et al. 2017). This is in contrast to the 1.7 million SNPs (0.63%) with closely related accession HM058. In addition to the SNPs, R108 also has 1.46 million of short Indels (<50 bp), 0.11 million of large Indels, 0.17 million of copy number variations, and 12,788 translocations in comparison with A17 (Zhou et al. 2017). As two distinct *M. truncatula* accessions, A17 and R108 not only differ genetically but also have differences in response to drought stress (Luo et al. 2016), salt and aluminum stress (Wang et al. 2014), and iron deficiency (Wang et al. 2014). Moreover, A17 and R108 have different compatible interacting rhizobia strains during their symbiotic nitrogen fixation (Kazmierczak et al. 2017). In contrast to the recalcitrance to genetic transformation observed in A17, R108 has much higher transformation efficiency, making it attractive for genetic studies (Tadege et al. 2005). Recently, R108 becomes more and more popular in the legume research community because of the generation of the near-saturated tobacco retrotransposon *Tnt1* insertion population, which is widely used in nearly all areas of legume functional analysis (Tadege et al. 2008; Sun et al. 2019) Recent advances in the understanding of nitrogen fixation, nodulation, compound leaf, inflorescence and plant architecture, floral development, pod formation, biosynthesis of secondary metabolites, and other developmental and metabolic aspects are all achieved based on the analyses of different *Tnt1* mutants (Chen et al. 2010; Tadege et al. 2011; Cheng et al. 2012; Lin et al. 2013; Berrabah et al. 2014; Huault et al. 2014; Zhou et al. 2014; Li et al. 2016; Niu et al. 2016; Cai et al. 2018; Gibelin-Viala et al. 2019; Zhao et al. 2019; He et al. 2020; Zhu et al. 2020).

3.3 Insertional Mutagenesis in plants

Insertional mutagenesis is a method of disrupting gene function by introducing a foreign DNA fragment (T-DNA, transposon, or retrotransposon) into the genome. This approach typically leads to a loss-of-function mutation, or with modification, it can be used for activation tagging, which typically leads to a gain-of-function mutation by over-expressing the target gene instead of inactivation. Insertional mutagenesis is a malleable genetic system in which a DNA sequence is used to mutate and tag the genome (Tadege et al. 2008). All efficacy of insertional mutagenesis rely on the genome size of the plant (Tadege et al. 2005).

One of the most popular insertion mutagenesis in plants is the *Agrobacterium* T-DNA-based mutagenesis. With the exception of Arabidopsis, T-DNA-tagged populations are generated by transformation, and the transformation process requires a tissue culture step that can cause many mutations. This creates extra effort to separate the tagged mutation from background mutations. Moreover, each mutant line requires an independent transformation event, which makes large-scale mutagenesis tedious, if not impossible. Though appealing in its usefulness, often T-DNA-based mutagenesis approaches have many disadvantages. Some of the issues with the T-DNA approaches are the T-DNA integration in tandem copies and re-arrangement of the insertion sites. A modified T-DNA can be used for activation tagging to identify gain-of-function mutants. The frequency of identifying dominant morphological mutants by activation tagging in Arabidopsis is 1 in 1000 (Weigel et al. 2000). The low percentage (0.1%) of gain-of-function mutant identification is mainly due to less number (average of 1.5 insertions per line) of T-DNA insertions in a given line. T-DNA-based insertion mutagenesis is popularly used in plants such as *Arabidopsis* (Alonso et al. 2003), rice (Jeon et al. 2000), and more recently in *Brachypodium distachyon* (An et al. 2016; Hsia et al. 2017).

An alternative to T-DNA-based mutagenesis is to use transposons. There are two major types of transposable elements that are studied, DNA transposons and retrotransposons (Tadege et al. 2005). The maize DNA transposons, *Ac/Ds*, *En/Spm*, and *Mu*, have been used extensively in maize and *Arabidopsis*, and to some extent in rice and tomato (Tissier et al. 1999). DNA transposons, once introduced into a genome, transpose into new locations by a "cut-and-paste" mechanism causing unstable mutations. DNA transposons like *Ac/Ds* and *En/Spm* generally jump to locations near the original insertion site, often to linked positions (Parinov et al. 1999; Ito and Meyerowitz 2000). Therefore, for a genome-wide coverage, the original insertions should be dispersed throughout the genome or *cis* transpositions should be counter selected. DNA transposons have been modified to use for activation tagging to obtain gain-of-function mutants (Marsch-Martinez et al. 2002). Based on dominant or semi-dominant visible mutant frequency in the transposon-based activation tagging, the frequency of mutant recovery was about 1%, which is ten times higher than T-DNA-based activation tagging (Marsch-Martinez et al. 2002). The second class of transposons, retrotransposons, are the most widespread eukaryotic mobile elements that transpose in a genome by a "copy-and-paste" mechanism (Kumar and Bennetzen 1999). Autonomous long-terminal repeat (LTR) retrotransposons are retrovirus-like elements that encode proteins required for their own replication and transposition. Retrotransposition involves an RNA intermediate in which the retroelement is first transcribed into a single mRNA and reverse transcribed into a cDNA by self-encoded reverse transcriptase. The new retroelements then insert into new locations unlinked to the original insertion site (Kumar and Bennetzen 1999). Since there is no excision during replicative transposition, mutations generated by retrotransposon insertions are stable. Moreover, since several independent insertions are obtained in one generation, multiple crossing and selfing is not required to increase the copy number per genome. There are three well-characterized actively transposing plant LTR retrotransposons, *Tos17* from rice, *Tnt1* and *Tto1* from tobacco, and *LORE1* from *Lotus japonicas,* all of which have been used as insertional mutagens (Okamoto and Hirochika 2000; Agrawal et al. 2001; Urbanski et al. 2012; Malolepszy et al. 2016; Mun et al. 2017). Tobacco *Tnt1,* which is among well-characterized plant autonomous LTR retrotransposons (Grandbastien et al. 1989; Grandbastien 1998), was used in *M. truncatula* (d'Erfurth et al. 2003; Tadege et al. 2008). A genome-wide mutagenesis approach using LTR retrotransposons has been used in many plant species including *L. japonicus* (Urbanski et al. 2012), *M. truncatula* (Tadege et al. 2008), rice (Miyao et al. 2003), *B. distachyon* (Nandety et al. 2020), lettuce (Mazier et al. 2007), soybean (Cui et al. 2013), and potato (Duangpan et al. 2013). Following the identification of the mutant phenotype, it is relatively easy to clone the gene using PCR-based techniques. Reverse genetic screening of DNA pools can also be done rapidly and efficiently using a combination of mutagen-specific and gene-specific primers (Cheng et al. 2014). *Tnt1* has been successfully used to develop a large collection ($\sim$21,700 lines representing >1,000,000 insertions) of insertion lines in *M. truncatula* (Tadege et al. 2008; Cheng et al. 2014; Sun et al. 2019). Below, we describe the generation of *M. truncatula Tnt1* insertion lines and its utilization to study gene function.

3.4 Generation of *Tnt1* Tagged *M. truncatula* Insertion Lines

Generation of *M. truncatula Tnt1* insertion lines has been described earlier (Tadege et al. 2008; Lee et al. 2018). Briefly, the complete *Tnt1* retroelement was cloned into a binary vector and transformed into *M. truncatula* via *Agrobacterium*-mediated transformation (d'Erfurth et al. 2003). A low-copy *Tnt1*-containing single transgenic line (with five *Tnt1*copies) was used as the starting material to generate mutants (Tadege et al. 2008). Leaf explants were cultured on Schenk and Hildebrandt medium supplemented with auxin and cytokinin to form calli (4–6 weeks). Calli were subsequently transferred to embryogenesis medium and then to hormone-free medium to ultimately regenerate new plants (Tadege et al. 2008). The initial number of new inserts per line showed an average of 25 inserts as confirmed by Southern blot analysis and flanking sequencing tags (FSTs) identified by Thermal Asymmetric Interlaced PCR (TAIL-PCR) (Tadege et al. 2008). All the *M. truncatula Tnt1* insertion lines generated (21, 741) were subjected to forward genetics screening for visible phenotypes as described (Yarce et al. 2013). Photographs of all the mutant phenotypes and their phenotypic descriptions along with the FST analysis information are publicly available through a database generated by the Noble Research Institute (https://medicago-mutant.dasnr.okstate.edu/mutant/index.php). Researchers can request the seeds of the *Tnt1* mutants after searching through the *Tnt1* mutant database.

3.5 Different Approaches Used for *Tnt1* FST Recovery

Three different methods were used to identify FSTs. The three methods include TAIL-PCR, whole genome sequencing (WGS), and sequence capture (SeqCap).

TAIL-PCR: Genomic DNA isolated from *M. truncatula* mature leaves was used for identification of FSTs. TAIL-PCR was developed as an efficient tool for the recovery of DNA fragments adjacent to known sequences (Liu et al. 2005; Liu and Chen 2007). TAIL-PCR uses three nested specific primers in consecutive reactions together with an arbitrary degenerate (AD) primer having a lower T_m (melting temperature), so that the relative amplification efficiencies of specific and non-specific products can be thermally controlled (Liu et al. 2005). Approximately, 33,000 FSTs were recovered from 2650 *Tnt1* insertion lines using TAIL-PCR, cloned into plasmid vectors, and sequenced by Sanger sequencing (Tadege et al. 2008; Sun et al. 2019). In addition, a two-dimensional DNA pooling strategy along with TAIL-PCR and next-generation sequencing (Illumina) was used to accelerate the FST recovery process from the rest of the regenerated *Medicago Tnt1* lines resulting in identification of $\sim$380,000 FSTs (Cheng et al. 2017). The TAIL-PCR approach though helpful is slow and cannot recover 100% of all FSTs present in a particular line. Therefore, to recover all the FSTs in a given *Tnt1* line, WGS was used.

Whole genome sequencing (WGS): WGS approach to identify the FSTs was used initially for two *Tnt1* symbiotic mutants, NF11217 and

NF10547, with defects in nodulation (Veerappan et al. 2016). For WGS, Illumina HiSeq (2000) platform was used, and it provided 188 and 180 million 90 bp paired-end (PE) clean reads from 500 bp insert libraries representing 44X and 40X total genome coverages of NF11217 and NF10547 mutants, respectively (Veerappan et al. 2016). Paired-end (PE) sequencing resulted in reads from both ends of a DNA fragment that mapped completely to the reference A17 genome (genomic) or *Tnt1* sequences only (*Tnt1*) or mapped to a combination of parts of *Tnt1* element and R108 genome, thus representing the insertion site (hybrid) (Veerappan et al. 2016; Nandety et al. 2020). Complete data analysis comparing TAIL-PCR and WGS shows the identification of substantially more *Tnt1* insertions in the WGS method (Veerappan et al. 2016; Nandety et al. 2020). Later, a similar WGS approach was used to identify FSTs from four additional *Tnt1* lines in *M. truncatula* (Sun et al. 2019). Out of the four lines, two of them NF1962 and NF11044 were also sequenced by SeqCap for comparing the sequencing methodologies (Sun et al. 2019). WGS approach was shown to identify approximately 100–150% higher insertions compared to TAIL-PCR, and that resulted in an average of 85 *Tnt1* insertions per line compared to ~27 insertions recovered by TAIL-PCR (Sun et al. 2019). Significant numbers of *Tnt1* insertions were identified in all the *Tnt1* insertion lines sequenced by WGS; 65 *Tnt1* insertions in NF0054, 164 in NF1962, 51 in NF2801, 65 in NF10547, 17 in NF11044, 97 in NF11217, and 136 in NF20764 (Jiang et al. 2015; Veerappan et al. 2016; Sun et al. 2019).

Sequence capture (SeqCap): SeqCap is similar to exome capture or target capture that is designed to capture the entire high-value genomic regions in the genome representing ~1–3% of the entire genome (Samorodnitsky et al. 2015; Kaur and Gaikwad 2017). This can be one of the important ways to achieve the reduction of genome complexity to focus on the targets of interest rather than on the genome in its entirety (Kaur and Gaikwad 2017). One of the first successful approaches was used in maize to enrich for a set of 43 dispersed genes in a 2.3 Mb chromosome interval (Fu et al. 2010). Similarly, this has been extended for identifying natural variation in wheat (Saintenac et al. 2011), barley (Mascher et al. 2013), soybean (Haun et al. 2011), black cottonwood (Zhou and Holliday 2012), and pine (Neves et al. 2013). In this method, akin to target capture or exon capture, a total of four 5′-biotinylated xGen Lockdown Probes (Integrated DNA Technologies, Skokie, IL, USA) of 120 bp were synthesized using the end sequence of *Tnt1*-LTR region (Sun et al. 2019; Nandety et al. 2020). Illumina libraries were pooled before hybridization-based capture of *Tnt1*-specific fragments. SeqCap exome libraries synthesized by various vendors have different bait densities and use short overlapping baits to cover the target regions (Clark et al. 2011). Captured DNA fragments were sequenced using Illumina NextSeq or HiSeq platform in *M. truncatula Tnt1* mutants (Sun et al. 2019). These captured DNA fragments processed through a bioinformatics approaches for trimming and reads were mapped to the reference genome. *Tnt1* insertions were identified in the mapping process when at least ten supporting reads support the insertion (Sun et al. 2019). Compared to WGS, SeqCap is relatively inexpensive and less laborious in terms of mapping the reads to the reference genome and to analyze the *Tnt1* FSTs (Sun et al. 2019; Nandety et al. 2020). Ten *M. truncatula Tnt1* lines were sequenced using SeqCap approach, and a subset of two lines was also sequenced used WGS to compare the efficiency (Sun et al. 2019). Comparative sequencing approaches of the two *M. truncatula Tnt1* lines NF1962 and NF11044 displayed the superiority of both the methods in accurately identifying all the 164 *Tnt1* insertions in NF1962 and 17 *Tnt1* insertions in NF11044 (Sun et al. 2019). In the *M. truncatula Tnt1* mutant line NF11044, one insertion was identified by WGS that was not recovered by the SeqCap, and the insertion identified through SeqCap was not identified in WGS (Sun et al. 2019). Based on the number of FSTs identified for each method, it is probably estimated that TAIL-PCR could only recover 50% or less *Tnt1* insertions in a given line (Veerappan et al. 2016; Sun et al. 2019). On an average, based on the reports from Sun et al. TAIL-PCR identified ~27 FSTs per mutant line, WGS identified ~85 FSTs per line, and SeqCap identified ~100 FSTs per *Tnt1* mutant line in *M. truncatula* (Sun et al. 2019). A comparison of all the three methods of FST recovery performed also concludes that both WGS and SeqCap identified most of the insertions in a given *Tnt1* mutant line.

3.6 Distribution of *Tnt1* in the Individual *M. truncatula* Lines

Though the overall distribution of *Tnt1* insertions in the *M. truncatula* genome is random (Tadege et al. 2008), this does not necessarily mean that the distribution of *Tnt1* insertions is random in individual lines (Sun et al. 2019). Earlier estimations placed the average number of insertions per line at 25 based on Southern analysis and also through recovery of insertions by TAIL-PCR (Tadege et al. 2005, 2008; Cheng et al. 2014, 2017). The average number of 25 inserts estimated per line in *M. truncatula Tnt1* insertion lines is significantly higher than that of other retrotransposons like *LORE1* (average 4.76 insertions) (Malolepszy et al. 2016) and *Tos17* (average 3.37 insertions) in rice cultivar Nipponbare (Piffanelli et al. 2007). WGS and *Tnt1*-SeqCap approaches revealed a significantly more number of *Tnt1* insertions per line (80) in *M. truncatula,* suggesting that the previous estimate of 25 inserts per line is an underestimation (Sun et al. 2019). Based on the average number of 80 *Tnt1* insertions per line, Sun et al. estimated that 21,741 *Tnt1* insertion lines might contain ~ 1.7 million insertions (Sun et al. 2019).

3.7 Distribution of *Tnt1* insertion in the *M. truncatula* genome

The target site specificity of *Tnt1* insertions in *M. truncatula* genome was analyzed for preference of any hotspots and was shown to have no specificity for their target site locations or any preferential insertion throughout the genome (Tadege et al. 2008; Sun et al. 2019). This was shown to be true with smaller dataset as well as with the more extensive FSTs recently identified (Tadege et al. 2008; Sun et al. 2019). This is consistent with *Tnt1* insertions in other plant species studied (Miyao et al. 2003; Cui et al. 2013; Duangpan et al. 2013; Nandety et al. 2020). Interestingly, it was shown that FST insertions in *M. truncatula* genome occurred in exon, intron + UTR, upstream/downstream regions of a gene, and intergenic regions at 33.2%, 22.6%, 20.1%, and 23.7%, respectively, suggesting that *Tnt1* preferentially inserts into exons and gene-rich regions when compared to intergenic regions.

3.8 Genes Containing *Tnt1* Insertions

Of the predicted 50,894 genes in the *M. truncatula* A17 Mt4.0 reference genome, 19,583 (38%) genes were found to have at least one *Tnt1* insertion and 6,105 genes with at least four insertions (Sun et al. 2019). *Tnt1* insertions analyzed in *M. truncatula* suggest an average of 3.43 insertions per gene. Maximum number of *Tnt1* insertions in a single gene occurred in *Medtr5g099060* with 60 *Tnt1* insertions (Sun et al. 2019). Further analysis of frequently inserted genes conducted by Sun et al. showed that genes mainly targeted were in functional categories of "stress," "signaling," "secondary metabolism," "transport," and "nucleotide metabolism" (Sun et al. 2019). Significant enrichments of gene groups that were targeted by *Tnt1* during mutagenesis led to correlation studies between the insertion frequency and the underlying methylation status of the gene groups suggesting that *Tnt1* insertion frequency positively correlates with the methylation frequency in gene groups (Sun et al. 2019). Furthermore, it was also observed that lowly expressed genes had more *Tnt1* insertions, whereas highly expressed genes had less *Tnt1* insertions (Sun et al. 2019).

Acknowledgements *Medicago truncatula*-related projects in the Mysore laboratory were supported by Noble Research Institute, LLC, and grants from National Science Foundation, USA (DBI 0703285, IOS-1127155, and IOS-1733470).

References

Agrawal GK, Yamazaki M, Kobayashi M, Hirochika R, Miyao A, Hirochika H (2001) Screening of the rice viviparous mutants generated by endogenous retrotransposon Tos17 insertion. Tagging of a zeaxanthin epoxidase gene and a novel ostatc gene. Plant Physiol 125:1248–1257

Alonso JM, Stepanova AN, Leisse TJ, Kim CJ, Chen H, Shinn P, Stevenson DK, Zimmerman J, Barajas P, Cheuk R, Gadrinab C, Heller C, Jeske A, Koesema E, Meyers CC, Parker H, Prednis L, Ansari Y, Choy N, Deen H, Geralt M, Hazari N, Hom E, Karnes M, Mulholland C, Ndubaku R, Schmidt I, Guzman P, Aguilar-Henonin L, Schmid M, Weigel D, Carter DE, Marchand T, Risseeuw E, Brogden D, Zeko A, Crosby WL, Berry CC, Ecker JR (2003) Genome-wide insertional mutagenesis of Arabidopsis thaliana. Science 301:653–657

An T, Cai Y, Zhao S, Zhou J, Song B, Bux H, Qi X (2016) Brachypodium distachyon T-DNA insertion lines: a model pathosystem to study nonhost resistance to wheat stripe rust. Sci Rep 6:25510

Bell CJ, Dixon RA, Farmer AD et al. (2001) The medicago genome initiative: a model legume database. Nucleic Acids Res 29:114–117

Benedito VA, Torres-Jerez I, Murray JD, Andriankaja A, Allen S, Kakar K, Wandrey M, Verdier J, Zuber H, Ott T, Moreau S, Niebel A, Frickey T, Weiller G, He J, Dai X, Zhao PX, Tang Y, Udvardi MK (2008) A gene expression atlas of the model legume Medicago truncatula. Plant J 55:504–513

Berrabah F, Bourcy M, Eschstruth A, Cayrel A, Guefrachi I, Mergaert P, Wen J, Jean V, Mysore KS, Gourion B, Ratet P (2014) A nonRD receptor-like kinase prevents nodule early senescence and defense-like reactions during symbiosis. New Phytol 203:1305–1314

Branca A, Paape TD, Zhou P, Briskine R, Farmer AD, Mudge J, Bharti AK, Woodward JE, May GD, Gentzbittel L, Ben C, Denny R, Sadowsky MJ, Ronfort J, Bataillon T, Young ND, Tiffin P (2011) Whole-genome nucleotide diversity, recombination, and linkage disequilibrium in the model legume Medicago truncatula. Proc Natl Acad Sci U S A 108:E864-870

Burks D, Azad R, Wen J, Dickstein R (2018) The Medicago truncatula genome: genomic data availability. Methods Mol Biol 1822:39–59

Cai K, Yin J, Chao H, Ren Y, Jin L, Cao Y, Duanmu D, Zhang Z (2018) A C3HC4-type RING finger protein regulates rhizobial infection and nodule organogenesis in Lotus japonicus. J Integr Plant Biol 60:878–896

Chen J, Yu J, Ge L, Wang H, Berbel A, Liu Y, Chen Y, Li G, Tadege M, Wen J, Cosson V, Mysore KS, Ratet P, Madueno F, Bai G, Chen R (2010) Control of dissected leaf morphology by a Cys(2)His(2) zinc finger transcription factor in the model legume Medicago truncatula. Proc Natl Acad Sci USA 107:10754–10759

Cheng X, Peng J, Ma J, Tang Y, Chen R, Mysore KS, Wen J (2012) NO APICAL MERISTEM (MtNAM) regulates floral organ identity and lateral organ separation in Medicago truncatula. New Phytol 195:71–84

Cheng X, Wang M, Lee HK, Tadege M, Ratet P, Udvardi M, Mysore KS, Wen J (2014) An efficient reverse genetics platform in the model legume Medicago truncatula. New Phytol 201:1065–1076

Cheng X, Krom N, Zhang S, Mysore KS, Udvardi M, Wen J (2017) Enabling reverse genetics in Medicago truncatula using high-throughput sequencing for Tnt1 flanking sequence recovery. Methods Mol Biol 1610:25–37

Clark MJ, Chen R, Lam HY, Karczewski KJ, Chen R, Euskirchen G, Butte AJ, Snyder M (2011) Performance comparison of exome DNA sequencing technologies. Nat Biotechnol 29:908–914

Cui Y, Barampuram S, Stacey MG, Hancock CN, Findley S, Mathieu M, Zhang Z, Parrott WA, Stacey G (2013) Tnt1 retrotransposon mutagenesis: a tool for soybean functional genomics. Plant Physiol 161:36–47

d'Erfurth I, Cosson V, Eschstruth A, Lucas H, Kondorosi A, Ratet P (2003) Efficient transposition of the Tnt1 tobacco retrotransposon in the model legume Medicago truncatula. Plant J 34:95–106

Duangpan S, Zhang W, Wu Y, Jansky SH, Jiang J (2013) Insertional mutagenesis using Tnt1 retrotransposon in potato. Plant Physiol 163:21–29

Fu Y, Springer NM, Gerhardt DJ, Ying K, Yeh CT, Wu W, Swanson-Wagner R, D'Ascenzo M, Millard T, Freeberg L, Aoyama N, Kitzman J, Burgess D, Richmond T, Albert TJ, Barbazuk WB, Jeddeloh JA, Schnable PS (2010) Repeat subtraction-mediated sequence capture from a complex genome. Plant J 62:898–909

Gepts P, Beavis WD, Brummer EC, Shoemaker RC, Stalker HT, Weeden NF, Young ND (2005) Legumes as a model plant family. Genomics for food and feed report of the cross-legume advances through genomics conference. Plant Physiol 137:1228–1235

Gibelin-Viala C, Amblard E, Puech-Pages V, Bonhomme M, Garcia M, Bascaules-Bedin A, Fliegmann J, Wen J, Mysore KS, le Signor C, Jacquet C, Gough C (2019) The Medicago truncatula LysM receptor-like kinase LYK9 plays a dual role in immunity and the arbuscular mycorrhizal symbiosis. New Phytol 223:1516–1529

Graham PH, Vance CP (2003) Legumes: importance and constraints to greater use. Plant Physiol 131:872–877

Grandbastien MA (1998) Activation of plant retrotransposons under stress conditions. Trends Plant Sci 3:181–187

Grandbastien MA, Spielmann A, Caboche M (1989) Tnt1, a mobile retroviral-like transposable element of tobacco isolated by plant cell genetics. Nature 337:376–380

Haun WJ, Hyten DL, Xu WW, Gerhardt DJ, Albert TJ, Richmond T, Jeddeloh JA, Jia G, Springer NM, Vance CP, Stupar RM (2011) The composition and origins of genomic variation among individuals of the soybean reference cultivar Williams 82. Plant Physiol 155:645–655

He J, Benedito VA, Wang M, Murray JD, Zhao PX, Tang Y, Udvardi MK (2009) The Medicago truncatula gene expression atlas web server. BMC Bioinform 10:441

He L, Liu Y, He H, Liu Y, Qi J, Zhang X, Li Y, Mao Y, Zhou S, Zheng X, Bai Q, Zhao B, Wang D, Wen J, Mysore KS, Tadege M, Xia Y, Chen J (2020) A molecular framework underlying the compound leaf pattern of Medicago truncatula. Nat Plants 6:511–521

Hsia MM, O'Malley R, Cartwright A, Nieu R, Gordon SP, Kelly S, Williams TG, Wood DF, Zhao Y, Bragg J, Jordan M, Pauly M, Ecker JR, Gu Y, Vogel JP (2017) Sequencing and functional validation of the JGI Brachypodium distachyon T-DNA collection. Plant J 91:361–370

Huault E, Laffont C, Wen J, Mysore KS, Ratet P, Duc G, Frugier F (2014) Local and systemic regulation of plant root system architecture and symbiotic nodulation by a receptor-like kinase. PLoS Genet 10: e1004891

Ito T, Meyerowitz EM (2000) Overexpression of a gene encoding a cytochrome P450, CYP78A9, induces large and seedless fruit in arabidopsis. Plant Cell 12:1541–1550

Jeon JS, Lee S, Jung KH, Jun SH, Jeong DH, Lee J, Kim C, Jang S, Yang K, Nam J, An K, Han MJ, Sung RJ, Choi HS, Yu JH, Choi JH, Cho SY, Cha SS, Kim SI, An G (2000) T-DNA insertional mutagenesis for functional genomics in rice. Plant J 22:561–570

Jiang C, Chen C, Huang Z, Liu R, Verdier J (2015) ITIS, a bioinformatics tool for accurate identification of transposon insertion sites using next-generation sequencing data. BMC Bioinform 16:72

Kamphuis LG, Williams AH, D'Souza NK, Pfaff T, Ellwood SR, Groves EJ, Singh KB, Oliver RP, Lichtenzveig J (2007) The Medicago truncatula reference accession A17 has an aberrant chromosomal configuration. New Phytol 174:299–303

Kaur P, Gaikwad K (2017) From genomes to GENE-omes: exome sequencing concept and applications in crop improvement. Front Plant Sci 8:2164

Kazmierczak T, Nagymihaly M, Lamouche F, Barriere Q, Guefrachi I, Alunni B, Ouadghiri M, Ibijbijen J, Kondorosi E, Mergaert P, Gruber V (2017) Specific host-responsive associations between Medicago truncatula accessions and sinorhizobium strains. Mol Plant Microbe Interact 30:399–409

Kumar A, Bennetzen JL (1999) Plant retrotransposons. Ann Rev Genet 33:479–532

Lee HK, Mysore KS, Wen J (2018) Tnt1 insertional mutagenesis in Medicago truncatula. Methods Mol Biol 1822:107–114

Li P, Chen B, Zhang G, Chen L, Dong Q, Wen J, Mysore KS, Zhao J (2016) Regulation of anthocyanin and proanthocyanidin biosynthesis by Medicago truncatula bHLH transcription factor MtTT8. New Phytol 210:905–921

Lin H, Niu L, Tadege M (2013) STENOFOLIA acts as a repressor in regulating leaf blade outgrowth. Plant Signal Behav 8:e24464

Liu YG, Chen Y (2007) High-efficiency thermal asymmetric interlaced PCR for amplification of unknown flanking sequences. Biotechniques 43:649–650, 652, 654 passim

Liu YG, Chen Y, Zhang Q (2005) Amplification of genomic sequences flanking T-DNA insertions by thermal asymmetric interlaced polymerase chain reaction. Methods Mol Biol 286:341–348

Luo SS, Sun YN, Zhou X, Zhu T, Zhu LS, Arfan M, Zou LJ, Lin HH (2016) Medicago truncatula genotypes Jemalong A17 and R108 show contrasting variations under drought stress. Plant Physiol Biochem 109:190–198

Malolepszy A, Mun T, Sandal N, Gupta V, Dubin M, Urbanski D, Shah N, Bachmann A, Fukai E, Hirakawa H, Tabata S, Nadzieja M, Markmann K, Su J, Umehara Y, Soyano T, Miyahara A, Sato S, Hayashi M, Stougaard J, Andersen SU (2016) The LORE1 insertion mutant resource. Plant J 88:306–317

Marsch-Martinez N, Greco R, Van Arkel G, Herrera-Estrella L, Pereira A (2002) Activation tagging using the En-I maize transposon system in Arabidopsis. Plant Physiol 129:1544–1556

Mascher M, Richmond TA, Gerhardt DJ, Himmelbach A, Clissold L, Sampath D, Ayling S, Steuernagel B, Pfeifer M, D'Ascenzo M, Akhunov ED, Hedley PE, Gonzales AM, Morrell PL, Kilian B, Blattner FR, Scholz U, Mayer KF, Flavell AJ, Muehlbauer GJ, Waugh R, Jeddeloh JA, Stein N (2013) Barley whole exome capture: a tool for genomic research in the genus Hordeum and beyond. Plant J 76:494–505

Matsushima R, Hu Y, Toyoda K, Sodmergen, Sakamoto W (2008) The model plant Medicago truncatula exhibits biparental plastid inheritance. Plant Cell Physiol 49:81–91

Mazier M, Botton E, Flamain F, Bouchet JP, Courtial B, Chupeau MC, Chupeau Y, Maisonneuve B, Lucas H (2007) Successful gene tagging in lettuce using the Tnt1 retrotransposon from tobacco. Plant Physiol 144:18–31

Miyao A, Tanaka K, Murata K, Sawaki H, Takeda S, Abe K, Shinozuka Y, Onosato K, Hirochika H (2003) Target site specificity of the Tos17 retrotransposon shows a preference for insertion within genes and against insertion in retrotransposon-rich regions of the genome. Plant Cell 15:1771–1780

Moll KM, Zhou P, Ramaraj T, Fajardo D, Devitt NP, Sadowsky MJ, Stupar RM, Tiffin P, Miller JR, Young ND, Silverstein KAT, Mudge J (2017) Strategies for optimizing BioNano and Dovetail explored through a second reference quality assembly for the legume model, Medicago Truncatula. BMC Genom 18:578

Mun T, Bachmann A, Gupta V, Stougaard J, Andersen SU (2016) Lotus base: an integrated information portal for the model legume Lotus japonicus. Sci Rep 6:39447

Mun T, Malolepszy A, Sandal N, Stougaard J, Andersen SU (2017) User guide for the LORE1 insertion mutant resource. Methods Mol Biol 1610:13–23

Nandety RS, Serrani-Yarce JC, Gill US, Oh S, Lee HK, Zhang X, Dai X, Zhang W, Krom N, Wen J, Zhao PX, Mysore KS (2020) Insertional mutagenesis of Brachypodium distachyon using Tnt1 retrotransposable element. Plant J 103(5):1924–1936

Neves LG, Davis JM, Barbazuk WB, Kirst M (2013) Whole-exome targeted sequencing of the uncharacterized pine genome. Plant J 75:146–156

Niu L, Fu C, Lin H, Wolabu TW, Wu Y, Wang ZY, Tadege M (2016) Control of floral transition in the bioenergy crop switchgrass. Plant Cell Environ 39:2158–2171

Okamoto H, Hirochika H (2000) Efficient insertion mutagenesis of Arabidopsis by tissue culture-induced activation of the tobacco retrotransposon Tto1. Plant J 23:291–304

Parinov S, Sevugan M, De Y, Yang WC, Kumaran M, Sundaresan V (1999) Analysis of flanking sequences from dissociation insertion lines: a database for reverse genetics in *Arabidopsis*. Plant Cell 11:2263–2270

Pecrix Y, Staton SE, Sallet E, Lelandais-Brière C, Moreau S, Carrère S, Blein T, Jardinaud M-F, Latrasse D, Zouine M, Zahm M, Kreplak J, Mayjonade B, Satgé C, Perez M, Cauet S, Marande W, Chantry-Darmon C, Lopez-Roques C, Bouchez O, Bérard A, Debellé F, Muños S, Bendahmane A, Bergès H, Niebel A, Buitink J, Frugier F, Benhamed M, Crespi M, Gouzy J, Gamas P (2018) Whole-genome landscape of Medicago truncatula symbiotic genes. Nature Plants 4:1017–1025

Piffanelli P, Droc G, Mieulet D, Lanau N, Bes M, Bourgeois E, Rouviere C, Gavory F, Cruaud C, Ghesquiere A, Guiderdoni E (2007) Large-scale characterization of Tos17 insertion sites in a rice T-DNA mutant library. Plant Mol Biol 65:587–601

Roy S, Liu W, Nandety RS, Crook A, Mysore KS, Pislariu CI, Frugoli J, Dickstein R, Udvardi MK (2020) Celebrating 20 years of genetic discoveries in legume nodulation and symbiotic nitrogen fixation. Plant Cell 32:15–41

Saintenac C, Jiang D, Akhunov ED (2011) Targeted analysis of nucleotide and copy number variation by exon capture in allotetraploid wheat genome. Genome Biol 12:R88

Samorodnitsky E, Jewell BM, Hagopian R, Miya J, Wing MR, Lyon E, Damodaran S, Bhatt D, Reeser JW, Datta J, Roychowdhury S (2015) Evaluation of hybridization capture versus amplicon-based methods for whole-exome sequencing. Hum Mutat 36:903–914

Stanton-Geddes J, Paape T, Epstein B, Briskine R, Yoder J, Mudge J, Bharti AK, Farmer AD, Zhou P, Denny R, May GD, Erlandson S, Yakub M, Sugawara M, Sadowsky MJ, Young ND, Tiffin P (2013) Candidate genes and genetic architecture of symbiotic and agronomic traits revealed by whole-genome, sequence-based association genetics in Medicago truncatula. PLoS One 8:e65688

Sun L, Gill US, Nandety RS, Kwon S, Mehta P, Dickstein R, Udvardi MK, Mysore KS, Wen J (2019) Genome-wide analysis of flanking sequences reveals that Tnt1 insertion is positively correlated with gene methylation in Medicago truncatula. Plant J 98:1106–1119

Tadege M, Ratet P, Mysore KS (2005) Insertional mutagenesis: a Swiss Army knife for functional genomics of Medicago truncatula. Trends Plant Sci 10:229–235

Tadege M, Wen J, He J, Tu H, Kwak Y, Eschstruth A, Cayrel A, Endre G, Zhao PX, Chabaud M, Ratet P, Mysore KS (2008) Large-scale insertional mutagenesis using the Tnt1 retrotransposon in the model legume Medicago truncatula. Plant J 54:335–347

Tadege M, Lin H, Bedair M, Berbel A, Wen J, Rojas CM, Niu L, Tang Y, Sumner L, Ratet P, McHale NA, Madueno F, Mysore KS (2011) STENOFOLIA regulates blade outgrowth and leaf vascular patterning in Medicago truncatula and Nicotiana sylvestris. Plant Cell 23:2125–2142

Tang H, Krishnakumar V, Bidwell S, Rosen B, Chan A, Zhou S, Gentzbittel L, Childs KL, Yandell M, Gundlach H, Mayer KFX, Schwartz DC, Town CD (2014) An improved genome release (version Mt4.0) for the model legume Medicago truncatula. BMC Genoms 15:312

Tissier AF, Marillonnet S, Klimyuk V, Patel K, Torres MA, Murphy G, Jones JD (1999) Multiple independent defective suppressor-mutator transposon insertions in *Arabidopsis*: a tool for functional genomics. Plant Cell 11:1841–1852

Urbanski DF, Malolepszy A, Stougaard J, Andersen SU (2012) Genome-wide LORE1 retrotransposon mutagenesis and high-throughput insertion detection in Lotus japonicus. Plant J 69:731–741

Veerappan V, Jani M, Kadel K, Troiani T, Gale R, Mayes T, Shulaev E, Wen J, Mysore KS, Azad RK, Dickstein R (2016) Rapid identification of causative insertions underlying Medicago truncatula Tnt1 mutants defective in symbiotic nitrogen fixation from a forward genetic screen by whole genome sequencing. BMC Genom 17:141

Wang TZ, Tian QY, Wang BL, Zhao MG, Zhang WH (2014) Genome variations account for different response to three mineral elements between Medicago truncatula ecotypes Jemalong A17 and R108. BMC Plant Biol 14:122

Weigel D, Ahn JH, Blazquez MA, Borevitz JO, Christensen SK, Fankhauser C, Ferrandiz C, Kardailsky I, Malancharuvil EJ, Neff MM, Nguyen JT, Sato S, Wang ZY, Xia Y, Dixon RA, Harrison MJ, Lamb CJ, Yanofsky MF, Chory J (2000) Activation tagging in Arabidopsis. Plant Physiol 122:1003–1013

Yarce JC, Lee HK, Tadege M, Ratet P, Mysore KS (2013) Forward genetics screening of Medicago truncatula Tnt1 insertion lines. Methods Mol Biol 1069:93–100

Young ND, Bharti AK (2012) Genome-enabled insights into legume biology. Annu Rev Plant Biol 63:283–305

Young ND, Udvardi M (2009) Translating Medicago truncatula genomics to crop legumes. Curr Opin Plant Biol 12:193–201

Young ND, Debelle F, Oldroyd GE, Geurts R, Cannon SB, Udvardi MK, Benedito VA, Mayer KF, Gouzy J, Schoof H, Van de Peer Y, Proost S, Cook DR, Meyers BC, Spannagl M, Cheung F, De Mita S, Krishnakumar V, Gundlach H, Zhou S, Mudge J, Bharti AK, Murray JD, Naoumkina MA, Rosen B, Silverstein KA, Tang H, Rombauts S, Zhao PX, Zhou P, Barbe V, Bardou P, Bechner M, Bellec A, Berger A, Berges H, Bidwell S, Bisseling T, Choisne N, Couloux A, Denny R, Deshpande S, Dai X, Doyle JJ, Dudez AM, Farmer AD, Fouteau S, Franken C, Gibelin C, Gish J, Goldstein S, Gonzalez AJ, Green PJ, Hallab A, Hartog M, Hua A, Humphray SJ, Jeong DH, Jing Y, Jocker A, Kenton SM, Kim DJ, Klee K, Lai H, Lang C, Lin S, Macmil SL, Magdelenat G, Matthews L, McCorrison J, Monaghan EL, Mun JH, Najar FZ, Nicholson C, Noirot C, O'Bleness M, Paule CR, Poulain J, Prion F, Qin B, Qu C, Retzel EF, Riddle C, Sallet E, Samain S, Samson N, Sanders I, Saurat O, Scarpelli C, Schiex T, Segurens B, Severin AJ, Sherrier DJ, Shi R, Sims S, Singer SR, Sinharoy S, Sterck L, Viollet A, Wang BB, Wang K, Wang M, Wang X, Warfsmann J, Weissenbach J, White DD, White JD, Wiley GB, Wincker P, Xing Y, Yang L, Yao Z, Ying F, Zhai J, Zhou L, Zuber A, Denarie J, Dixon RA, May GD, Schwartz DC, Rogers J, Quetier F, Town CD, Roe BA (2011) The Medicago genome provides insight into the evolution of rhizobial symbioses. Nature 480:520–524

Zhao Y, Liu R, Xu Y, Wang M, Zhang J, Bai M, Han C, Xiang F, Wang ZY, Mysore KS, Wen J, Zhou C (2019) AGLF provides C-function in floral organ identity through transcriptional regulation of AGAMOUS in Medicago truncatula. Proc Natl Acad Sci USA 116:5176–5181

Zhou L, Holliday JA (2012) Targeted enrichment of the black cottonwood (Populus trichocarpa) gene space using sequence capture. BMC Genom 13:703

Zhou C, Han L, Li G, Chai M, Fu C, Cheng X, Wen J, Tang Y, Wang ZY (2014) STM/BP-like KNOXI is uncoupled from ARP in the regulation of compound leaf development in Medicago truncatula. Plant Cell 26:1464–1479

Zhou P, Silverstein KA, Ramaraj T, Guhlin J, Denny R, Liu J, Farmer AD, Steele KP, Stupar RM, Miller JR, Tiffin P, Mudge J, Young ND (2017) Exploring structural variation and gene family architecture with De Novo assemblies of 15 Medicago genomes. BMC Genom 18:261

Zhu F, Deng J, Chen H, Liu P, Zheng L, Ye Q, Li R, Brault M, Wen J, Frugier F, Dong J, Wang T (2020) A CEP peptide receptor-like kinase regulates auxin biosynthesis and ethylene signaling to coordinate root growth and symbiotic nodulation in Medicago truncatula. Plant Cell 32:2855–2877

Fast Neutron Bombardment (FNB)-Induced Mutant Resources in *Medicago truncatula*

4

Yuhui Chen and Rujin Chen

Abstract

Genetic mutant resources are very useful and necessary tools to elucidate functions of the very large number of genes present in plant genomes. Here, we report the large mutant resources that we have developed in the model legume, *Medicago truncatula* cv. Jemalong A17, using a physical mutagen, fast neutron. We first discuss the scale of the mutant resources with an aim of saturation mutagenesis. We then discuss methods that have been used for high-throughput molecular cloning of underlying genes in fast neutron bombardment (FNB)-induced mutants in *M. truncatula*. And lastly, we briefly summarize some FNB mutants that have been studied to facilitate our understanding of the molecular mechanisms that underlie a broad range of developmental processes such as compound leaf and seed development, root gravitropism, and symbiotic nitrogen fixation (SNF) in legumes.

Y. Chen · R. Chen (✉)
School of Life Sciences, Lanzhou University, Lanzhou 30000, China
e-mail: rjchen@lzu.edu.cn

R. Chen
MOE Key Laboratory of Cell Activities and Stress Adaptations, Lanzhou University, Lanzhou 730000, China

4.1 Introduction

Medicago truncatula has been selected as one of the model legume species for gene functional studies, mainly because 1. It has a relatively small genome; 2. It is a self-pollinating diploid species; 3. It can be stably transformed; 4. Genome sequences for the reference plant (*M. truncatula* cv. Jemalong A17) and a large number of ecotypes have been completed; and 5. Comprehensive mutant collections and genomic resources (i.e., in silico gene expression) have been developed. Using chemical (i.e., ethyl methane sulfonate or EMS), physical (i.e., ionization radiation such as fast neutron bombardment or FNB), and biological (i.e., transfer DNA or T-DNA and tobacco *Tnt1* retrotransposon) mutagens, several large mutant collections have been developed for gene functional studies in *M. truncatula*. Mutants generated by different mutagens carry different types of mutations. EMS mutagenesis generates a large number of single nucleotide changes in the genome. FNB mutagenesis typically induces deletions of different sizes ranging from a single nucleotide to thousands of base pairs, among other types of mutations. *Tnt1* mutagenesis gives rise to multiple copies of insertions of the retrotransposon in the genome. Because different types of mutations have different mutational effects on the associated genes, phenotypes of mutants derived from different mutagenesis approaches are very useful in understanding gene functions.

S. Sinharoy et al. (eds.), *The Medicago truncatula Genome*, Compendium of Plant Genomes,
https://doi.org/10.1007/978-3-030-90757-0_4

Interestingly, genome sequence analyses reveal that many genes are members of tandemly repeated gene families. For example, in *Arabidopsis thaliana* (Arabidopsis), about 19% of gene families have members in tandem repeats. In rice (*Oryza sativa*), the number goes to as high as 29% (Goff et al. 2002). In *M. truncatula*, about 33% of gene families have members in tandem repeats (Young et al. 2011). Large chromosomal deletions removing tandemly repeated genes can be induced by physical mutagens such as ionizing radiations. High-energy ionizing radiations such as fast neutrons have been shown to induce a broad range of deletions and other types of chromosomal alterations in several organisms including Arabidopsis (Alonso et al. 2003; Silverstone et al. 1998), rice (*O. sativa*) (Li et al. 2002), tomato (*Solanum lycopersicum*), soybean (*Glycine soja*; *G. max*) (Bolon et al. 2011; Men et al. 2002), barley (*Hordeum vulgare*), and *Lotus japonicus* (Hoffmann et al. 2007). In plant genomes, the number of genes that encode small peptides or regulatory small RNAs is also very high. EMS and FNB mutagenesis are effective in generating mutations in genes with short sequences. On the other hand, characterization of FNB mutants in different organisms has contributed to our understanding of the nature and range of FNB-induced mutations in plant genomes for the purpose of molecular breeding of new cultivars.

4.2 Development of FNB-Induced Mutant Resources in *Medicago Truncatula*

To facilitate functional studies of genes, a comprehensive collection of mutants in the range of over 150,000 M_1 lines has been generated using FNB mutagenesis in *M. truncatula* cv. Jemalong A17 by our group at the Noble Research Institute and Giles Oydrod's group at the John Innes Center (Wang et al. 2006; Rogers et al. 2009).

For FNB mutagenesis, wild-type seeds (M_1 seeds) are treated with fast neutron irradiation. The M_1 plants are chimeric and unstable in nature due to some mutations in non-germline cells in M_1 seeds. Dominant mutations would result in visible phenotypes in M_1 plants. Only mutations in germline cells would be transmitted to the M_2 generation. M_2 plants segregate for recessive mutations and are subject to phenotypic screens.

For a given plant species under studies, a small-scale mutagenesis experiment is usually carried out as the first step to determine the optimal dosage level for efficient mutagenesis. This is mainly because high-dosage levels would usually cause a high degree of sterility and lethality, while low-dosage levels result in low mutation frequencies, and different plant species have different radiation tolerance. It has been reported that FNB at 60 Gy worked well for Arabidopsis (Kodym and Afza 2003). For rice, efficient mutagenesis was carried out with 18–20 Gy FNB (Li et al. 2001). In our studies, we tested several FNB dosage levels for mutagenesis of *M. truncatula*. The segregation of albino phenotypes in the M_2 progeny of mutagenized seeds has also been used as an indicator of mutagenic rate. For a fast neutron-mutagenized population of Arabidopsis an albino frequency of about 2% has been equated to around ten induced mutations per line (Koornneeffa et al. 1982). We observed that FNB at 40 Gy results in 1.85% albino M_2 mutants in *M. truncatula* (Table 4.1) (Wang et al. 2006). Studies have shown that the mutation frequency per locus caused by FNB mutagenesis is independent of the genome size (Koornneeffa et al. 1982), and mutant populations required to reach the same mutation coverage are comparable among different plant species with different genome sizes. It has been estimated that a population of 84,825 and 130, 397 would have 95 and 99% of mutation coverages, respectively, to recover mutants of any target genes (Koornneeffa et al. 1982). Similarly, we estimated that a population of 150,000 M_1 lines would be sufficient to identify mutants of nearly any target genes in *M. truncatula* (Wang et al. 2006; Chen and Chen 2018).

Table 4.1 Categories and frequencies of mutants observed in an M2 population of *M. truncatula* cv. Jemalong A17 mutagenized by 40 Gy fast neutron radiation (modified from Wang et al. 2006)

Phenotypes	Number (frequency) of mutants in M2 population
Albino mutants	40 (1.85%)
Dwarf mutants	15 (0.7%)
Yellow-leaf mutants	35 (1.62%)
Mutants with increased anthocyanin spots	4 (0.19%)
Leaf-form mutants	10 (0.46%)
Bonsai mutants	11 (0.51%)
Short-root mutants	2 (0.93%)
Total	115 (5.3%)

4.3 *M. truncatula* FNB Mutants with Distinct Phenotypes

An initial survey of visual mutant phenotypes was carried out in a population of 2500 independent *M. truncatula* FNB M_2 lines, and about 5.3% of the M_2 lines exhibited visual aboveground phenotypes (Table 4.1) (Wang et al. 2006). In the past ten years or so, we organized multiple communal mutant screening workshops and screened over 10,000 pools of *M. truncatula* FNB M_2 lines for phenotypes in the following categories: nodulation, root, leaf, stem, and others such as trichomes, early and late flowering. Figure 4.1 shows some representative *M. truncatula* FNB mutants with distinct phenotypes, including increased anthocyanin accumulation in leaves (Fig. 4.1a), palmate-like pentafoliata leaves (Fig. 4.1b), fused compound leaves (Fig. 4.1c), and serrated leaves (Fig. 4.1d), compared with wild-type leaves (Fig. 4.1e). Figure 4.1 also shows mutants with striking root and nodule development phenotypes such as short and bushy roots (Fig. 4.1f), defective symbiotic nitrogen fixation (fix^-; Fig. 4.1g), an

Fig. 4.1. Representative *M. truncatula* FNB mutants with visual phenotypes. **a** Mutant with increased anthocyanin accumulation (red spots) in leaflets; **b** *Palmate-like pentafoliata1* (*palm1*) mutant; **c** *Fused compound leaf1* (*fcl1*) mutant; **d** Mutant with serrated leaflets; **e** Wild-type plant (*M. truncatula* cv Jemalong A17); **f** Mutant with short and bushy roots (left, wild type; middle and right, mutants); **g** Mutant with defective symbiotic nitrogen fixation; **h** Mutant with an increased number of nodules; **i** Mutant with no nodules; **j** Wild-type plant (A17) with elongated and pink functional nodules

increased number of nodules (nod^{++}; Fig. 4.1h), and absence of nodules (nod^{-}; Fig. 4.1i), compared with elongated, pink nodules developed on wild-type roots (Fig. 4.1j).

4.4 Genomic Analysis of *M. truncatula* FNB Mutants Using Genome Array-Based Comparative Genomic Hybridization (aCGH)

Traditionally, molecular cloning of underlying genes from FNB mutants relies on a time-consuming map-based approach. With the availability of annotated whole genome sequences of *M. truncatula*, several complimentary approaches have been employed to facilitate high-throughput cloning of underlying genes from FNB mutants, including transcript-based, whole genome array-based comparative genomic hybridization (aCGH) and whole genome resequencing methods (Chen and Chen 2018; Mitra et al. 2004; Ge et al. 2016; Kim et al. 2015; Chen et al. 2017; Sun et al. 2018; Du et al. 2021).

To develop a whole genome array-based comparative genomic hybridization (aCGH) platform, we first designed an Agilent 1 × 1 M (million) *M. truncatula* whole genome array with a total of 971,041 unique probes targeting all annotated genes, with a spacing of approximately 150 bp in exonic regions and 261 bp in intronic and 5′- and 3′-untranslated regions in the *M. truncatula* genome. The design of the current genome array is to maximize the coverage for detection of copy number variations (CNVs) in gene sequences in the genome (Chen et al. 2017). Once detected by the aCGH method, CNVs such as deletions are confirmed by PCR and gel electrophoresis, and deletion junctions are identified by sequencing of deletion borders.

We then tested the effectiveness of the array-based CGH method in detecting CNVs between wild type and the FNB mutant, *dmi1-4* (*does not make infections1-4*; Jemalong A17 background) (Catoira et al. 2000). The *dmi1-4* mutant is defective in symbiotic interactions with the soil bacterium, *Sinorhizobium meliloti*, due to a large deletion that removes the promoter region and the 5′ portion of the *DMI1* gene (Ane et al. 2002, 2004). The coding region of *DMI1* is located between 363,876 and 370,020 bases on chromosome 2 (Ane et al. 2002, 2004). Our aCGH data revealed segments corresponding to regions between 353,865 and 366,551 bases on Chromosome 2 as deletions. A closer inspection revealed that a total of 42 probes located between 353,865 and 366,551 bases on chromosome 2 have the mean $\log_2$ ratios of −2.5, indicating that a deletion event likely occurred in this region in the *dmi1-4* mutant (Fig. 4.2a, b). Segments in other regions on chromosome 2 and other chromosomes were identified as normal by aCGH data. PCR amplification and subsequent Sanger-sequencing of deletion borders indicated a deletion of 18,550 bp between 347,884 and 366,434 bases on chromosome 2 in the *dmi1-4* mutant (Fig. 4.2c). Included within the deleted region is Medtr2g005870.1 (*DMI1*). Taken together, the results demonstrate that our aCGH method is effective in detecting chromosomal deletions in FNB mutants, and when combined with PCR amplification and sequencing, it can pinpoint the precise borders of deletions in FNB mutants.

The aCGH method coupled with genetic linkage analyses of F_2 backcrossed populations can be used to rapidly identify candidate genes in FNB mutants. This analysis platform has been used to identify and characterize a large number of genes from *M. truncatula* FNB mutants (Table 4.2) (Rogers et al. 2009; Mitra et al. 2004; Ge et al. 2016; Kim et al. 2015; Chen et al. 2017; Du et al. 2021; Vernié et al. 2008; Chen et al. 2010; Wang et al. 2010; Velde et al. 2010; Peng et al. 2011; Murray et al. 2011; Horváth et al. 2015; Ge and Chen 2016; Liu et al. 2019; Ng et al. 2020). The current high-density whole genome array, however, covers mostly exonic, intronic, and 5′- and 3′-UTR of genes in the genome with spacing. This limits the detection of

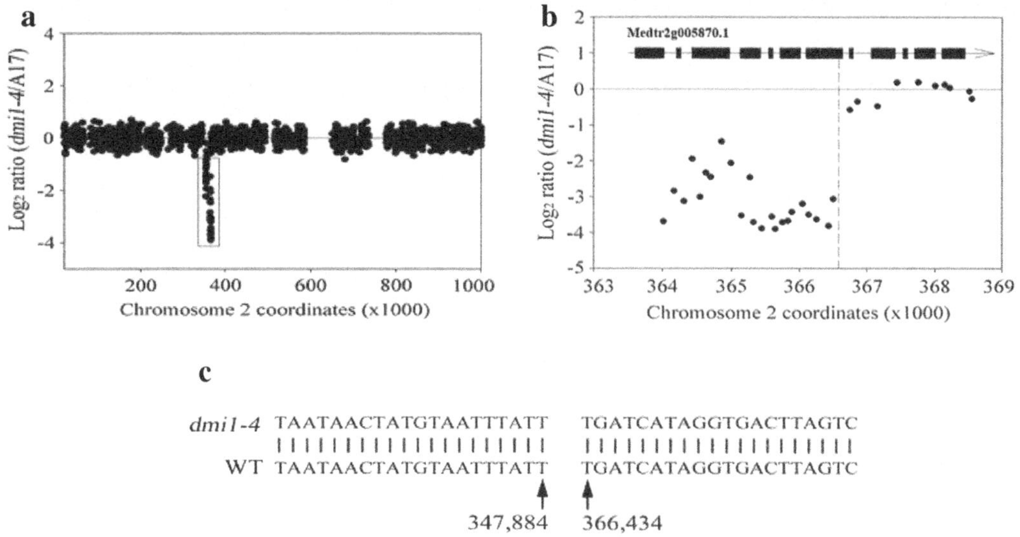

Fig. 4.2 aCGH-based detection of 18 kb deletion in the *M. truncatula dmi1-4* mutant. **a** Normalized log2 ratios of *dmi1-4* and WT (Jemalong A17) signals for probes spanning the first 1 Mb on chromosome 2. Note that, besides the deleted probes that encompass the *DMI1* locus (rectangle), no other probes in the chromosomal region were detected as being deleted. **b** Close-up view of **a**, together with the exon and intron structure of the *DMI1* gene (Medtr2g005870.1), highlighting the deletion borders. A vertical dashed line depicts the actual deletion breakpoint in the seventh exon of the *DMI1* gene, which coincides with the deletion border detected by aCGH approach. **c** Deletion border sequences from the *dmi1-4* mutant compared with the corresponding sequence of the WT. Arrows indicate the deletion borders, and numbers underneath denote the coordinates of nucleotide sequences on chromosome 2

CNVs to regions only covered by the array probes. In addition, the current genome array has a detection resolution limited to relatively larger Indels and is not designed to detect single nucleotide polymorphisms (SNPs). With the latest annotation of the *M. truncatula* genome sequences (Mt5.0; Pecrix et al. 2018), a combination of aCGH and whole genome resequencing approaches would allow us to detect SNPs, Indels, or CNVs with an improved detection sensitivity in *M. truncatula* FNB mutants. Furthermore, combining with bulked segregant analysis (BSA) of F_2 populations would allow us to identify SNPs, Indel, or CNVs that are tightly linked to mutant phenotypes.

In summary, we have developed a large population of *M. truncatula* FNB mutant lines in collaboration with the John Innes Center. In the past ten years, we have organized multiple communal mutant screening workshops at the Noble Research Institute and uncovered a large number of mutants with defects in the development of leaves, roots, nodules, and seeds, among others. We discussed several complementary methods for high-throughput molecular cloning of underlying genes in *M. truncatula* FNB mutants. And, we discussed the effectiveness of FNB mutagenesis in generating mutations in homologous genes in tandem repeats and genes encoding small peptides such as NCRs.

Table 4.2 Analyses of selected *Medicago truncatula* mutants generated by fast neutron bombardment (FNB)

Mutant	Cloning methods	Gene location (Mt4.0)	Gene (Mt4.0)	Citations
DMI3	Transcript-based	Chr8: 16,814,558.0.16819041	Medtr8g043970	Mitra et al. (2004)
EFD	De-tilling	Chr4: 1725344.0.1727163	Medtr4g008860	Vernie et al. (2008)
MCA8	De-tilling	Chr7: 47229504.0.47236330	Medtr7g100110	Rogers et al. (2009)
PALM1	Mapping	Chr5: 4830733.0.4831488	Medtr5g014400	Chen et al. (2010)
DNF1	Transcript-based	Chr3: 8796837.0.8801535	Medtr3g027890	Wang et al. (2010), Van de Velde et al. (2010)
FCL1	Mapping	Chr6: 26316380.0.26320112	Medtr6g071190	Peng et al. (2011)
Vapyrin	Transcript-based	Chr6: 9767853.0.9770137	Medtr6g027840	Murray et al. (2011)
DNF4	Genome resequencing; aCGH	Chr4: 12109474.0.12429570	Medtr4g035705	Kim et al. (2015)
DNF7	Mapping; transcript-based; aCGH	Chr7: 14135847.0.14324851	Medtr7g029760	Horváth et al. (2015)
BS1	Mapping; transcript-based	Chr1: 46507198.0.46513810	Medtr1g102900	Ge et al. (2016)
NGR	aCGH	Chr8: 7629444.0.7632442	Medtr8g021237	Ge and Chen (2016)
SUNN	aCGH	Chr4: 26730299.0.26756862	Medtr4g070970	Chen et al. (2017)
NIN	aCGH	Chr5: 43407481.0.43411238	Medtr5g099060	Liu et al. (2019)
PIN2	De-tilling	Chr4: 52715977.0.52718935	Medtr4g127100	Ng et al. (2020)
SymCRK	Genome resequencing	Chr3: 36065815.0.36068995	Medtr3g079850	Du et al. (2021)
MtEGY1	Genome resequencing	Chr7: 33548272.0.33553669	Medtr7g086430	Du et al. (2021)
DNF2	aCGH	Chr4: 33517981.0.33569847	Medtr4g085800	This study
DMI1	aCGH	Chr2: 347884.0.366434	Medtr2g005870	This study
NSP2	aCGH	Chr3: 32725078.0.32727063	Medtr3g072710	This study
EIN2	aCGH	Scaffold0041: 13660.0.21631	Medtr0041s0030	This study

Acknowledgements This work was supported by the Fundamental Research Fund for the Central Universities of China (lzujbky-2020-sp04) and Lanzhou University. Funding for generating *M. truncatula* FNB mutant resources was provided by the National Science Foundation (IOS-1127155) and Noble Research Institute (to RC).

References

Alonso JM, Stepanova AN, Solano R, Wisman E, Ferrari S, Ausubel FM, Ecker JR (2003) Five components of the ethylene-response pathway identified in a screen for weak ethylene-insensitive mutants in Arabidopsis. Proc Natl Acad Sci USA 100 (5):2992–2997

Ane JM, Levy J, Thoquet P, Kulikova O, de Billy F, Penmetsa V, Kim DJ, Debelle F, Rosenberg C, Cook DR, Bisseling T, Huguet T, Denarie J (2002) Genetic and cytogenetic mapping of *DMI1*, *DMI2*, and *DMI3* genes of *Medicago truncatula* involved in Nod factor transduction, nodulation, and mycorrhization. Mol Plant Microbe Interact 15:1108–1118

Ane J-M, Kiss GB, Riely BK, Penmetsa RV, Oldroyd GED, Ayax C, Levy J, Debelle F, Baek J-M, Kalo P, Rosenberg C, Roe BA, Long SR, Denarie J, Cook DR (2004) *Medicago truncatula DMI1* required for bacterial and fungal symbioses in legumes. Science 303:1364–1367

Bolon YT, Haun WJ, Xu WW, Grant D, Stacey MG, Nelson RT, Gerhardt DJ, Jeddeloh JA, Stacey G, Muehlbauer GJ, Orf JH, Naeve SL, Stupar RM, Vance CP (2011) Phenotypic and genomic analyses of a fast neutron mutant population resource in soybean. Plant Physiol 156(1):240–253

Catoira R, Galera C, de Billy F, Penmetsa RV, Journet E-P, Maillet F, Rosenberg C, Cook D, Gough C, Denarie J (2000) Four genes of *Medicago truncatula* controlling components of a nod factor transduction pathway. Plant Cell 12:1647–1666

Chen Y, Chen R (2018) Physical mutagenesis in M. truncatula using Fast Neutron Bombardment (FNB) for symbiosis and developmental biology studies. Methods Mol Biol 1822:61–69

Chen Y, Wang X, Lu S, Li S, Chen R (2017) An array-based comparative genomic hybridization (CGH) platform for efficient detection of copy number variations (CNVs) in fast neutron induced *Medicago truncatula* mutants. J Vis Exp (JOVE), 129

Chen J, Yu J, Ge L, Wang H, Berbel A, Liu Y, Chen Y, Li G, Tadege M, Wen J, Cosson V, Mysore KS, Ratet P, Madueño F, Bai G, Chen R (2010) Control of dissected leaf morphology by a Cys(2)His(2) zinc finger transcription factor in the model legume *Medicago truncatula*. Proc Natl Acad Sci USA 107 (23):10754–10759

Du H, Jiao Z, Liu J, Huang W, Ge L (2021) Rapid identification of mutations caused by fast neutron bombardment in Medicago truncatula. Plant Methods 17:62

Ge L, Chen R (2016) Negative gravitropism in plant roots. Nat Plants 2:16155

Ge L, Yu J, Wang H, Luth D, Bai G, Wang K, Chen R (2016) Increasing seed size and quality by manipulating BIG SEEDS1 in legume species. Proc Natl Acad Sci USA 113(44):12414–12419

Goff SA, Ricke D, Lan TH, Presting G, Wang RL, Dunn M, Glazebrook J, Sessions A, Oeller P, Varma H, Hadley D, Hutchinson D, Martin C, Katagiri F, Lange BM, Moughamer T, Xia Y, Budworth P, Zhong JP, Miguel T, Paszkowski U, Zhang SP, Colbert M, Sun WL, Chen LL, Cooper B, Park S, Wood TC, Mao L, Quail P, Wing R, Dean R, Yu YS, Zharkikh A, Shen R, Sahasrabudhe S, Thomas A, Cannings R, Gutin A, Pruss D, Reid J, Tavtigian S, Mitchell J, Eldredge G, Scholl T, Miller RM, Bhatnagar S, Adey N, Rubano T, Tusneem N, Robinson R, Feldhaus J, Macalma T, Oliphant A, Briggs S (2002) A draft sequence of the rice genome (Oryza sativa L. ssp. japonica). Science 296(5565):92–100

Hoffmann D, Jiang Q, Men A, Kinkema M, Gresshoff PM (2007) Nodulation deficiency caused by fast neutron mutagenesis of the model legume *Lotus japonicus*. J Plant Physiol 164(4):460–469

Horváth B, Domonkos Á, Kereszt A, Szűcs A, Ábrahám E, Ayaydin F, Bóka K, Chen Y, Chen R, Murray JD, Udvardi MK, Kondorosi É, Kaló P (2015) Loss of the nodule-specific cysteine rich peptide, NCR169, abolishes symbiotic nitrogen fixation in the Medicago truncatula dnf7 mutant. Proc Natl Acad Sci USA 112 (49):15232–15237

Kim M, Chen Y, Xi J, Waters C, Chen R, Wang D (2015) An antimicrobial peptide essential for bacterial survival in the nitrogen-fixing symbiosis. Proc Natl Acad Sci USA 112(49):15238–15243

Kodym A, Afza R (2003) Physical and chemical mutagenesis. Methods Mol Biol 236:189–204

Koornneeffa M, Dellaert LWM, van der Veena JH (1982) EMS- and relation-induced mutation frequencies at individual loci in Arabidopsis thaliana (L.) Heynh. Mutation Res/Fund Molecul Mech Mutagenesis 93(1)

Li X, Lassner M, Zhang Y (2002) Deleteagene: a fast neutron deletion mutagenesis-based gene knockout system for plants. Comp Funct Genom 3(2):158–160

Li X, Song YJ, Century K, Straight S, Ronald P, Dong XN, Lassner M, Zhang YL (2001) A fast neutron deletion mutagenesis-based reverse genetics system for plants. Plant J: Cell Molecul Biol 27 (3):235–242

Liu J, Rutten L, Limpens E, van der Molen T, van Velzen R, Chen R, Chen Y, Geurts R, Kohlen W, Kulikova O, Bisseling T (2019) A remote *cis*-regulatory region is required for *NIN* expression in the pericycle to initiate nodule primordium formation in *Medicago truncatula*. Plant Cell 31(1):68–83

Men AE, Laniya TS, Searle IR, Iturbe-Ormaetxe I, Gresshoff I, Jiang QY, Carrol BJ, Gresshoff PM (2002) Fast neutron mutagenesis of Soybean (Glycine soja L.) Produces a supernodulating mutant containing a large deletion in Linkage Group H. Genome Lett 1 (3):147–155

Mitra R, Gleason C, Edwards A, Hadfield J, Downie A, Oldroyd G, Long S (2004) A Ca^{2+}/calmodulin-dependent protein kinase required for symbiotic nodule development: gene identification by transcript-based cloning. Proc Natl Acad Sci USA 101(13):4701–4705

Murray JD, Muni RR, Torres-Jerez I, Tang Y, Allen S, Andriankaja M, Li G, Laxmi A, Cheng X, Wen J, Vaughan D, Schultze M, Sun J, Charpentier M, Oldroyd G, Tadege M, Ratet P, Mysore KS, Chen R, Udvardi MK (2011) *Vapyrin*, a gene essential for intracellular progression of arbuscular mycorrhizal symbiosis, is also essential for infection by rhizobia in the nodule symbiosis of *Medicago truncatula*. Plant J 65(2):244–252

Ng J, Welvaert A, Wen J, Chen R, Mathesius U (2020) The *Medicago truncatula* PIN2 auxin transporter mediates basipetal auxin transport but is not necessary for nodulation. J Exp Bot 71(4):1562–1573

Pecrix Y, Staton SE, Sallet E, Lelandais-Brere C, Moreau S, Carrere S, Blein T, Jardinaud MF, Latrasse D, Zouine M, Zahm M, Kreplak J, Mayjonade B, Satge C, Perez M, Cauet S, Marande W, Chantry-Darmon C, Lopez-Roques C, Bouchez O, Berard A, Debelle F, Munos S, Bendahmane A, Berges H, Niebel A, Buitink J, Frugier F, Benhamed M, Crespi M, Gouzy J, Gamas P (2018) Whole-genome landscape of Medicago truncatula symbiotic genes. Nat Plants 4(12):1017–1025

Peng J, Yu J, Wang H, Guo Y, Li G, Bai G, Chen R (2011) Regulation of compound leaf development in *Medicago truncatula* by fused compound leaf1, a class M *KNOX* gene. Plant Cell 23(11):3929–3943

Rogers C, Wen J, Chen R, Oldroyd G (2009) Deletion-based reverse genetics in Medicago truncatula. Plant Physiol 151(3):1077–1086

Silverstone AL, Ciampaglio CN, Sun T (1998) The Arabidopsis RGA gene encodes a transcriptional regulator repressing the gibberellin signal transduction pathway. Plant Cell 10(2):155–169

Sun L, Ge Y, Bancroft AC, Cheng X, Wen J (2018) FNBtools: a software to identify homozygous lesions in deletion mutant populations. Front Plant Sci 9:976

Van de Velde W, Zehirov G, Szatmari A, Debreczeny M, Ishihara H, Kevei Z, Farkas A, Mikulass K, Nagy A, Tiricz H (2010) Plant peptides govern terminal differentiation of bacteria in symbiosis. Science 327:1122–1126

Vernié T, Moreau S, De Billy F, Plet J, Combier J-P, Rogers C, Oldroyd G, Frugier F, Niebel A, Gamas P (2008) EFD is an ERF transcription factor involved in the control of nodule number and differentiation in *Medicago truncatula*. Plant Cell 20:2696–2713

Wang H, Li G, Chen R (2006) Fast neutron bombardment (FNB) induced deletion mutagenesis for forward and reverse genetic studies in plants. In: da Silva JT (ed) Floriculture, ornamental and plant biotechnology: advances and topical issues, 1st ed. Global Science Books, Isleworth, UK, pp 629–639

Wang D, Griffitts J, Starker C, Fedorova E, Limpens E, Ivanov S, Bisseling T, Long S (2010) A nodule-specific protein secretory pathway required for nitrogen-fixing symbiosis. Science 327:1126–1129

Young ND, Debelle F, Oldroyd GD, Geurts R, Cannon SB, Udvardi MK, Benedito VA, Mayer KFX, Gouzy J, Schoof H, Van de Peer Y, Proost S, Cook DR, Meyers BC, Spannagl M, Cheung F, De Mita S, Krishnakumar V, Gundlach H, Zhou SG, Mudge J, Bharti AK, Murray JD, Naoumkina MA, Rosen B, Silverstein KAT, Tang HB, Rombauts S, Zhao PX, Zhou P, Barbe V, Bardou P, Bechner M, Bellec A, Berger A, Berges H, Bidwell S, Bisseling T, Choisne N, Couloux A, Denny R., Deshpande S, Dai XB, Doyle JJ, Dudez AM, Farmer AD, Fouteau S, Franken C, Gibelin C, Gish J, Goldstein S, Gonzalez AJ, Green PJ, Hallab A, Hartog M, Hua A, Humphray SJ, Jeong DH, Jing Y, Jocker A, Kenton SM, Kim DJ, Klee K, Lai HS, Lang CT, Lin SP, Macmil SL, Magdelenat G, Matthews L, McCorrison J, Monaghan EL, Mun JH., Najar FZ, Nicholson C, Noirot C O'Bleness M, Paule CR, Poulain J, Prion F, Qin BF, Qu CM, Retzel EF, Riddle C, Sallet E, Samain S, Samson N, Sanders I, Saurat O, Scarpelli C, Schiex T, Segurens B, Severin AJ, Sherrier DJ, Shi RH, Sims S, Singer SR, Sinharoy S, Sterck L, Viollet A, Wang BB, Wang KQ, Wang MY, Wang XH, Warfsmann J, Weissenbach J, White DD, White JD, Wiley GB, Wincker P, Xing YB, Yang LM, Yao ZY, Ying F, Zhai JX, Zhou LP, Zuber A, Denarie J, Dixon RA, May GD, Schwartz DC, Rogers J, Quetier F, Town CD, Roe BA (2011) The Medicago genome provides insight into the evolution of rhizobial symbioses. Nature 480(7378):520–524

Medicago truncatula as a Model to Decipher Powdery Mildew Resistance in Legumes

5

Megha Gupta, Arunima Gupta, and Divya Chandran

Abstract

Powdery mildew, caused by fungal pathogens of the order Erysiphales, is a serious disease that destroys grain and forage legumes worldwide. Disease control strategies based on the application of chemical fungicides, though effective, are often uneconomical and hazardous to the environment. Moreover, the risk of developing fungicide-resistant fungal strains is also high. The use of genetic means for developing resistant legume cultivars is considered to be a superior strategy for powdery mildew disease management due to its eco-friendly and sustainable nature. However, genetic characterization of crop legumes has been challenging because of their large and complex genomes. In the early 2000s, *Medicago truncatula* (barrel medic) emerged as a powerful model to study legume biology because of its relatively small diploid genome, short generation time, and genetically diverse germplasm. Furthermore, the availability of *M. truncatula* genotypes spanning a broad range of powdery mildew resistance phenotypes has not only made it an excellent system for deciphering resistance mechanisms but also a reservoir for the identification of novel resistance loci. In this chapter, we summarize our current knowledge of the genetic and molecular basis of powdery mildew resistance in *M. truncatula* and review the genomic and genetic tools that are available for the identification of new resistance genes that could potentially be used to develop legume cultivars with enhanced powdery mildew resistance. We anticipate that the high conservation of synteny between the genomes of *M. truncatula* and other legumes will facilitate the cross-species translation of genomic information obtained from this model species.

M. Gupta, A. Gupta—These authors contributed equally.

M. Gupta · A. Gupta · D. Chandran (✉)
Laboratory of Plant–Microbe Interactions, Regional Centre for Biotechnology, NCR Biotech Science Cluster, Faridabad, Haryana 121001, India
e-mail: divya.chandran@rcb.res.in

M. Gupta
e-mail: megha.gupta@rcb.res.in

A. Gupta
e-mail: arunima.gupta@rcb.res.in

M. Gupta
Kalinga Institute of Industrial Technology, Bhubaneswar, Orissa, India

5.1 Introduction

Grain and forage legumes contribute significantly to human and animal health as well as agricultural sustainability owing to their high nutritional value and environmental benefits (Graham and Vance 2003; Polak et al. 2015; Stagnari et al. 2017). In recent years, several countries have

S. Sinharoy et al. (eds.), *The Medicago truncatula Genome*, Compendium of Plant Genomes,
https://doi.org/10.1007/978-3-030-90757-0_5

developed food-based dietary guidelines in partnership with the Food and Agriculture Organization (FAO), which recommend an increase in the daily intake of pulses (Rawal et al. 2019). Yet global pulse production has not kept pace with the growing demand due to various socio-economic and environmental factors. Among the environmental factors, fungal diseases constitute one of the major constraints to legume cultivation (Kankanala et al. 2019), resulting in substantial yield losses per year (Mahmoud 2021). Powdery mildew, caused by filamentous fungi of the order Erysiphales and division Ascomycota, is one of the more common diseases of legume crops, including pea, mung bean, common bean, lentils, and annual medics (Khunti et al. 2005; Martins et al. 2020; Rubiales et al. 2015). It is particularly devastating on pea, causing 25–50% annual yield losses through a reduction in total biomass and pod/seed number and quality (Fondevilla and Rubiales 2012).

5.1.1 The Pathogen, Host Range, and Disease Symptoms

Powdery mildew in legumes is generally caused by various *Erysiphe* species (Fondevilla et al. 2013) although *Oidium* sp., *Microsphaera diffusa*, and *Podosphaera phaseoli* have also been reported on annual medics, soybean, and cowpea, respectively (Graves et al. 1999; Sillero et al. 2006; Wang et al. 2013). Pea powdery mildew is mainly caused by *E. pisi* and to a lesser extent by *E. trifolii* and *E. baeumleri* (Fondevilla and Rubiales 2012). Unlike the monocot-adapted powdery mildews that can infect only one host species (Wyand and Brown 2003), the legume-adapted species have a broader host range and can typically infect more than one legume crop. For example, *E. pisi* also infects faba bean and annual medics (Prats et al. 2007; Sillero et al. 2006; Yang et al. 2013), *E. trifolii* infects lentil and *Lathyrus cicero* (chickling pea) (Attanayake et al. 2009, 2010), and *E. polygonii* infects common bean and mung bean (Chankaew et al. 2013; Campa and Ferreira 2017). More recently, *E. viciae-unijugae* was reported to cause powdery mildew on *Vicia sativa* (common vetch) (Lee and Nguyen 2021).

Powdery mildews are true obligate biotrophs, requiring living host plant tissue for their survival (Panstruga 2003). They infect only the aerial parts of the plant and produce characteristic white or gray powder-like symptoms, which represent a mixture of epiphytic fungal mycelia and asexual reproductive structures (conidiophores bearing conidia). The powdery mildew disease cycle is typically completed in 5–7 days and consists of three main stages. The early infection stage includes conidia germination, host cell penetration, and primary haustorium formation; the intermediate stage includes colony expansion; and the final stage includes asexual reproduction (Smith et al. 1996). Upon landing on a host, the fungal conidium germinates within 2 h and produces a germ tube that differentiates into a multilobed appressorium. Between 6 and 12 h post-infection (hpi), the fungus penetrates the host epidermal cell wall via the appressorium and invaginates the plasma membrane to form a specialized feeding structure known as the haustorium. The primary haustorium fully develops by ~24 hpi and is the primary structure through which the pathogen secretes virulence molecules termed 'effectors' into the host cell to suppress immunity, including cell death (Hückelhoven 2005; Sharma et al. 2019). It is also the primary feeding structure through which host nutrients are diverted into the pathogen. The formation of a functional primary haustorium leads to colony expansion between 24 and 72 hpi, which includes the development of an extensive epiphytic hyphal network and secondary haustoria. Between 5 and 7 dpi, the fungal colony transitions from vegetative to reproductive growth, resulting in the production of asexual reproductive structures known as conidiophores that bear single conidia at their tips. These conidia are readily wind-disseminated to initiate new infection cycles. Although the sexual stage (cleistothecia) of this pathogen is rarely observed under laboratory/greenhouse conditions, in the field, small black cleistothecia are often found within the mycelial network on senescing host tissues, which ensures their survival during the winter months (Tiwari et al. 1999).

5.1.2 Powdery Mildew Disease Management: Chemical and Biocontrol Agents

Several fungicides, including hexaconazole, tebuconazole, propiconazole, azoxystrobin, and chemical formulations, such as wettable sulfur and zinc, are effective in reducing legume powdery mildew disease severity under field conditions and recommended for powdery mildew disease management (Khunti et al. 2005; Ransom et al. 1991; Singh et al. 2013; Singh and Shekar 2020; Warkentin et al. 1996). However, their repeated usage is harmful to the environment and costly, often being economically beneficial only under conditions of high disease pressure (Warkentin et al. 1996). Further, the ability of powdery mildew fungi to develop resistance to these fungicides is problematic (Vielba-Fernández et al. 2020).

Environmental concerns have fueled the evaluation of natural compounds as potential antifungal agents against legume powdery mildews. Natural bioactive plant products, such as azadirachtin, inuloxin C, norsecurinine and chitosan, fungal metabolites, such as cavoxin and sphaeropsidin, and extracts from cashewnut shell and brown algae were found to be effective in controlling powdery mildew disease severity on pea (Bahadur et al. 2008; Barilli et al. 2019; Patel et al. 2020; Sahni et al. 2005; Singh and Prithiviraj 1997). However, the efficacy of these treatments under field conditions is yet to be demonstrated. More recently, crop diversification via intercropping pea with barley has been proposed as an alternative strategy to reduce powdery mildew in pea (Villegas-Fernández et al. 2021).

5.1.3 Breeding for Genetic Resistance Against Powdery Mildew in Legumes

Due to the apprehensions surrounding fungicide resistance and environmental safety, the introduction of powdery mildew resistance through genetic means is considered to be a more efficient and eco-friendly strategy for disease management in legumes (Fondevilla and Rubiales 2012). So far, few powdery mildew resistance gene(s) and/or quantitative trait loci (QTL) have been identified in different legumes and even fewer have been employed in breeding programs to develop powdery mildew-resistant legume cultivars. Early genetic studies performed on resistant and susceptible pea cultivars identified recessive (*er1* and *er2*) and dominant (*Er3*) powdery mildew resistance genes (Fondevilla et al. 2007; Heringa et al. 1969; Tiwari et al. 1997). Pea cultivars carrying the *er1* gene provide complete to moderate resistance to most *E. pisi* isolates and have therefore been extensively used in breeding programs (Fondevilla and Rubiales 2012). Candidate powdery mildew resistance genes or QTLs have also been identified in common bean (Campa and Ferreira 2017; Pérez-Vega et al. 2013), soybean (Wang et al. 2013), mung bean (Chankaew et al. 2013), and chickling pea (Santos et al. 2020) that are available for use in breeding programs. However, the continuous cultivation of legume varieties harboring the same resistance gene could lead to a breakdown in resistance as a result of pathogen counter-evolution. For example, a breakdown in *er1*-based resistance has been reported in pea against an *E. pisi* Italian isolate (Lahoz et al. 2013) and *E. trifolii* (Fondevilla et al. 2013). Additionally, the resistance provided by *er2* is influenced by temperature and leaf age (Fondevilla et al. 2006). Therefore, there is a need to identify new sources of resistance to powdery mildews in legumes for continuous influx into breeding programs.

The forage legume, *Medicago truncatula* (barrel medic), is an attractive model to study powdery mildew resistance in legumes owing to its relatively small ($\sim$450 Mb), diploid, fully sequenced and annotated genome, short life cycle, amenability to transformation, availability of a genetically diverse germplasm collection and mutant populations, and synteny with agriculturally relevant legume crops (Krishnakumar 2020; Rispail et al. 2019). In the following sections, we summarize key findings from previous investigations on powdery mildew resistance in *M. truncatula* and present a set of genomic and genetic tools that can be exploited for the identification of novel resistance genes and the underlying molecular mechanisms of resistance against powdery mildew disease in legumes.

5.2 *Medicago truncatula* Germplasm Exhibits Natural Variation in Powdery Mildew Resistance

The genetic variability in the core *M. truncatula* germplasm collections of the U.S. National Plant Germplasm System and INRA Montpellier, France has served as an excellent resource for the identification of powdery mildew-resistant genotypes by several labs (Table 5.1). One of the earliest screening studies involved the evaluation of nine *M. truncatula* accessions from the core annual *Medicago* germplasm collection for susceptibility to an alfalfa isolate of *E. pisi* (Yaege and Stuteville 2002). Macroscopic assessment of disease symptoms identified accessions PI292436 and PI566892 as *E. pisi* resistant, with 100% and 85% of the plants, respectively, remaining mildew-free in controlled environment growth chambers until 21 days after inoculation. Subsequently, a larger-scale assessment was performed on the same collection by Prats et al. (2007), in which 277 *M. truncatula* accessions were screened for resistance to an *M. truncatula* isolate of *E. pisi*. Macroscopic assessment of symptom development identified 4% of the accessions as resistant or moderately resistant (<40% of leaf area covered by mildew) to *E. pisi* f. sp. *medicaginis*. Similar screening studies performed by other research groups against alfalfa, *M. truncatula* or pea isolates of *E. pisi* have identified the *M. truncatula* genotypes A17, A20, R108, and DZA315.16 (DZA) as powdery mildew resistant, moderately resistant, moderately susceptible, and susceptible, respectively (Ameline-Torregrosa et al. 2007; Foster-Hartnett et al. 2007; Gupta et al. 2020, 2021; Yang et al. 2013). At the macroscopic level, powdery mildew symptoms are visible on leaves of the susceptible genotype DZA and, to a lesser extent, on the moderately susceptible genotype R018 at 12 dpi but not on the resistant genotype A17 (Fig. 5.1a). On a microscopic level, pathogen growth is arrested at the appressorial stage on A17, whereas a mycelial mesh containing asexual reproductive structures (conidiophores) develops on R108 and DZA by 5–7 dpi (Fig. 5.1b). However, since DZA is highly susceptible, it supports denser mycelial growth and a larger number of conidiophores compared to the moderately susceptible R108 (Fig. 5.1b). Susceptibility to *E. trifolii* was also evaluated in 10 cultivars of *M. truncatula* under field and greenhouse conditions at two different locations in Southern Australia (Ballard et al. 2012). The study found that the cultivar Paraggio had the lowest powdery mildew disease score and the slowest rate of disease development.

Histological studies on the different resistant genotypes indicate that distinct defense mechanisms are triggered in host plants, that act either alone or in combination, to restrict pathogen development at different infection stages, ranging from spore germination to colony expansion [(-summarized in Table 5.1); Foster-Hartnett et al. 2007; Gupta et al. 2020; Prats et al. 2007; Samac et al. 2011]. Based on these studies, three major powdery mildew resistance mechanisms appear to be operational in *M. truncatula*: (1) pre-penetration resistance, in which plant cell wall appositions, also known as papillae, formed at the site of attempted fungal entry limit pathogen penetration, (2) penetration resistance, in which the fungal penetration pegs that overcome papilla formation are restricted from further development by a localized cell death response known as

Table 5.1 Powdery mildew resistance/susceptibility phenotypes of *Medicago truncatula* accessions identified through germplasm screening studies

Powdery mildew species (isolate)	*M. truncatula* accession/cultivar	Powdery mildew phenotype	Resistance mechanism (if known)	References
Erysiphe pisi PM1 (alfalfa)	PI292436	Resistant	–	Yaege and Stuteville (2002)
	PI566892	Resistant	–	
	PI566887	Moderately resistant	–	
	PI566888	Moderately susceptible	–	
	PI566890	Moderately susceptible	–	
	PI566889	Susceptible	–	
	PI384648	Susceptible	–	
	PI566886	Susceptible	–	
	PI566891	Susceptible	–	
E. pisi (pea)	A17	Resistant	Early HR; Pre-penetration resistance	Foster-Hartnett et al. (2007), Samac et al. (2011)
	A20	Moderately resistant	Delayed HR; Post-penetration resistance	
	R108	Moderately susceptible	–	
	DZA315.16	Susceptible	–	
E. pisi f.sp. *medicaginis* CO05 *(M. truncatula)*	PI283661	Resistant	Inhibition of spore germination	Prats et al. (2007) Curto et al. (2014)
	PI564941	Resistant	Inhibition of spore germination; Pre-penetration/papilla resistance; Post-haustorial resistance	
	PI190084	Resistant	Pre-penetration/papilla resistance	
	PI464816	Resistant	Pre-penetration/papilla resistance; HR	
	W65999	Resistant	Inhibition of spore germination; Pre-penetration/papilla resistance; Post-haustorial resistance	
	W66018	Resistant	Inhibition of spore germination; HR	
	W66026	Resistant	Inhibition of spore germination; Post-haustorial resistance	
	SA1306	Resistant	HR	
	PI199257	Moderately resistant	HR	
	PI384647	Moderately resistant	Post-haustorial resistance	
	PI190086	Moderately resistant	Post-haustorial resistance	
	Parabinga	Susceptible	–	
	PI292436	Susceptible	–	
	PI384648	Susceptible	–	
	PI566886	Susceptible	–	

(continued)

Table 5.1 (continued)

Powdery mildew species (isolate)	*M. truncatula* accession/cultivar	Powdery mildew phenotype	Resistance mechanism (if known)	References
E. pisi Ep-p (pea) *Ep-a* (alfalfa)	A17	Resistant	Localized HR with production of ROS and phenolics	Ameline-Torregrosa et al. (2008)
	F83005.5	Susceptible	–	
	DZA315.16	Susceptible	–	
E. pisi race 1 (*M. truncatula*)	A17	Resistant	–	Yang et al. (2013)
	F83005.5	Susceptible	–	
E. pisi Palampur-1 (pea)	A17	Resistant	HR	Gupta et al. (2020, 2021)
	R108	Moderately resistant	–	
	DZA	Susceptible	–	
E. trifolii (medic)	Paraggio	Moderately susceptible	–	Ballard et al. (2012)
	Sephi	Moderately susceptible	–	
	Hannaford	Susceptible	–	
	Mogul	Susceptible	–	
	Caliph	Susceptible	–	
	Jemlaong	Susceptible	–	
	Jester	Susceptible	–	
	Borung	Susceptible	–	
	Cyprus	Susceptible	–	
	Parabinga	Susceptible	–	

hypersensitive response (HR) that includes the accumulation of autofluorescent compounds and hydrogen peroxide and cytoplasmic disorganization in the infected epidermal cell, and (3) post-haustorial resistance, in which fungal colony development is restricted through a delayed HR response.

5.3 Genetic and Molecular Basis of Powdery Mildew Resistance in *M. truncatula*

Plants have evolved a multitiered innate immune system to defend themselves against pathogen infection (Cui et al. 2015). Pattern recognition receptors (PRRs) present at the plant surface act as primary sensors that detect pathogen-associated molecular patterns (PAMPs) to induce a basal resistance response called pattern-triggered immunity (PTI). Pathogens, in turn, have evolved an arsenal of effectors, some of which are secreted into host cells to interfere with PTI and dampen basal defenses. To block pathogen effectors, plants have evolved a diverse group of intracellular nucleotide-binding/leucine-rich repeat (NLR) receptors that, upon activation by effectors/effector activities, transcriptionally induce a robust form of resistance known as effector-triggered immunity (ETI) that is often accompanied by localized HR and cell death (Mur et al. 2008). Recent studies indicate that these two immune pathways are not distinct but function synergistically to activate resistance against pathogens (Yuan et al. 2021). Genetic and comparative transcriptomics studies on resistant

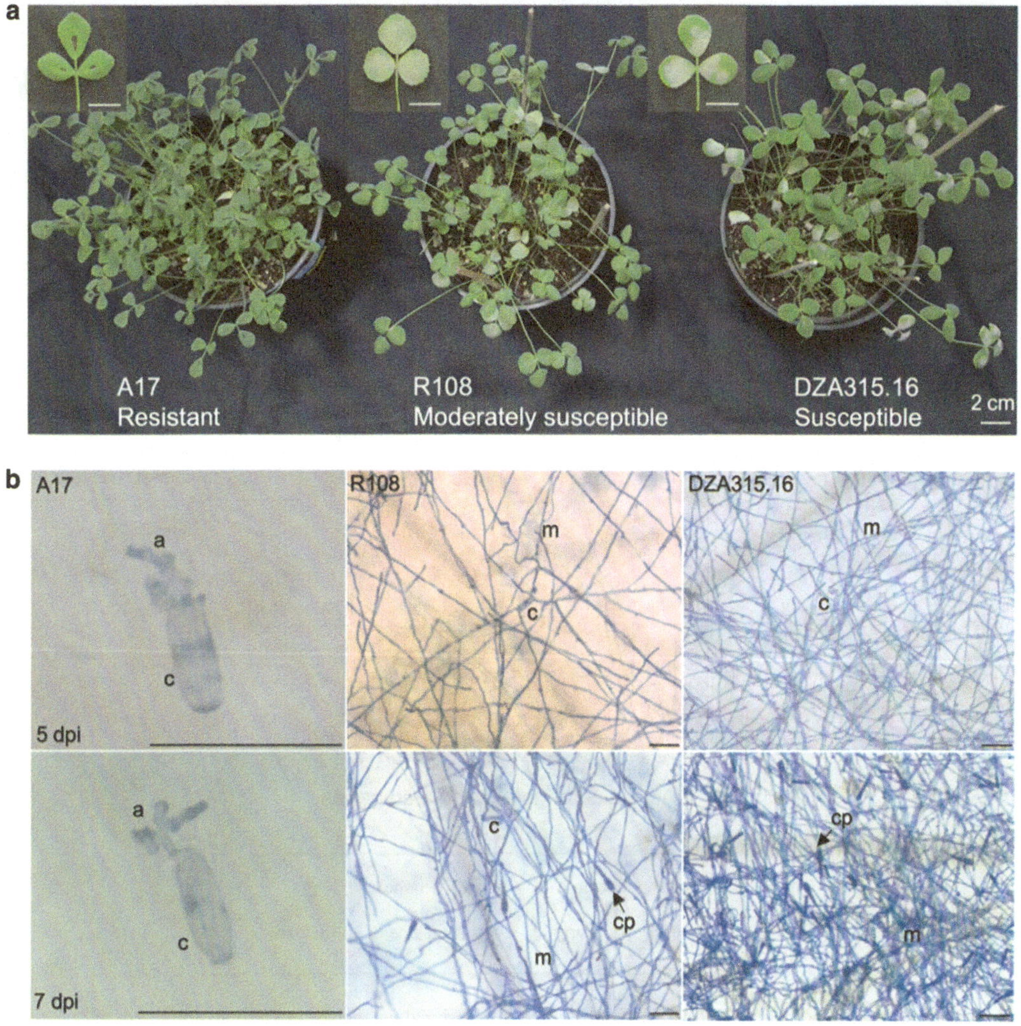

Fig. 5.1 *M. truncatula* genotypes exhibiting natural variation in powdery mildew resistance. **a** Whole plants at 12 days post inoculation (dpi) with *E. pisi* Palampur-1. White powdery symptoms were visible on inoculated leaves of the moderately susceptible host R108 and susceptible host DZA, whereas no symptoms were observed on comparable leaves of the resistant host A17. Inset shows a single trifoliate of each genotype at 12 dpi. **b** Representative microscopic images showing *E. pisi* Palampur-1 conidia arrested at the appressorium stage on resistant A17 leaves at 5 and 7 dpi, and colonies forming mycelial networks and conidiophores on moderately susceptible R108 and susceptible DZA at 5 dpi and 7 dpi, respectively. c, conidia; a, appressorium; m, mycelium; cp, conidiophores. Scale bar, 100 μm

and susceptible *M. truncatula* genotypes have identified candidate powdery mildew resistance loci/*NLR* genes (Ameline-Torregrosa et al. 2007; Yang et al. 2013) and provided insights into the signaling components and defense pathways activated in response to infection (Curto et al. 2015; Foster-Hartnett et al. 2007; Gupta et al. 2020) (summarized as a model in Fig. 5.2).

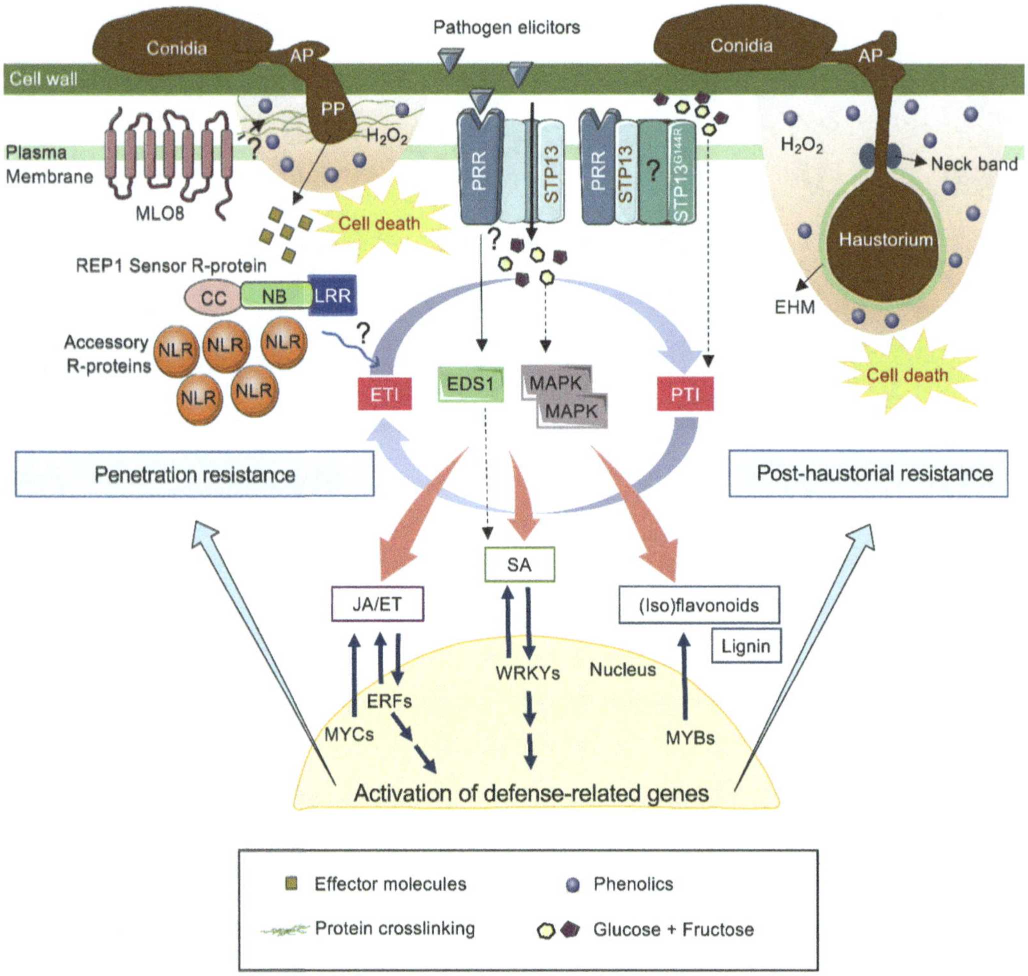

Fig. 5.2 Proposed hypothetical model depicting the molecular mechanisms contributing to penetration and/or post-haustorial resistance against powdery mildew infection in *M. truncatula*. PTI is initiated when *M. truncatula* PRRs recognize fungal elicitors (e.g., chitosan), triggering a signaling cascade via EDS1, MAPKs, and phytohormones SA and JA that lead to transcriptional reprogramming via transcription factors and activation of defense-related genes (e.g., *PR*s and (iso)flavonoid and lignin pathway genes). Signaling events mediated by the pathogen-induced hexose sugar transporter STP13, via interaction with PRRs and/or intracellular hexose accumulation, can also lead to the induction of defense-related genes. The outcome of PTI is localized accumulation of hydrogen peroxide (H_2O_2) and antifungal phenolics (isoflavonoids/lignin) at the infection site that eventually leads to host cell death and powdery mildew resistance. ETI is initiated when pathogen effectors/effector activities are recognized by a sensor NLR protein like MtREP1. The signal is then relayed and amplified by accessory NLR proteins, largely through the same pathways triggered during PTI, resulting in a more rapid and robust induction of the same defense responses. A functional synergism between ETI and PTI appears to be involved in penetration resistance, whereas post-haustorial resistance may be mainly dependent on PTI. Penetration resistance may also be mediated by MLO8-based protein crosslinking in the infected epidermal cell. In addition, overexpression of the transport-deficient MtSTP13^{G144R} variant results in the induction of defense-related genes (possibly via impairment of wildtype STP13 function through heterodimer formation) and reduced susceptibility to powdery mildew in pea, demonstrating the applicability of using *M. truncatula* genes for developing resistance in crop legumes. AP, appressorium; PP, penetration peg; EHM, extra-haustorial membrane; PTI, PAMP-triggered immunity; ETI, effector triggered immunity

5.3.1 Powdery Mildew Resistance Loci and *R* Gene

Genetic analysis of an F7 RIL population generated from an A17 (resistant) x DZA (susceptible) cross led to the identification of three distinct major quantitative trait loci (QTL) for resistance to *E. pisi* (*Epp1, Epa1 and Epa2*). *Epp1* was found to confer ~66% resistance against a pea isolate *Ep-p*, whereas *Epa1* and *Epa2* were responsible for 26 and 23% of the resistance against an alfalfa isolate, *Ep-a* (Ameline-Torregrosa et al. 2007). The identification of distinct resistance loci against different isolates of the pathogen implies that host defense responses are triggered in an isolate-specific manner. The segregation pattern of the resistance phenotypes in the F7 RIL population suggested the involvement of one and two major resistance genes against the pea and alfalfa isolates, respectively (Ameline-Torregrosa et al. 2007). To identify the resistance gene(s) responsible for powdery mildew resistance, Yang et al. (2013) performed a genetic analysis on an F2 mapping population derived from an A17 (resistant) x F83005.5 (susceptible) cross using cleaved amplified polymorphic sequences (CAPS)-based markers. This analysis discovered a resistance locus designated as *MtREP1* (resistance to *E. pisi* race 1) on *M. truncatula* chromosome 5, occupying the same position as one of the two *Epa* loci (Ameline-Torregrosa et al. 2007). Within the *MtREP1* locus, the gene responsible for powdery mildew resistance was cloned and characterized by Yang et al. (2013) and shown to encode a coiled-coil (CC) type of NLR protein (Medtr5g072340). Introduction of *Medtr5g072340,* designated as *MtREP1,* into alfalfa (*Medicago sativa*) conferred full resistance against *E. pisi*, demonstrating the potential of using *M. truncatula* genes for genetic improvement of alfalfa (Yang et al. 2013).

The higher basal expression (without pathogen infection) of *MtREP1* in resistant (A17) versus susceptible (F83005.5, DZA) *M. truncatula* genotypes was found to correlate with early penetration resistance against different *E. pisi* isolates (Gupta et al. 2020; Yang et al. 2013). Further, an examination of the methylation status of *MtREP1* promoter regions in resistant and susceptible genotypes revealed lower cytosine methylation rates in the promoter of the resistant allele, indicating that *MtREP1* expression is epigenetically regulated (Yang et al. 2013). Although the downstream defense pathways activated by MtREP1 have not yet been entirely elucidated, expression data from a recent genome-wide transcriptomics study on compatible and incompatible *M. truncatula–E. pisi* interactions suggest that MtREP1 may serve as the initial sensor R-protein that recognizes the pathogen signal to initiate ETI, while the pathogen-dependent induction of other NLRs may be required for boosting defense via PTI amplification (Gupta et al. 2020). In line with this, genes encoding PRRs and early PTI signaling components, including calmodulin-related proteins, oxidative burst generating peroxidases and MAP kinases were found to be induced by powdery mildew infection to a greater extent in resistant genotypes compared to susceptible ones (Curto et al. 2014; Foster-Hartnett et al. 2007; Gupta et al. 2020).

5.3.2 Role of Plant Secondary Metabolites in Powdery Mildew Resistance

In response to biotic stress, plants synthesize a wide variety of secondary metabolites with antimicrobial properties (Jan et al. 2021). Legumes are rich in phenolic compounds, such as flavonoids, isoflavonoids, and lignin, which are synthesized via the phenylpropanoid pathway (Dixon et al. 2002). Many of these compounds are considered to act as phytoalexins, i.e., antimicrobial compounds that are rapidly synthesized in response to (a)biotic elicitors (Jiménez-González et al. 2008; Samac and Graham 2007). For example, Ishiga et al. (2015) showed that the phytoalexin medicarpin accumulates in *M. truncatula* on infection with the Asian soybean rust fungus *Phakospora phachyrhizi* and that treatment with medicarpin inhibits urediniospore germination and differentiation in vitro.

In addition to their role as defense compounds, (iso)flavonoids serve as signaling molecules during plant–symbiont interactions (Stacey et al. 2006).

Comparative transcriptomic analyses of resistant (A17, A20, SA1306) and susceptible (DZA, Parabinga) *M. truncatula* genotypes have revealed that the (iso)flavonoid and lignin branches of the phenylpropanoid biosynthetic pathway are up-regulated during both incompatible and compatible *M. truncatula–E. pisi* interactions (Curto et al. 2014; Foster-Hartnett et al. 2007; Gupta et al. 2020). However, in general, lower levels of pathway gene induction are observed in the susceptible genotypes as compared to resistant ones, particularly at early infection time points [12 and 24 h post-inoculation (hpi)], supporting a role for these metabolites in defense (Foster-Hartnett et al. 2007; Gupta et al. 2020). Additionally, *M. truncatula* orthologs of transcription factors known to positively regulate (iso)flavonoid biosynthesis in Arabidopsis or soybean are significantly induced by *E. pisi* in A17 but not in DZA (Gupta et al. 2020), suggesting that induction of this pathway correlates with resistance. To test the role of this pathway in defense, Samac et al. (2011) silenced chalcone synthase (CHS), the first committed reaction in (iso) flavonoid biosynthesis, through an RNA interference (RNAi) approach. Stable expression of the *CHS* hairpin RNA construct in the moderately susceptible R108 genotype reduced *CHS* transcript accumulation by 30% but did not increase susceptibility to *E. pisi*. A higher level of silencing may likely be required to significantly perturb this pathway and produce a disease phenotype. Alternatively, *CHS* silencing may not necessarily affect (iso)flavonoid biosynthesis since substrates for the successive steps of the pathway can be replenished through the hydrolysis of their glucoside storage forms. Such availability of substrates from the glucoside pool has been reported for medicarpin synthesis on elicitation with methyl jasmonate (MeJA) in *M. truncatula* cell suspensions (Naoumkina et al. 2007).

The localized accumulation of de novo synthesized autofluorescent phenolic compound(s) at the site of pathogen penetration also appears to be important for powdery mildew resistance. For example, treatment of barley leaves with inhibitors of key enzymes of the phenylpropanoid pathway reduced the frequency and intensity of autofluorescence associated with the barley powdery mildew germ tube (Carver et al. 1994). Similar infection-localized autofluorescence has been observed as a part of the HR response in resistant and moderately resistant *M. truncatula* genotypes (Foster-Hartnett et al. 2007; Prats et al. 2007) and in *er1*, *er2*, or *Er3* containing pea accessions that exhibit pre-penetration or post-haustorial HR (Barilli et al. 2014). This suggests that localized accumulation of phenolics during incompatible legume–powdery mildew interactions is important for resistance, but the exact composition of these compounds is not known. Collectively, these studies provide correlative evidence that the (iso)flavonoid pathway may function in legume powdery mildew resistance; however, more direct evidence needs to be generated to unequivocally prove their role in defense. *M. truncatula*, with its high isoflavonoid content (Barreira et al. 2015), can be exploited as a model to study the spatiotemporal accumulation of the different pathway intermediates and phytoalexin end product in response to powdery mildew infection as well as to evaluate their antifungal activity and ability to activate defense signaling in plants.

5.3.3 Role of Phytohormones in Powdery Mildew Resistance

Signaling mediated by the phytohormones salicylic acid (SA) and jasmonate (JA)/ethylene (ET) is principally required for the two modes of plant immunity, PTI, and ETI (Mine et al. 2018). In general, the SA signaling network is important for immunity against biotrophic pathogens, whereas the JA/ET signaling sector is considered to be essential for resistance against necrotrophic pathogens (Thaler et al. 2012). Comparative transcriptomics analyses of resistant and susceptible *M. truncatula* genotypes suggest that both SA and JA/ET-mediated defenses are involved in powdery mildew resistance in legumes, but as discussed below, the two signaling pathways may contribute differently to the two layers of immunity.

Known regulators of SA biosynthesis and components of the SA signaling network are activated by *E. pisi* infection in resistant and susceptible *M. truncatula* genotypes, but often earlier and/or to a greater magnitude in the resistant genotypes (Foster-Hartnett et al. 2007; Gupta et al. 2020), indicating that SA is involved in basal defense (PTI) and ETI against powdery mildew in legumes. For example, a *M. truncatula* homolog of the Arabidopsis *enhanced disease susceptibility 1* (*EDS1*), a well-known positive regulator of induced SA accumulation and signaling (Dempsey et al. 2011; Wiermer et al. 2005), was comparably induced in A17, A20, and DZA genotypes at 12 hpi with *E. pisi* (Foster-Hartnett et al. 2007) but only in the resistant genotype A17 at 24 hpi (Gupta et al. 2020). Similarly, *M. truncatula* homologs of WRKY transcription factors, known to function as positive regulators of SA biosynthesis (*WRKY75*) and signaling (*WRKY51*) in Arabidopsis (Gao et al. 2011; Guo et al. 2017), were induced in both A17 and DZA in response to *E. pisi* infection; however, several additional WRKYs were significantly induced only in resistant *M. truncatula* genotypes (Curto et al. 2015; Gupta et al. 2020). In addition, known markers of SA signaling such as the *pathogenesis-related* (*PR*) genes *PR2*, *PR4*, *PR5*, and *PR10* were significantly up-regulated in resistant and susceptible *M. truncatula* genotypes at early infection time points, but frequently to a greater degree in resistant genotypes (Curto et al. 2014; Foster-Hartnett et al. 2007; Gupta et al. 2020). Plant PR proteins exhibit diverse functions, including β-1,3-glucanase (PR2), chitinase (PR4), thaumatin-like (PR5), and ribonuclease (PR10), and are reported to function as antifungal agents (Ali et al. 2018); nevertheless, their exact contribution to legume powdery mildew resistance is yet to be demonstrated. Strong support for SA's role in defense against legume powdery mildew comes from a recent study in which pre-established nodulation of susceptible *M. truncatula* (DZA) and pea plants by *Sinohizobium meliloti* was found to systemically induce SA accumulation, SA-dependent *PR* gene expression, and *E. pisi* resistance (Smigielski et al. 2019). Whether this SA-mediated resistance in legumes is dependent on the EDS1 pathway remains to be determined.

By contrast, *M. truncatula* homologs of known positive regulators of the JA/ET signaling pathway in Arabidopsis, such as *ERF1*, *ERF59*, *ANACO19*, and other *HD-like* and *AP2/EREBP* family transcription factors, were significantly up-regulated in response to *E. pisi* infection in resistant *M. truncatula* genotypes while other known positive regulators of the JA/ET pathway, including the master regulator *MYC2*, were repressed in susceptible genotypes (Curto et al. 2015; Gupta et al. 2020). Collectively, these findings suggest that in addition to SA, JA/ET-mediated defense responses may contribute substantially to ETI against powdery mildews, a hypothesis that is supported by previous studies in which coordinated accumulation of SA and JA was reported to occur during ETI (Liu et al. 2016; Zhu et al. 2014). It has been speculated that the induction of JA-mediated defenses during incompatible host–pathogen interactions may enable the host plant to utilize cell death as a major defense mechanism against biotrophic pathogens without making them susceptible to necrotrophic pathogens (Liu et al. 2016), or offset any pathogen-mediated interference in SA signaling (Tsuda et al. 2009). Further insights on the role of these phytohormones in legume powdery mildew resistance can be obtained through a systematic quantification of these metabolites in resistant and susceptible *M. truncatula* genotypes at different time points of infection, and elucidating whether the activation of these defense signaling pathways is dependent on pathogen effector recognition by an NLR protein like MtREP1.

5.3.4 Role of Sugar Transporters in Powdery Mildew Resistance

Besides the use of conventional *NLR* genes in resistance breeding programs, modulation of the sugar transport machinery has emerged as an innovative and alternate approach to develop broad-spectrum and durable resistance against biotrophic pathogens in crops such as wheat and barley (Milne et al. 2019; Moore et al. 2015). The infectivity of biotrophic pathogens is largely determined by their ability to tap nutrients, especially sugars, from host cells through haustoria (Hückelhoven 2005). In this context, members of the *M. truncatula* SWEET (sugars will be eventually exported transporters) sugar uniporter family, that is implicated in pathogen nutrition (Chen 2014), have been reported to be induced by diverse microbes (Chandran 2015) but, in general, the role of plant SWEETs in powdery mildew nutrition is not known. Contrary to this, there is considerable evidence to suggest that members of the sugar transport protein (STP) family of plant transporters restrict carbon allocation toward pathogens via apoplastic hexose retrieval and consequently influence the rate of disease development (Lemonnier et al. 2014; Yamada et al. 2016). However, since sugars can also serve as the resource or signal for plants to mount a successful defense response (Morkunas and Ratajczak 2014), alterations in sugar transport, particularly at infection sites, may also lead to resistance via an induction of sugar-mediated defense signaling (Moore et al. 2015). This has been observed in wheat, where an STP13 variant encoding *Lr67* gene provides partial resistance to rust and powdery mildew fungi in adult plants (Moore et al. 2015). The variant STP13 hexose transporter (Lr67res) differs from the susceptible form of the same protein (Lr67sus) by two amino acids (G144R and V387L) (Moore et al. 2015). Functional studies in a hexose transport deficient yeast mutant strain revealed that the Lr67 resistant allele (Lr67res) is incapable of transporting a broad range of hexose substrates, including glucose and fructose (Moore et al. 2015). Since Lr67res forms heterodimers with Lr67sus in planta, it was hypothesized that the inhibition of Lr67sus transport function via Lr67res-sus heterodimerization increases the apoplastic hexose/sucrose ratio, which in turn induces a sugar signaling-mediated defense response and disease resistance (Moore et al. 2015). In support of this hypothesis, the introduction of the wheat *Lr67res* gene into barley plants resulted in higher basal and pathogen-inducible *PR* gene expression seedling resistance against rust and powdery mildew fungi (Milne et al. 2019).

Gupta et al. (2021) recently demonstrated that the Lr67res-mediated resistance is transferable to legumes as the transient expression of an *M. truncatula* STP13 ortholog carrying one of the Lr67res-specific mutations (G144R) provided penetration resistance to *E. pisi* in pea. In addition to the G144R mutant, the wildtype MtSTP13 was also found to confer basal resistance against *E. pisi* in *M. truncatula* and pea (Gupta et al. 2021). Among the 30 *STP* genes in *M. truncatula*, *MtSTP13.1* (one of the two *MtSTP13* paralogs) displayed the highest fold change in expression at 72 and 120 hpi with *E. pisi* and was found to be rapidly induced by treatment with the fungal PAMP chitosan. Virus-induced gene silencing of *MtSTP13* enhanced the powdery mildew susceptibility of the moderately susceptible *M. truncatula* R108 genotype, whereas transient overexpression of *MtSTP13.1* provided partial resistance to *E. pisi* in leaves of the susceptible AP-3 pea genotype. The resistance phenotypes were attributed to an MtSTP13.1/MtSTP13.1^{G144R}-mediated activation or priming of defense responses as defense-associated *PR* and/or isoflavonoid pathway genes were repressed in *MtSTP13*-silenced R108 leaves but up-regulated in *MtSTP13.1* or *MtSTP13.1*^{G144R} overexpressing pea leaves (Gupta et al. 2021). This study demonstrates the applicability of *M. truncatula* genes for the improvement of powdery mildew resistance in crop legumes such as pea. Future investigations into the spatiotemporal distribution of sugars in *mtstp13* mutants and stable *MtSTP13.1* or

MtSTP13.1^{G144R} overexpression lines will shed light on the precise role of sugar signaling in defense induction against powdery mildews.

5.4 Harnessing the *M. truncatula* Genome for New Sources of Powdery Mildew Resistance

The availability of high-quality *M. truncatula* reference genome assemblies has proven to be important for the comprehensive identification of structural elements (genes and repeat elements), large-scale analysis of gene expression, and cross-legume comparisons [e.g., Gupta et al. (2020), Lee et al. (2017), Tang et al. (2014)]. Sequencing of the first *M. truncatula* genome commenced in 2003, and the initial strategy employed a combination of Sanger and Illumina-based sequencing to build a BAC-based draft assembly (Mt3.5) of the euchromatic portion of the A17 genome (Young et al. 2011). This was later refined by incorporating de novo assembled Illumina-derived whole genome shotgun (WGS) sequences (Mt4.0) into the already existing assembly to produce a more comprehensive and accurate genome (Tang et al. 2014). The Mt4.0 A17 pseudomolecules encompasses ~360 Mb of actual sequences spanning 390 Mb plus 28.4 Mb of unanchored scaffolds. Further improvements to the Mt4.0 assembly were achieved through long-read PacBio sequencing and advanced assembly protocols, yielding a highly contiguous reference genome assembly of 430 Mb (Mt5.0) (Pecrix et al. 2018). A total of 50,894 genes were predicted from the A17 Mt4.0 assembly (Tang et al. 2014) which was further refined to 44,623 protein-coding genes in the Mt5.0 version (Pecrix et al. 2018). The genotype R108 is more attractive for genetic studies because of its higher transformation efficiency compared to A17 and the availability of a large-scale *Tnt1*-transposon insertion mutant population for functional studies. Therefore, the R108 genome was the second high-quality *M. truncatula* reference assembly to be released (Moll et al. 2017; Kaur et al. 2021). In addition, as a part of the Medicago Hapmap project, the genomes of 288 *M. truncatula* accessions were sequenced, providing a platform for the discovery of single nucleotide variations (SNPs and indels) and structural variants associated with complex traits (Stanton-Geddes et al. 2013; Zhou et al. 2017). The availability of these reference genome assemblies also enables genome-wide expression studies in *M. truncatula* via RNA-Sequencing.

The different *M. truncatula* genome assemblies are available for download from Genbank (https://www.ncbi.nlm.nih.gov/genbank/) or the *M. truncatula* Hapmap project website (http://www.medicagohapmap.org/) (Fig. 5.3). The genomes and their associated resources are also hosted in specific databases that include several bioinformatics tools for data analysis (Fig. 5.3). The A17 Mt4.0v1 data is available at the Phytozome13 website (https://phytozome-next.jgi.doe.gov/info/Mtruncatula_Mt4_0v1) while the Mt5.0 data is hosted on the INRA MtrunA17r5.0-ANR website (https://medicago.toulouse.inra.fr/MtrunA17r5.0-ANR/). Both websites include several resources and bioinformatics services for data mining from the respective genome versions, including a keyword search for a gene of interest, detailed gene reports with sequence information, functional annotations, protein homologs, and expression data, a BLAST tool that allows searches against the genome and proteome, a genome browser for navigation and comparative analyses, and a data portal from which the assemblies and annotations can be downloaded. In addition, the reference-quality MtR108v1.0 sequence and de novo assemblies of 20 *M. truncatula* accessions and the associated SNP genotype files are available at the Medicago Hapmap project website. These resources can be readily used for genome-wide association studies (GWAS) to detect genetic variants related to powdery mildew resistance in *M. truncatula*, as recently shown in common bean (Binagwa et al. 2021) and identify candidate genes for powdery mildew resistance

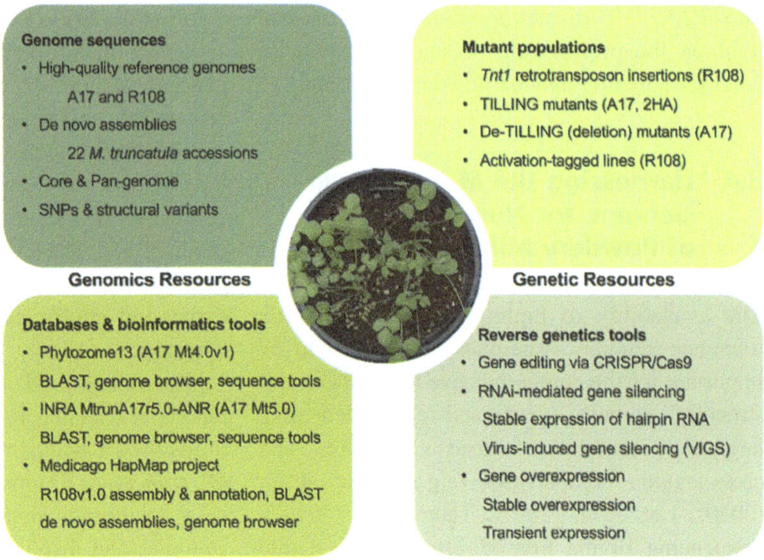

Fig. 5.3 *M. truncatula* genomic and genetic resources for the identification of powdery mildew resistance in legumes

breeding in crop legumes. The high level of synteny between the genomes of *M. truncatula* and other legumes, such as pea, chickpea, lentil, and alfalfa (Gujaria-Verma et al. 2014; Lee et al. 2017; Varshney et al. 2013) will aid in the cross-species translation of genomic information.

5.4.1 New *NLRs* as a Potential Source of Powdery Mildew Resistance

To date, only one *NLR* resistance gene (*MtREP1*) and three resistance loci have been identified in *M. truncatula* against powdery mildew (Ameline-Torregrosa et al. 2007; Yang et al. 2013). Since resistance mediated via *NLR* genes is often based on recognition of a single effector/effector activity, the risk of resistance breakdown is generally high. Therefore, a constant inflow of novel resistance genes in breeding programs is required. Among the sequenced legume species, *M. truncatula* harbors the largest number of *NLR* resistance genes (Varshney et al. 2013). Analysis of the *Mt3.5* A17 genome assembly identified over 750 putative toll-interleukin receptor (TIR)-NLRs (TNLs) and non-TIR NLRs that include the coiled-coil (CC) and resistance to powdery mildew 8 (RPW8) domain-containing proteins (Young et al. 2011). The majority of the NLRs exhibit a high degree of clustering in the genome with the TNL clusters mainly found on chromosomes 4, 6, and 8 and CNL clusters on chromosomes 3 and 5 (Ameline-Torregrosa et al. 2008; Nepal et al. 2017; Young et al. 2011). Some of the CNL-dense chromosomes show large blocks of synteny with chromosomes in common bean and soybean; however, these syntenic regions were not found to contain CNL genes (Nepal et al. 2017). Examination of orthologous gene families across 15 de novo *M. truncatula* genome assemblies revealed that the *NLR* gene family exhibits high levels of SNP diversity, copy number variability, and protein domain diversity (Zhou et al. 2017). In addition, many members of this gene family were found to be accession-specific, present in some but absent in others (Zhou et al. 2017). This large reservoir of genetic diversity can be utilized for the discovery of new sources of powdery mildew resistance in legumes. Notably, nine CNL genes with similarities to 'Arabidopsis Resistance to powdery mildew 8' (AtRPW8) have been identified in *M. truncatula* so far (Young et al. 2011). AtRPW8 is a well-studied broad-spectrum resistance gene that confers full resistance to powdery mildew in Arabidopsis (Wang et al.

2009; Xiao et al. 2001). It will be interesting to test whether these MtRPW8-like NLRs also confer powdery mildew resistance in legumes.

5.4.2 Target *Mildew Locus 0 (mlo)* Genes for Reduced Powdery Mildew Susceptibility

The *MLO (Mildew resistance locus O)* gene family encodes a set of plant-specific, plasma membrane-localized, transmembrane domain-containing proteins that are ubiquitously found in monocots and dicots (Appiano et al. 2015). Loss-of-function *mlo* alleles have been reported to confer non-race specific penetration resistance against powdery mildew in barley, Arabidopsis, and tomato [reviewed in Acevedo-Garcia et al. 2014], indicating their role as powdery mildew susceptibility factors. Phylogenetic analysis indicates that MLO proteins cluster into 6 to 8 clades with members related to powdery mildew susceptibility categorized under clade IV in monocots (barley HvMLO, and wheat TaMLO_A1 and TaMLO_B1) and clade V in dicots (tomato SlMLO1 and Arabidopsis AtMLO1, AtMLO6 and AtMLO12, and pea PsMLO1) (Acevedo-Garcia et al. 2014; Appiano et al. 2015). In legumes, the *er1*-driven powdery mildew resistance that is extensively used in pea breeding programs is caused by recessive loss-of-function mutations in the *PsMLO1* gene (Humphry et al. 2011; Pavan et al. 2011; Sun et al. 2016, 2019). *Psmlo1*-based penetration resistance against *E. pisi* is largely associated with protein cross-linking at sites of attempted penetration rather than callose deposition or HR (Iglesias-García et al. 2015) and can be complemented by heterologous expression of orthologous *MLO1* genes from *M. truncatula* and *Lotus japonicus*, indicating functional conservation of these genes across legumes (Humphry et al. 2011).

Although *Psmlo1*-based pea powdery mildew resistance is considered to be highly durable under field conditions, a breakdown in *er1*-mediated resistance has been reported against certain isolates of the pathogen (Lahoz et al. 2013; Fondevilla et al. 2013). Hence, the functional characterization of additional members of the legume *MLO* family may enable the identification of new targets for utilization in resistance breeding programs. Genome-wide analyses of the *MLO* gene family in legumes identified 16 *MLOs* in *M. truncatula* but only 4 in pea (Rispail and Rubiales 2016; Deshmukh et al. 2017), highlighting the usefulness of the *M. truncatula* genome in terms of the number of candidates available for loss-of-function studies. A recent study showed that *Mtmlo8* knockout mutants exhibit enhanced resistance to *E. pisi*, indicating that *MtMLO8* is required for powdery mildew colonization in *M. truncatula* (Jacott et al. 2020). Notably, MtMLO8 clusters with clade IV MLOs (Rispail and Rubiales 2016), which are associated with powdery mildew susceptibility in monocots. This suggests that MLOs, in addition to those present in clade V, may function as powdery mildew susceptibility factors in legumes. Of the remaining MtMLOs, MtMLO3 is a promising candidate for reverse genetics studies as it clusters with MtMLO1 in clade V. Another likely candidate is *MtMLO10,* which was reported to be induced in response to powdery mildew infection only in a resistant *M. truncatula* genotype (Gupta et al. 2020).

5.5 Forward and Reverse Genetics Tools for Dissecting Powdery Mildew Resistance Mechanisms

Large-scale comparative transcriptomics studies using microarrays and/or high-throughput next-generation sequencing platforms (e.g., Illumina) have identified a sizable number of differentially expressed genes and uncovered distinct host transcriptional signatures during incompatible and compatible *M. truncatula-E. pisi* interactions (Curto et al. 2014, 2015; Foster-Hartnett et al. 2007; Gupta et al. 2020). Surprisingly, however, very few *M. truncatula* genes have been functionally characterized for their role in powdery mildew resistance (Gupta et al. 2021; Jacott et al. 2020; Samac et al. 2011). Over the last decade, an extensive genetic toolbox has been developed for the model legume *M. truncatula*, which has predominantly been used to characterize genes involved in symbiosis [e.g., Pislariu et al. (2012), Veerappan et al. (2016)]. However, many of these same tools can be exploited for the identification of novel powdery mildew resistance/susceptibility factors and their underlying mechanism of action. In the following subsections, we review the various forward and reverse genetics tools available in *M. truncatula* for the functional characterization of genes involved in legume powdery mildew interactions (Fig. 5.3).

5.5.1 *Tnt1* Retrotransposon-Tagged and Other Mutant Populations

Unlike in Arabidopsis, T-DNA insertional mutagenesis is not an efficient and practical method for genome-wide mutagenesis in *M. truncatula* (Tadege et al. 2005). Transposon-tagging using the tobacco-derived *Tnt1* retrotransposon offers an alternative method for large-scale insertional mutagenesis in plant species that rely on tissue culture-based methods for transformation, as *Tnt1* transposition is activated during the tissue culture step (D'Erfurth et al. 2003; Tadege et al. 2005). Therefore, a near saturation insertional *Tnt1* mutant population was developed in the *M. truncatula* R108 background through a collaborative effort by researchers at the Noble Research Institute (Oklahoma, USA) and CNRS (France) (D'Erfurth et al. 2003; Porceddu et al. 2008; Tadege et al. 2008). This population consists of roughly 21,000 regenerated lines carrying more than 1 million *Tnt1* insertions and covers about 90% of the genome with each line harboring on average 25 insertions (Lee et al. 2018; Tadege et al. 2008). An efficient PCR-based screening method was developed for the identification of knockout mutants and a BLAST-enabled database containing flanking sequence tag (FST) information has been made available to the scientific community for the easy identification of insertion lines for genes of interest (Cheng et al. 2011). This database is currently hosted on the Oklahoma State University (OSU) server (https://medicago-mutant.dasnr.okstate.edu/mutant/index.php) and contains information on more than 390,000 high confidence FSTs. This mutant population is particularly well-suited for powdery mildew-related forward and reverse genetic screens as it was developed in the moderately susceptible R108 genotype, allowing for the assessment of both reduced and enhanced powdery mildew susceptibility phenotypes. In addition, *Tnt1* insertions were found to be frequently present in genes involved in stress, signaling, defense, secondary metabolism, and transport (Kaur et al. 2021), all pathways that are known to be affected by powdery mildews. So far, only one powdery mildew susceptibility gene, i.e., *MtMLO8* (discussed in Sect. 1.4.2) has been identified through the analysis of *M. truncatula Tnt1* knockout lines (Jacott et al. 2020).

One major disadvantage of the *Tnt1* insertional mutagenesis method is that the tissue culture process may induce additional potential phenotypic background mutations that might be unrelated to the *Tnt1* insertion (Tadege et al. 2005). Therefore, as an alternative, researchers at the John Innes Centre, UK and the Noble Research Institute generated deletion TILLING (De-TILLING) mutant populations consisting of ~ 150,000 M1 lines of the *M. truncatula* A17 genotype in which genomic deletions were physically induced using fast neutron bombardment (FNB) (Rogers et al. 2009). In general, De-TILLING lines have far fewer background mutations, and complete knockout mutants can be efficiently recovered at a low cost by combining a 3D-pooling strategy with PCR-based screening (Rogers et al. 2009), making them ideal for reverse genetics studies. The list of FNB-induced deletion mutants along with phenotypic descriptions of developmental and symbiotic mutants identified in the M2 generation (Chen and Chen 2018) are available at the OSU *Medicago truncatula* mutant database (https://medicago-mutant.dasnr.okstate.edu/mutant/index.php).

In addition to the de-TILLING and *Tnt1* mutant populations, TILLING (Targeting-Induced Local Lesions IN Genomes) collections generated via traditional chemical (EMS) mutagenesis are available for high-throughput screening studies. These include a population of 3,162 individual lines of the A17 genotype developed at the John Innes Centre, UK in collaboration with INRA, France [le Signor et al. (2009); https://www.jic.ac.uk/research-impact/technology-research-platforms/reverse-genetics/medicago-truncatula/] and a collection of 2,400 lines of the 2HA genotype created at CNR IBBR Perugia, in collaboration with CRA FLC Lodi, Italy (Carelli et al. 2013; https://www.ibbr.cnr.it/ibbr/resources//medicago-truncatula-mutant-collections). However, like the *Tnt1* insertional mutants, TILLING mutants typically contain a high number of background mutations and recovery of mutations in small genes (<1 kb) is inefficient (Rogers et al. 2009).

Activation tagging using enhancer sequences has been used as a complementary approach for the creation of a small population of gain-of-function mutants in the *M. truncatula* R108 genotype (Panara et al. 2018; Porceddu et al. 2008). This population is especially useful for the functional characterization of functionally redundant genes or essential genes that cannot be studied via loss-of-function mutations (Panara et al. 2018).

5.5.2 CRISPR/Cas9 Based Gene Editing

Developed as an alternative to traditional mutagenesis approaches, the CRISPR/Cas9 (clustered regularly interspaced palindromic repeats system/CRISPR-associated protein 9) technology has revolutionized reverse genetics studies in plant systems, including legumes, through targeted gene editing (Bhowmik et al. 2021). This system enables the programmable targeting and editing of DNA by an RNA-guided nuclease complex and uses three major components: (1) the CAS9 protein, (2) a guide RNA (gRNA), which consists of a 20 nucleotide RNA sequence (crRNA) that is complementary to the target DNA and a short transactivating RNA that binds to the crRNA and forms a stable complex with Cas9, and (3) a short sequence motif adjacent to the target sequence termed protospacer adjacent motif (PAM), which is required for the correct recognition of the target site (Heintze et al. 2013; Schiml and Puchta 2016). Recognition of the target DNA by the gRNA-Cas9 nuclease complex induces a double-stranded break at the target locus, which leads to the activation of the error-prone host repair system known as non-homologous end joining (NHEJ). This introduces insertions, deletions or substitutions, resulting in the generation of functional knock-outs or gene variants that are stably inherited (Curtin et al. 2018). Moreover, the CRISPR transgene can often be removed from the mutant lines by genetic segregation to obtain non-transgenic mutant plants.

CRISPR/Cas9 reagents which employ a binary vector for the co-expression of Cas9 and gRNA have been optimized for targeted mutagenesis in *M. truncatula* (Curtin 2018; Meng et al. 2020). The *M. truncatula U6* promoter is used for the optimal expression of the gRNA while the 2 × CaMV 35S promoter or the soybean ubiquitin promoter drives optimal expression of Cas9. Using this system, Meng et al. (2017) successfully generated monoallelic and biallelic mutants of an *M. truncatula* gene involved in carotenoid biosynthesis (*Phytoene desaturase; PDS*) with ~10% mutation efficiency in the T0 generation. Additionally, Curtin et al. (2017) used this targeted mutagenesis platform to validate the role of three *M. truncatula* genes previously predicted to control quantitative variation in nodulation. In terms of its applicability for improving powdery mildew resistance in legumes, this technology can be deployed for the targeted disruption of *M. truncatula* genes controlling powdery mildew susceptibility, such as the *MLO*s. The CRISPR/Cas9 approach has been successfully used to target *MLO* genes in tomato (*SlMlo1*), wheat (*TaMlo*), and grapevine (*VvMlo3*) (Pramanik et al. 2021; Wan et al. 2020; Wang et al. 2014). *Mlo*-edited lines showed either complete or enhanced resistance to powdery mildew in these plant species. In addition, components of the plant defense system, such as sugar transporters (e.g., *STP13*), *NLR* genes, and pathogen effector-targets can be targeted by CRISPR/Cas9-based editing as a mechanism to enhance powdery mildew disease resistance in legumes (Andolfo et al. 2016).

5.5.3 Virus-Induced Gene Silencing (VIGS)

Screening large populations of *M. truncatula* mutants for the identification of a single gene mutation is laborious while analysis of genes having a lethal phenotype in *Tnt1*-tagged mutants is problematic. RNAi-based gene silencing of target genes (Saurabh et al. 2014) can address some of these issues, and protocols based on the stable expression of hairpin RNA constructs are available in *M. truncatula* [e.g., Limpens et al. (2004), Samac et al. (2011)]. The hairpin RNA technology is especially useful for silencing multiple genes within a family where gene redundancy might mask phenotypic effects and for genes that are homozygous lethal (Curtin 2018). However, the lengthy transformation protocol makes the process of generating stable hairpin RNA-expressing transgenics tedious and time-consuming, or as in the case of *Agrobacterium rhizogenes*-mediated hairy root transformation, inapplicable for functional characterization of genes expressed in above-ground organs. Hence, virus-induced gene silencing (VIGS) via *Pea early browning virus* (PEBV)-based vectors has emerged as an attractive reverse genetics tool for the systematic and high-throughput identification of gene function in several legumes, including *M. truncatula* (Constantin et al. 2004; Grønlund et al. 2008). VIGS is a transient posttranscriptional gene silencing method that does not require a rigorous regeneration protocol, and the interval between cloning of the gene target into the VIGS vector to phenotype analysis is relatively short (Constantin et al. 2004). Silencing based on VIGS vectors is induced in plants as a part of the RNA-mediated viral defense system, which results in the degradation of endogenous mRNA with high sequence identity to the target gene fragment present in the recombinant virus (Baulcombe 1999; Ruiz et al. 1998).

The PEBV-based VIGS vectors, pCAPE1 and pCAPE2, were generated by exploiting the bipartite RNA genome of the virus (Constantin et al. 2004). pCAPE1 carries the genes that encode viral replication and movement proteins, whereas pCAPE2 carries the coat protein-encoding gene and a fragment of the gene that is targeted for silencing. Infiltration of these vectors in a 1:1 ratio through the *Agrobacterium tumefaciens* strain GV3101 can efficiently silence target genes in the *M. truncatula* genotypes R108, SA1326, and SA1335 (Grønlund et al. 2008), and other genotypes that are systemically infected by PEBV (Serwatowska et al. 2018). This technique was recently used to validate the role of the *M. truncatula* sugar transporter STP13 in legume powdery mildew resistance (Gupta et al. 2021). VIGS of *MtSTP13* in the moderately susceptible R108 genotype reduced *STP13* transcript levels by ~ 80% in leaves of PEBV:*MtSTP13*-infiltrated plants compared to PEBV:*GFP*-infiltrated controls. The appearance of photobleaching symptoms in similar-aged leaves of PEBV: *MtPDS*-infiltrated R108 control plants was used as a visual marker to monitor the rate of symptom development and silencing efficiency. On average, *PDS* silencing symptoms were visible on 4–5 newly developed non-infiltrated trifoliate leaves approximately 24–30 days post agroinfiltration. At this time point, powdery mildew inoculations were performed on *MtSTP13*-silenced and *GFP*-silenced control plants followed by pathogen phenotype assessment at 72 hpi. Higher fungal growth was observed on PEBV:*MtSTP13*-silenced leaves relative to control leaves, indicating that knockdown of *MtSTP13* increases susceptibility to powdery mildew.

5.5.4 Overexpression Through Stable and Transient Approaches

In addition to knockout and knockdown approaches, constitutive/inducible overexpression of candidate genes is another strategy that is routinely used for the functional characterization of candidate genes. Several Agrobacterium-mediated transformation and regeneration protocols have been developed for the stable expression of candidate genes in *M. truncatula* because of differences in callus induction, embryogenesis, and regenerative capacities among genotypes (Chabaud et al. 2003; Cosson et al. 2015; Wright and Wang 2015; Jiang et al. 2019; Wen et al. 2019). The genotypes R108 followed by 2HA (derived from cultivar Jemalong) were found to be best suited for the development of transgenics because of their high in vitro regenerative capacities. Transformation efficiencies of ~50–60% can be achieved with R108 (Cosson et al. 2015) and ~24% with 2HA (Chabaud et al. 2003). Depending on the type of explant used, transgenic plants can be regenerated in 6–8 months (leaf explants) or 3–4 months (cotyledonary explants) after transformation (Cosson et al. 2015; Jiang et al. 2019; Wright and Wang 2015). However, the transformation efficiency drastically drops when cotyledonary explants are used [12% for R108; Wright and Wang (2015)] as compared to leaf explants [50% for R108; Cosson et al. 2015]. Therefore, it is important to consider these points while selecting which protocol to use for the development of *M. truncatula* transgenics.

The production of stable *M. truncatula* transgenics is a slow and laborious process. Hence, transient overexpression in leaves using Agrobacterium-mediated infiltration offers a rapid and high-throughput alternative for gene characterization in *M. truncatula* (Picard et al. 2013; Bond et al. 2016). This method is particularly useful for the functional characterization of transcription factors as it can be readily combined with quantitative reverse transcription-PCR (qRT-PCR) (Picard et al. 2013) or RNA-Seq (Bond et al. 2016) for the identification of downstream targets.

5.6 Conclusions and Future Perspectives

Twenty years ago, *M. truncatula* emerged as the model organism of choice to study legume biology because of its short life cycle, reasonably small diploid genome, and ease of transformation. Since then, major advances in the development of genomic and genetic tools for this model have propelled its use in symbiotic and pathogenic plant–microbe interactions research. *M. truncatula* is well-suited for investigating powdery mildew resistance in legumes because of the inherent genetic diversity in its germplasm and its ability to be infected by some of the same fungal species that cause powdery mildew disease in agriculturally important legumes. Despite this, to date, very few genetic determinants of powdery mildew resistance have been identified in this model legume.

Much of the knowledge of the molecular basis of powdery mildew resistance in *M. truncatula* has been based on comparative transcriptomics analyses of resistant and susceptible genotypes during powdery mildew infection. These large-scale expression studies indicate that (iso)flavonoids, SA, and JA are key components of the defense response against this pathogen. However, further work is required to gain holistic insights into this response at the different levels of biological organization, such as proteins and metabolites. In this regard, an integrated multi-omics approach will provide a systems level understanding of the molecular networks underlying powdery mildew resistance. In addition, further studies are essential to determine the signaling and regulatory factors controlling these resistance responses. For instance, it is still not known whether isoflavonoid metabolites and phytohormones act independently or synergistically to confer protection against this disease in legumes. The molecular basis for defense activation via altered host sugar transport at the powdery mildew infection site is also poorly understood although modulation in the expression and/or activity of a sugar transporter correlates with resistance. It is critical to understand how these different pathways are integrated to mount an effective defense response in *M. truncatula* against powdery mildew.

The *M. truncatula* genome contains a large number of *NLR* and *Mlo* genes that can be exploited for use in legume resistance breeding programs. To this end, several forward and reverse genetics tools have been developed in *M. truncatula* for the functional characterization of candidate genes, each having its own advantages and disadvantages. The *Tnt1* insertion mutant population is useful for performing knockout studies, but unwanted background mutations are problematic, whereas VIGS provides a high-throughput but transient method for conducting knockdown studies. The most promising among these tools is the CRISPR/Cas9 technology, which provides an efficient and precise method for modifying genes through the addition or deletion of genetic material. This technology can be used to identify novel powdery mildew resistance/susceptibility genes and, through their targeted insertion/deletion, potentially introduce durable and broad-spectrum resistance in legumes. In addition, with the continued release of draft genome assemblies of genetically diverse *M. truncatula* accessions through the Medicago HapMap initiative, the use of advanced techniques, such as genome-wide association studies (GWAS), to discover new genetic loci for powdery mildew resistance is possible.

Acknowledgements This work was supported by an Innovative Young Biotechnologist Award [BT/09/IYBA/2015/12] from the Department of Biotechnology, Govt. of India and intramural funds from the Regional Centre for Biotechnology to DC. MG is supported by a University Grants Commission-National Eligibility Test (UGC-NET) Research Fellowship.

References

Acevedo-Garcia J, Kusch S, Panstruga R (2014) Magical mystery tour: MLO proteins in plant immunity and beyond. New Phytol 204:273–281. https://doi.org/10.1111/nph.12889

Ali S, Ganai BA, Kamili AN, Bhat AA, Mir ZA, Bhat JA, Tyagi A, Islam ST, Mushtaq M, Yadav P, Rawat S, Grover A (2018) Pathogenesis-related proteins and peptides as promising tools for engineering plants with multiple stress tolerance. Microbiol Res 212–213:29–37. https://doi.org/10.1016/j.micres.2018.04.008

Ameline-Torregrosa C, Cazaux M, Danesh D, Chardon F, Cannon SB, Esquerre-Tugaye MT, Dumas B, Young ND, Samac DA, Huguet T, Jacquet C, Esquerré-Tugayé M-T, Dumas B, Young ND, Samac DA, Huguet T, Jacquet C (2007) Genetic dissection of resistance to anthracnose and powdery mildew in *Medicago truncatula*. Mol Plant-Microbe Interact 21:61–69. https://doi.org/10.1094/MPMI-21-1-0061

Ameline-Torregrosa C, Wang BB, O'Bleness MS, Deshpande S, Zhu H, Roe B, Young ND, Cannon SB (2008) Identification and characterization of nucleotide-binding site-leucine-rich repeat genes in the model plant *Medicago truncatula*. Plant Physiol 146:5–21. https://doi.org/10.1104/pp.107.104588

Andolfo G, Iovieno P, Frusciante L, Ercolano MR (2016) Genome-editing technologies for enhancing plant disease resistance. Front Plant Sci 7:1813. https://doi.org/10.3389/fpls.2016.01813

Appiano M, Catalano D, Santillán Martínez M, Lotti C, Zheng Z, Visser RGF, Ricciardi L, Bai Y, Pavan S (2015) Monocot and dicot MLO powdery mildew susceptibility factors are functionally conserved in spite of the evolution of class-specific molecular features. BMC Plant Biol 15:1–10. https://doi.org/10.1186/s12870-015-0639-6

Attanayake RN, Glawe DA, Dugan FM, Chen W (2009) *Erysiphe trifolii* causing powdery mildew of lentil (*Lens culinaris*). Plant Dis 93:797–803. https://doi.org/10.1094/PDIS-93-8-0797

Attanayake RN, Glawe DA, McPhee KE, Dugan FM, Chen W (2010) *Erysiphe trifolii*—a newly recognized powdery mildew pathogen of pea. Plant Pathol 59:712–720. https://doi.org/10.1111/j.1365-3059.2010.02306.x

Bahadur A, Singh UP, Sign DP, Sarma BK, Singh KP, Singh A, Aust HJ (2008) Control of *Erysiphe pisi* causing powdery mildew of pea (*Pisum sativum*) by cashewnut (*Anacardium occidentale*) shell extract. Mycobiology 36:60–65. https://doi.org/10.4489/MYCO.2008.36.1.060

Ballard RA, Peck DM, Lloyd DL, Howie JH, Hughes SJ, Hutton RE, Morgan BA (2012) Susceptibility of annual medics (Medicago spp.) to powdery mildew (*Erysiphe trifolii*). In: Proceedings of the 16th Australian agronomy conference, October, pp 14–18

Barilli E, Rubiales D, Gjetting T, Lyngkjaer MF (2014) Differential gene transcript accumulation in peas in response to powdery mildew (*Erysiphe pisi*) attack. Euphytica 198:13–28. https://doi.org/10.1007/s10681-014-1062-z

Barilli E, González-Bernal MJ, Cimmino A, Agudo-Jurado FJ, Masi M, Rubiales D, Evidente A (2019) Impact of fungal and plant metabolites application on early development stages of pea powdery mildew. Pest Manag Sci 75:2464–2473. https://doi.org/10.1002/ps.5351

Barreira JCM, Visnevschi-Necrasov T, Nunes E, Cunha SC, Pereira G, Oliveira MBPP (2015) Medicago spp. as potential sources of bioactive isoflavones: characterization according to phylogenetic and phenologic factors. Phytochemistry 116:230–238. https://doi.org/10.1016/j.phytochem.2015.04.011

Baulcombe DC (1999) Fast forward genetics based on virus-induced gene silencing. Curr Opin Plant Biol

2:109–113. https://doi.org/10.1016/S1369-5266(99)80022-3

Bhowmik P, Konkin D, Polowick P, Hodgins CL, Subedi M, Xiang D, Yu B, Patterson N, Rajagopalan N, Babic V, Ro D, Tar'an B, Bandara M, Smyth SJ, Cui Y, Kagale S (2021) CRISPR/Cas9 gene editing in legume crops: opportunities and challenges. Legume Sci 1–16. https://doi.org/10.1002/leg3.96

Binagwa PH, Traore SM, Egnin M, Bernard GC, Ritte I, Mortley D, Kamfwa K, He G, Bonsi C (2021) Genome-wide identification of powdery mildew resistance in common bean (*Phaseolus vulgaris* L.). Front Genet 12:673069. https://doi.org/10.3389/fgene.2021.673069

Bond DM, Albert NW, Lee RH, Gillard GB, Brown CM, Hellens RP, Macknight RC (2016) Infiltration-RNAseq: transcriptome profiling of Agrobacterium-mediated infiltration of transcription factors to discover gene function and expression networks in plants. Plant Methods 12:1–11. https://doi.org/10.1186/s13007-016-0141-7

Campa A, Ferreira JJ (2017) Gene coding for an elongation factor is involved in resistance against powdery mildew in common bean. Theor Appl Genet 130:849–860. https://doi.org/10.1007/s00122-017-2864-x

Carelli M, Calderini O, Panara F, Porceddu A, Losini I, Piffanelli P, Arcioni S, Scotti C (2013) Reverse genetics in *Medicago truncatula* using a TILLING mutant collection. In: Legume genomics. Humana Press Inc., pp 101–118. https://doi.org/10.1007/978-1-62703-613-9_9

Carver T, Zeyen R, Bushnell W, Robbins M (1994) Inhibition of phenylalanine ammonia lyase and cinnamyl alcohol dehydrogenase increases quantitative susceptibility of barley to powdery mildew (*Erysiphe graminis* D.C.). Physiol Mol Plant Pathol 44:261–272. https://doi.org/10.1016/S0885-5765(05)80029-3

Chabaud M, de Carvalho-Niebel F, Barker DG (2003) Efficient transformation of *Medicago truncatula* cv. Jemalong using the hypervirulent *Agrobacterium tumefaciens* strain AGL1. Plant Cell Rep 22:46–51. https://doi.org/10.1007/s00299-003-0649-y

Chandran D (2015) Co-option of developmentally regulated plant SWEET transporters for pathogen nutrition and abiotic stress tolerance. IUBMB Life 67:461–471. https://doi.org/10.1002/iub.1394

Chankaew S, Somta P, Isemura T, Tomooka N, Kaga A, Vaughan DA, Srinives P (2013) Quantitative trait locus mapping reveals conservation of major and minor loci for powdery mildew resistance in four sources of resistance in mungbean [*Vigna radiata* (L.) Wilczek]. Mol Breed 32:121–130. https://doi.org/10.1007/s11032-013-9856-6

Chen LQ (2014) SWEET sugar transporters for phloem transport and pathogen nutrition. New Phytol 201:1150–1155. https://doi.org/10.1111/nph.12445

Chen Y, Chen R (2018) Physical mutagenesis in *Medicago truncatula* using fast neutron bombardment (FNB) for symbiosis and developmental biology studies. In: Functional genomics in *Medicago truncatula*. Humana Press, New York pp 61–69. https://doi.org/10.1007/978-1-4939-8633-0_4

Cheng X, Wen J, Tadege M, Ratet P, Mysore KS (2011) Reverse genetics in *Medicago truncatula* using Tnt1 insertion mutants. In: Plant reverse genetics, vol 678. Humana Press, Totowa, NJ, pp 179–190. https://doi.org/10.1007/978-1-60761-682-5_13

Constantin GD, Krath BN, MacFarlane SA, Nicolaisen M, Johansen E, Lund OS (2004) Virus-induced gene silencing as a tool for functional genomics in a legume species. Plant J 40:622–631. https://doi.org/10.1111/j.1365-313X.2004.02233.x

Cosson V, Eschstruth A, Ratet P (2015) *Medicago truncatula* transformation using leaf explants. In: Agrobacterium protocols, vol 1223. Springer, New York, pp 43–56. https://doi.org/10.1007/978-1-4939-1695-5_4

Cui H, Tsuda K, Parker JE (2015) Effector-triggered immunity: from pathogen perception to robust defense. Ann Rev Plant Biol 66:487–511. https://doi.org/10.1146/annurev-arplant-050213-040012

Curtin SJ, Tiffin P, Guhlin J, Trujillo D, Burghart L, Atkins P, Baltes NJ, Denny R, Voytas DF, Stupar RM, Young ND (2017) Validating genome-wide association candidates controlling quantitative variation in nodulation. Plant Physiol 173:921–931. https://doi.org/10.1104/pp.16.01923

Curtin SJ, Xiong Y, Michno JM, Campbell BW, Stec AO, Čermák T, Starker C, Voytas DF, Eamens AL, Stupar RM (2018) CRISPR/Cas9 and TALENs generate heritable mutations for genes involved in small RNA processing of *Glycine max* and *Medicago truncatula*. Plant Biotechnol J 16:1125–1137. https://doi.org/10.1111/pbi.12857

Curtin SJ (2018) Editing the *Medicago truncatula* genome: targeted mutagenesis using the CRISPR-Cas9 reagent. In: Functional genomics in *Medicago truncatula*. Humana Press, New York, pp 161–174. https://doi.org/10.1007/978-1-4939-8633-0_12

Curto M, Krajinski F, Küster H, Rubiales D (2014) Plant defense responses in *Medicago truncatula* unveiled by microarray analysis. Plant Mol Biol Rep 33:569–583. https://doi.org/10.1007/s11105-014-0770-9

Curto M, Krajinski F, Schlereth A, Rubiales D (2015) Transcriptional profiling of *Medicago truncatula* during *Erysiphe pisi* infection. Front Plant Sci 6. https://doi.org/10.3389/fpls.2015.00517

D'Erfurth I, Cosson V, Eschstruth A, Lucas H, Kondorosi A, Ratet P (2003) Efficient transposition of the *Tnt1* tobacco retrotransposon in the model legume *Medicago truncatula*. Plant J 34:95–106. https://doi.org/10.1046/j.1365-313X.2003.01701.x

Dempsey DA, Vlot AC, Wildermuth MC, Klessig DF (2011) Salicylic acid biosynthesis and metabolism. In: The arabidopsis book, vol 9. American Society of Plant Biologists, p e0156. https://doi.org/10.1199/tab.0156

Deshmukh R, Singh VK, Singh BD (2017) Mining the *Cicer arietinum* genome for the *mildew locus O* (*Mlo*)

gene family and comparative evolutionary analysis of the *Mlo* genes from *Medicago truncatula* and some other plant species. J Plant Res 130:239–253. https://doi.org/10.1007/s10265-016-0868-2

Dixon RA, Achnine L, Kota P, Liu CJ, Reddy MSS, Wang L (2002) The phenylpropanoid pathway and plant defence—a genomics perspective. Mol Plant Pathol 3:371–390. https://doi.org/10.1046/j.1364-3703.2002.00131.x

Fondevilla S, Rubiales D (2012) Powdery mildew control in pea. A review. Agron Sustain Dev 32:401–409. https://doi.org/10.1007/s13593-011-0033-1

Fondevilla S, Carver TLW, Moreno MT, Rubiales D (2006) Macroscopic and histological characterisation of genes *er1* and *er2* for powdery mildew resistance in pea. Eur J Plant Pathol 115:309–321. https://doi.org/10.1007/s10658-006-9015-6

Fondevilla S, Torres AM, Moreno T, Rubiales D (2007) Identification of a new gene for resistance to powdery mildew in *Pisum fulvum*, a wild relative of pea. Breed Sci 57:181–184. https://doi.org/10.1270/jsbbs.57.181

Fondevilla S, Chattopadhyay C, Khare N, Rubiales D (2013) *Erysiphe trifolii* is able to overcome *er1* and *Er3*, but not *er2*, resistance genes in pea. Eur J Plant Pathol 136:557–563. https://doi.org/10.1007/s10658-013-0187-6

Foster-Hartnett D, Danesh D, Peñuela S, Sharopova N, Endre G, Vandenbosch KA, Young ND, Samac DA (2007) Molecular and cytological responses of *Medicago truncatula* to *Erysiphe pisi*. Mol Plant Pathol 8:307–319. https://doi.org/10.1111/j.1364-3703.2007.00395.x

Gao QM, Venugopal S, Navarre D, Kachroo A (2011) Low oleic acid-derived repression of jasmonic acid-inducible defense responses requires the WRKY50 and WRKY51 proteins. Plant Physiol 155:464–476. https://doi.org/10.1104/pp.110.166876

Graham PH, Vance CP (2003) Legumes: importance and constraints to greater use. Plant Physiol 131:872–877. https://doi.org/10.1104/pp.017004

Graves WL, Stuteville DL, Johnson RC, Greene SL (1999) Powdery mildew caused by an Oidium sp. on twenty-one annual Medicago spp. in California. Plant Dis 83:1176–1176. https://doi.org/10.1094/pdis.1999.83.12.1176c

Grønlund M, Constantin G, Piednoir E, Kovacev J, Johansen IE, Lund OS (2008) Virus-induced gene silencing in *Medicago truncatula* and *Lathyrus odorata*. Virus Res 135:345–349. https://doi.org/10.1016/j.virusres.2008.04.005

Gujaria-Verma N, Vail SL, Carrasquilla-Garcia N, Penmetsa RV, Cook DR, Farmer AD, Vandenberg A, Bett KE (2014) Genetic mapping of legume orthologs reveals high conservation of synteny between lentil species and the sequenced genomes of Medicago and chickpea. Front Plant Sci 5. https://doi.org/10.3389/fpls.2014.00676

Guo P, Li Z, Huang P, Li B, Fang S, Chu J, Guo H (2017) A tripartite amplification loop involving the transcription factor WRKY75, salicylic acid, and reactive oxygen species accelerates leaf senescence. Plant Cell 29:2854–2870. https://doi.org/10.1105/tpc.17.00438

Gupta M, Sharma G, Saxena D, Budhwar R, Vasudevan M, Gupta V, Gupta A, Gupta R, Chandran D (2020) Dual RNA-Seq analysis of *Medicago truncatula* and the pea powdery mildew *Erysiphe pisi* uncovers distinct host transcriptional signatures during incompatible and compatible interactions and pathogen effector candidates. Genomics 112:2130–2145. https://doi.org/10.1016/j.ygeno.2019.12.007

Gupta M, Dubey S, Jain D, Chandran D (2021) The Medicago truncatula sugar transport protein 13 and its Lr67res-like variant confer powdery mildew resistance in legumes via defense modulation. Plant Cell Physiol. https://doi.org/10.1093/pcp/pcab021

Heintze J, Luft C, Ketteler R (2013) A CRISPR CASe for high-throughput silencing. Front Genet 4:193. https://doi.org/10.3389/fgene.2013.00193

Heringa R, van Norel A, Tazelaar M (1969) Resistance to powdery mildew (*Erysiphe polygoni* D.C.) in peas (*Pisum sativum* L.). Euphytica 18:163–169. https://doi.org/10.1007/BF00035687

Hückelhoven R (2005) Powdery mildew susceptibility and biotrophic infection strategies. FEMS Microbiol Lett 245:9–17. https://doi.org/10.1016/j.femsle.2005.03.001

Humphry M, Reinstädler A, Ivanov S, Bisseling T, Panstruga R (2011) Durable broad-spectrum powdery mildew resistance in pea *er1* plants is conferred by natural loss-of-function mutations in *PsMLO1*. Mol Plant Pathol 12:866–878. https://doi.org/10.1111/j.1364-3703.2011.00718.x

Iglesias-García R, Rubiales D, Fondevilla S (2015) Penetration resistance to Erysiphe pisi in pea mediated by er1 gene is associated with protein cross-linking but not with callose apposition or hypersensitive response. Euphytica 201:381–387. https://doi.org/10.1007/s10681-014-1221-2

Ishiga Y, Rao Uppalapati S, Gill US, Huhman D, Tang Y, Mysore KS (2015) Transcriptomic and metabolomic analyses identify a role for chlorophyll catabolism and phytoalexin during Medicago nonhost resistance against Asian soybean rust. Sci Rep 5:1–17. https://doi.org/10.1038/srep13061

Jacott CN, Charpentier M, Murray JD, Ridout CJ (2020) Mildew Locus O facilitates colonization by arbuscular mycorrhizal fungi in angiosperms. New Phytol 227:343–351. https://doi.org/10.1111/nph.16465

Jan R, Asaf S, Numan M, Lubna KKM (2021) Plant secondary metabolite biosynthesis and transcriptional regulation in response to biotic and abiotic stress conditions. Agronomy 11:938. https://doi.org/10.3390/agronomy11050968

Jiang Q, Fu C, Wang ZY (2019) A unified Agrobacterium-mediated transformation protocol for alfalfa (*Medicago sativa* L.) and *Medicago truncatula*. In: Transgenic plants. Humana Press, New York, pp 153–163. https://doi.org/10.1007/978-1-4939-8778-8_11

Jiménez-González L, Álvarez-Corral M, Muñoz-Dorado M, Rodríguez-García I (2008) Pterocarpans: interesting natural products with antifungal activity and other biological properties. Phytochem Rev 7:125–154. https://doi.org/10.1007/s11101-007-9059-z

Kankanala P, Nandety RS, Mysore KS (2019) Genomics of plant disease resistance in legumes. Front Plant Sci 10:1345. https://doi.org/10.3389/fpls.2019.01345

Kaur P, Lui C, Dudchenko O, Nandety RS, Hurgobin B, Pham M, Aiden EL, Wen J, Mysore K (2021) Delineating the tnt1 insertion landscape of the model legume *Medicago truncatula* cv. R108 at the hi-c resolution using a chromosome-length genome assembly. Int J Mol Sci 22:4326. https://doi.org/10.3390/ijms22094326

Khunti JP, Bhoraniya ME, Vora VD (2005) Management of powdery mildew and cercospora leaf spot of mung bean by some systemic fungicides. Legume Res 28:65–67

Krishnakumar V (2020) MTGD: the Medicago truncatula genome database. Plant Cell Physiol 56:e1–e1. https://doi.org/10.1093/pcp/pcu179

Lahoz E, Carrieri R, Parisi B, Pentangelo A, Raimo F (2013) Overcoming of the resistance in resistant genotypes of dry pea (*Pisum sativum*) by an isolate of *Erysiphe pisi* in Italy. J Plant Pathol 171–176. https://www.jstor.org/stable/23721750

le Signor C, Savois V, Aubert G, Verdier J, Nicolas M, Pagny G, Moussy F, Sanchez M, Baker D, Clarke J, Thompson R (2009) Optimizing TILLING populations for reverse genetics in *Medicago truncatula*. Plant Biotech J 7:430–441. https://doi.org/10.1111/j.1467-7652.2009.00410.x

Lee C, Yu D, Choi HK, Kim RW (2017) Reconstruction of a composite comparative map composed of ten legume genomes. Genes Genom 39:111–119. https://doi.org/10.1007/s13258-016-0481-8

Lee HB, Nguyen TTT (2021) First report of powdery mildew caused by *Erysiphe viciae-unijugae* on *Vicia sativa* subsp. nigra in Korea. Plant Dis 105:493. https://doi.org/10.1094/PDIS-08-20-1816-PDN

Lee HK, Mysore KS, Wen J (2018) Tnt1 insertional mutagenesis in *Medicago truncatula*. In: Functional genomics in Medicago truncatula. Humana Press, New York, pp 107–114. https://doi.org/10.1007/978-1-4939-8633-0_7

Lemonnier P, Gaillard C, Veillet F, Verbeke J, Lemoine R, Coutos-Thévenot P, la Camera S (2014) Expression of Arabidopsis sugar transport protein STP13 differentially affects glucose transport activity and basal resistance to *Botrytis cinerea*. Plant Mol Biol 85:473–484. https://doi.org/10.1007/s11103-014-0198-5

Limpens E, Ramos J, Franken C, Raz V, Compaan B, Franssen H, Bisseling T, Geurts R (2004) RNA interference in *Agrobacterium rhizogenes*-transformed roots of Arabidopsis and *Medicago truncatula*. J Exp Bot 55:983–992. https://doi.org/10.1093/jxb/erh122

Liu L, Sonbol FM, Huot B, Gu Y, Withers J, Mwimba M, Yao J, He SY, Dong X (2016) Salicylic acid receptors activate jasmonic acid signalling through a non-canonical pathway to promote effector-triggered immunity. Nat Commun 7:1–10. https://doi.org/10.1038/ncomms13099

Mahmoud GA-E (2021) Biotic stress to legumes: Fungal diseases as major biotic stress factor. In: Sustainable agriculture reviews 51: legume agriculture and biotechnology, vol 2, p 181. https://doi.org/10.1007/978-3-030-68828-8_7

Martins D, Araújo S de S, Rubiales D, Patto MCV (2020) Legume crops and biotrophic pathogen interactions: a continuous cross-talk of a multilayered array of defense mechanisms. Plants 9:1–24. https://doi.org/10.3390/plants9111460

Meng Y, Hou Y, Wang H, Ji R, Liu B, Wen J, Niu L, Lin H (2017) Targeted mutagenesis by CRISPR/Cas9 system in the model legume *Medicago truncatula*. Plant Cell Rep 36:371–374. https://doi.org/10.1007/s00299-016-2069-9

Meng Y, Wang C, Yin P, Zhu B, Zhang P, Niu L, Lin H (2020) Targeted mutagenesis by an optimized agrobacterium-delivered CRISPR/Cas9 system in the model legume *Medicago truncatula*. In: The model legume *Medicago truncatula*, pp 1015–1018. https://doi.org/10.1002/9781119409144.ch130

Milne RJ, Dibley KE, Schnippenkoetter W, Mascher M, Lui ACW, Wang L, Lo C, Ashton AR, Ryan PR, Lagudah ES (2019) The wheat LR67 gene from the sugar transport protein 13 family confers multi-pathogen resistance in barley. Plant Physiol 179:1285–1297. https://doi.org/10.1104/pp.18.00945

Mine A, Seyfferth C, Kracher B, Berens ML, Becker D, Tsuda K (2018) The defense phytohormone signaling network enables rapid, high-amplitude transcriptional reprogramming during effector-triggered immunity. Plant Cell 30:1199–1219. https://doi.org/10.1105/tpc.17.00970

Moll KM, Zhou P, Ramaraj T, Fajardo D, Devitt NP, Sadowsky MJ, Stupar RM, Tiffin P, Miller JR, Young ND, Silverstein KAT, Mudge J (2017) Strategies for optimizing BioNano and Dovetail explored through a second reference quality assembly for the legume model, *Medicago truncatula*. BMC Genom 18:1–16. https://doi.org/10.1186/s12864-017-3971-4

Moore JW, Herrera-Foessel S, Lan C, Schnippenkoetter W, Ayliffe M, Huerta-Espino J, Lillemo M, Viccars L, Milne R, Periyannan S, Kong X, Spielmeyer W, Talbot M, Bariana H, Patrick JW, Dodds P, Singh R, Lagudah E (2015) A recently evolved hexose transporter variant confers resistance to multiple pathogens in wheat. Nat Genet 47:1494–1498. https://doi.org/10.1038/ng.3439

Morkunas I, Ratajczak L (2014) The role of sugar signaling in plant defense responses against fungal pathogens. Acta Physiol Plant 36:1607–1619. https://doi.org/10.1007/s11738-014-1559-z

Mur LAJ, Kenton P, Lloyd AJ, Ougham H, Prats E (2008) The hypersensitive response; The centenary is upon us but how much do we know? J Exp Bot 59:501–520. https://doi.org/10.1093/jxb/erm239

Naoumkina M, Farag MA, Sumner LW, Tang Y, Liu CJ, Dixon RA (2007) Different mechanisms for phytoalexin induction by pathogen and wound signals in *Medicago truncatula*. Proc Natl Acad Sci U S A 104:17909–17915. https://doi.org/10.1073/pnas.0708697104

Nepal MP, Andersen EJ, Neupane S, Benson BV (2017) Comparative genomics of non-TNL disease resistance genes from six plant species. Genes 8:10. https://doi.org/10.3390/genes8100249

Panara F, Calderini O, Porceddu A (2018) T-DNA insertional mutagenesis and activation tagging in Medicago truncatula. In: Functional genomics in Medicago truncatula. Humana Press, New York, pp 83–105. https://doi.org/10.1007/978-1-4939-8633-0_6

Panstruga R (2003) Establishing compatibility between plants and obligate biotrophic pathogens. Curr Opin Plant Biol 6:320–326. https://doi.org/10.1016/S1369-5266(03)00043-8

Patel JS, Selvaraj V, Gunupuru LR, Rathor PK, Prithiviraj B (2020) Combined application of *Ascophyllum nodosum* extract and chitosan synergistically activates host-defense of peas against powdery mildew. BMC Plant Biol 20. https://doi.org/10.1186/s12870-020-2287-8

Pavan S, Schiavulli A, Appiano M, Marcotrigiano AR, Cillo F, Visser RGF, Bai Y, Lotti C, Ricciardi L (2011) Pea powdery mildew er1 resistance is associated to loss-of-function mutations at a MLO homologous locus. Theor Appl Genet 123:1425–1431. https://doi.org/10.1007/s00122-011-1677-6

Pecrix Y, Staton SE, Sallet E, Lelandais-Brière C, Moreau S, Carrère S, Blein T, Jardinaud MF, Latrasse D, Zouine M, Zahm M, Kreplak J, Mayjonade B, Satgé C, Perez M, Cauet S, Marande W, Chantry-Darmon C, Lopez-Roques C, Bouchez O, Bérard A, Debellé F, Muños S, Bendahmane A, Bergès H, Niebel A, Buitink J, Frugier F, Benhamed M, Crespi M, Gouzy J, Gamas P (2018) Whole-genome landscape of *Medicago truncatula* symbiotic genes. Nat Plants 4:1017–1025. https://doi.org/10.1038/s41477-018-0286-7

Pérez-Vega E, Trabanco N, Campa A, Ferreira JJ (2013) Genetic mapping of two genes conferring resistance to powdery mildew in common bean (*Phaseolus vulgaris* L.). Theor Appl Genet 126:1503–1512. https://doi.org/10.1007/s00122-013-2068-y

Picard K, Lee R, Hellens R, Macknight R (2013) Transient gene expression in *Medicago truncatula* leaves via agroinfiltration. In: Legume genomics, vol 1069. Humana Press, Totowa, NJ, pp 215–226. https://doi.org/10.1007/978-1-62703-613-9_15

Pislariu CI, Murray JD, Wen JQ, Cosson V, Muni RSRD, Wang M, Benedito VA, Andriankaja A, Cheng X, Jerez IT, Mondy S, Zhang S, Taylor ME, Tadege M, Ratet P, Mysore KS, Chen R, Udvardi MK (2012) A *Medicago truncatula* tobacco retrotransposon insertion mutant collection with defects in nodule development and symbiotic nitrogen fixation. Plant Physiol 159:1686–1699. https://doi.org/10.1104/pp.112.197061

Polak R, Phillips EM, Campbell A (2015) Legumes: health benefits and culinary approaches to increase intake. Clin Diabetes 33:198–205. https://doi.org/10.2337/diaclin.33.4.198

Porceddu A, Panara F, Calderini O, Molinari L, Taviani P, Lanfaloni L, Scotti C, Carelli M, Scaramelli L, Bruschi G, Cosson V, Ratet P, de Larembergue H, Duc G, Piano E, Arcioni S (2008) An Italian functional genomic resource for *Medicago truncatula*. BMC Res Notes 1:1–7. https://doi.org/10.1186/1756-0500-1-129

Pramanik D, Shelake RM, Park J, Kim MJ, Hwang I, Park Y, Kim JY (2021) CRISPR/Cas9-mediated generation of pathogen-resistant tomato against tomato yellow leaf curl virus and powdery mildew. Int J Mol Sci 22:1–18. https://doi.org/10.3390/ijms22041878

Prats E, Llamas MJ, Rubiales D (2007) Characterization of resistance mechanisms to *Erysiphe pisi* in *Medicago truncatula*. Phytopathol 97:1049–1053. https://doi.org/10.1094/PHYTO-97-9-1049

Ransom LM, O'Brien RG, Glass RJ (1991) Chemical control of powdery mildew in green peas. Australas Plant Pathol 20:16–20. https://doi.org/10.1071/APP9910016

Rawal V, Charrondiere R, Xipsiti M, Grande F (2019) Pulses: nutritional benefits and consumption pattern. In: Rawal V, Navarro D (eds) The global economy of pulses. Food and Agriculture Organization of the United Nations, Rome, pp 9–19

Rispail N, Prats E, Rubiales D (2019) *Medicago truncatula* as a model to study powdery mildew resistance. In: The model legume *Medicago truncatula*, pp 390–397. https://doi.org/10.1002/9781119409144.ch49

Rispail N, Rubiales D (2016) Genome-wide identification and comparison of legume MLO gene family. Sci Rep 6:1–12. https://doi.org/10.1038/srep32673

Rogers C, Wen J, Chen R, Oldroyd G (2009) Deletion-based reverse genetics in *Medicago truncatula*. Plant Physiol 151:1077–1086. https://doi.org/10.1104/pp.109.142919

Rubiales D, Fondevilla S, Chen W, Gentzbittel L, Higgins TJV, Castillejo MA, Singh KB, Rispail N (2015) Achievements and challenges in legume breeding for pest and disease resistance. Crit Rev Plant Sci 34:195–236. https://doi.org/10.1080/07352689.2014.898445

Ruiz MT, Voinnet O, Baulcombe DC (1998) Initiation and maintenance of virus-induced gene silencing. Plant Cell 10:937–946. https://doi.org/10.1105/tpc.10.6.937

Sahni S, Maurya S, Singh UP, Singh AK, Singh VP, Pandey VB (2005) Antifungal activity of Norsecurinine against some phytopathogenic fungi. Mycobiology 33:97–103. https://doi.org/10.4489/MYCO.2005.33.2.097

Samac DA, Graham MA (2007) Recent advances in legume-microbe interactions: recognition, defense response, and symbiosis from a genomic perspective.

Plant Physiol 144:582–587. https://doi.org/10.1104/pp.107.096503

Samac DA, Peñuela S, Schnurr JA, Hunt EN, Foster-Hartnett D, Vandenbosch KA, Gantt JS (2011) Expression of coordinately regulated defence response genes and analysis of their role in disease resistance in *Medicago truncatula*. Mol Plant Pathol 12:786–798. https://doi.org/10.1111/j.1364-3703.2011.00712.x

Santos C, Martins D, Rubiales D, Vaz Patto MC (2020) Partial resistance against *Erysiphe pisi* and *E. trifolii* under different genetic control in *Lathyrus cicera*: outcomes from a linkage mapping approach. Plant Dis 104:2875–2884. https://doi.org/10.1094/PDIS-03-20-0513-RE

Saurabh S, Vidyarthi AS, Prasad D (2014) RNA interference: concept to reality in crop improvement. Planta 239:543–564. https://doi.org/10.1007/s00425-013-2019-5

Schiml S, Puchta H (2016) Revolutionizing plant biology: multiple ways of genome engineering by CRISPR/Cas. Plant Methods 12:1–9. https://doi.org/10.1186/s13007-016-0103-0

Serwatowska J, Lund OS, Johansen IE (2018) Transient posttranscriptional gene silencing in *Medicago truncatula*: Virus-induced gene silencing (VIGS). Functional genomics in Medicago truncatula. Humana Press, New York, pp 115–122

Sharma G, Aminedi R, Saxena D, Gupta A, Banerjee P, Jain D, Chandran D (2019) Effector mining from the *Erysiphe pisi* haustorial transcriptome identifies novel candidates involved in pea powdery mildew pathogenesis. Mol Plant Pathol 20:1506–1522. https://doi.org/10.1111/mpp.12862

Sillero JC, Fondevilla S, Davidson J, Vaz Patto MC, Warkentin TD, Thomas J, Rubiales D (2006) Screening techniques and sources of resistance to rusts and mildews in grain legumes. Euphytica 147:255–272. https://doi.org/10.1007/s10681-006-6544-1

Singh UP, Prithiviraj B (1997) Neemazal, a product of neem (*Azadirachta indica*), induces resistance in pea (*Pisum sativum*) against *Erysiphe pisi*. Physiol Mol Plant Pathol 51:181–194. https://doi.org/10.1006/pmpp.1997.0112

Singh A, Bhatt B, Singh K, Kumar A, Manibhushan KU, Chandra N, Bharati R (2013) Dynamics of powdery mildew (Erysiphe trifolii) disease of lentil influenced by sulphur and zinc nutrition. Plant Pathol J 12:71–77. https://doi.org/10.3923/ppj.2013.71.77

Singh U, Shekar C (2020) Evaluation of fungicides management of powdery mildew (*Erysiphe Polygoni* DC) disease of field pea (*Pisum Sativum* L.) in Eastern U.P. Int J Chem Stud 8:2590–2593. https://doi.org/10.22271/chemi.2020.v8.i2an.9141

Smigielski L, Laubach EM, Pesch L, Glock JML, Albrecht F, Slusarenko A, Panstruga R, Kuhn H (2019) Nodulation induces systemic resistance of *Medicago truncatula* and *Pisum sativum* against *Erysiphe pisi* and primes for powdery mildew-triggered salicylic acid accumulation. Mol Plant-Microbe Interact 32:1243–1255. https://doi.org/10.1094/MPMI-11-18-0304-R

Smith PH, Foster EM, Boyd LA, Brown JKM (1996) The early development of *Erysiphe pisi* on *Pisum sativum* L. Plant Pathol 45:302–309. https://doi.org/10.1046/j.1365-3059.1996.d01-111.x

Stacey G, Libault M, Brechenmacher L, Wan J, May GD (2006) Genetics and functional genomics of legume nodulation. Curr Opin Plant Biol 9:110–121. https://doi.org/10.1016/j.pbi.2006.01.005

Stagnari F, Maggio A, Galieni A, Pisante M (2017) Multiple benefits of legumes for agriculture sustainability: an overview. Chem Biol Technol Agric 4:1–13. https://doi.org/10.1186/s40538-016-0085-1

Stanton-Geddes J, Paape T, Epstein B, Briskine R, Yoder J, Mudge J, Bharti AK, Farmer AD, Zhou P, Denny R, May GD, Erlandson S, Yakub M, Sugawara M, Sadowsky MJ, Young ND, Tiffin P (2013) Candidate genes and genetic architecture of symbiotic and agronomic traits revealed by whole-genome, sequence-based association genetics in *Medicago truncatula*. PLoS One 8. https://doi.org/10.1371/journal.pone.0065688

Sun S, Fu H, Wang Z, Duan C, Zong X, Zhu Z (2016) Discovery of a novel er1 allele conferring powdery mildew resistance in Chinese Pea (*Pisum sativum* L.) landraces. PLoS One 11:e0147624. https://doi.org/10.1371/journal.pone.0147624

Sun S, Deng D, Duan C, Zong X, Xu D, He Y, Zhu Z (2019) Two novel er1 alleles conferring powdery mildew (*Erysiphe pisi*) resistance identified in a worldwide collection of pea (*Pisum sativum* L.) germplasms. Int J Mol Sci 20:5071. https://doi.org/10.3390/ijms20205071

Tadege M, Ratet P, Mysore KS (2005) Insertional mutagenesis: a Swiss Army knife for functional genomics of *Medicago truncatula*. Trends Plant Sci 10:229–235. https://doi.org/10.1016/j.tplants.2005.03.009

Tadege M, Wen J, He J, Tu H, Kwak Y, Eschstruth A, Cayrel A, Endre G, Zhao PX, Chabaud M, Ratet P, Mysore KS (2008) Large-scale insertional mutagenesis using the Tnt1 retrotransposon in the model legume *Medicago truncatula*. Plant J 54:335–347. https://doi.org/10.1111/j.1365-313X.2008.03418.x

Tang H, Krishnakumar V, Bidwell S, Rosen B, Chan A, Zhou S, Gentzbittel L, Childs KL, Yandell M, Gundlach H, Mayer KFX, Schwartz DC, Town CD (2014) An improved genome release (version Mt4.0) for the model legume *Medicago truncatula*. BMC Genom 15:1–14. https://doi.org/10.1186/1471-2164-15-312

Thaler JS, Humphrey PT, Whiteman NK (2012) Evolution of jasmonate and salicylate signal crosstalk. Trends Plant Sci 17:260–270. https://doi.org/10.1016/j.tplants.2012.02.010

Tiwari KR, Penner GA, Warkentin TD (1997) Inheritance of powdery mildew resistance in pea. Can J Plant Sci 77:307–310. https://doi.org/10.4141/P96-157

Tiwari KR, Warkentin TD, Penner GA, Menzies JG (1999) Studies on winter survival strategies of *Erysiphe pisi* in Manitoba. Can J Plant Pathol 21:159–164. https://doi.org/10.1080/07060669909501207

Tsuda K, Sato M, Stoddard T, Glazebrook J, Katagiri F (2009) Network properties of robust immunity in plants. PLoS Genet 5:e1000772. https://doi.org/10.1371/journal.pgen.1000772

Varshney RK, Song C, Saxena RK, Azam S, Yu S, Sharpe AG, Cannon S, Baek J, Rosen BD, Tar'an B, Millan T, Zhang X, Ramsay LD, Iwata A, Wang Y, Nelson W, Farmer AD, Gaur PM, Soderlund C, Penmetsa RV, Xu C, Bharti AK, He W, Winter P, Zhao S, Hane JK, Carrasquilla-Garcia N, Condie JA, Padhyaya HD, Luo MC, Thudi M, Gowda CLL, Singh NP, Lichtenzveig J, Gali KK, Rubio J, Nadarajan N, Dolezel J, Bansal KC, Xu X, Edwards D, Zhang G, Kahl G, Gil J, Singh KB, Datta SK, Jackson SA, Wang J, Cook DR (2013) Draft genome sequence of chickpea (*Cicer arietinum*) provides a resource for trait improvement. Nat Biotech 31:240–246. https://doi.org/10.1038/nbt.2491

Veerappan V, Jani M, Kadel K, Troiani T, Gale R, Mayes T, Shulaev E, Wen J, Mysore KS, Azad RK, Dickstein R (2016) Rapid identification of causative insertions underlying *Medicago truncatula* Tnt1 mutants defective in symbiotic nitrogen fixation from a forward genetic screen by whole genome sequencing. BMC Genom 17:1–11. https://doi.org/10.1186/s12864-016-2452-5

Vielba-Fernández A, Polonio Á, Ruiz-Jiménez L, de Vicente A, Pérez-García A, Fernández-Ortuño D (2020) Fungicide resistance in powdery mildew fungi. Microorganisms 8:1–34. https://doi.org/10.3390/microorganisms8091431

Villegas-Fernández ÁM, Amarna AA, Moral J, Rubiales D (2021) Crop diversification to control powdery mildew in pea. Agronomy 11:690. https://doi.org/10.3390/AGRONOMY11040690

Wan DY, Guo Y, Cheng Y, Hu Y, Xiao S, Wang Y, Wen YQ (2020) CRISPR/Cas9-mediated mutagenesis of VvMLO3 results in enhanced resistance to powdery mildew in grapevine (*Vitis vinifera*). Hortic Res 7:1–14. https://doi.org/10.1038/s41438-020-0339-8

Wang W, Wen Y, Berkey R, Xiao S (2009) Specific targeting of the Arabidopsis resistance protein RPW8.2 to the interfacial membrane encasing the fungal haustorium renders broad-spectrum resistance to powdery mildew. Plant Cell 21:2898–2913. https://doi.org/10.1105/tpc.109.067587

Wang Y, Shi A, Zhang B, Chen P (2013) Mapping powdery mildew resistance gene in V97–3000 soybean. Plant Breed 132:625–629. https://doi.org/10.1111/pbr.12072

Wang Y, Cheng X, Shan Q, Zhang Y, Liu J, Gao C, Qiu JL (2014) Simultaneous editing of three homoeoalleles in hexaploid bread wheat confers heritable resistance to powdery mildew. Nat Biotech 32:947–951. https://doi.org/10.1038/nbt.2969

Warkentin TD, Rashid KY, Xue AG (1996) Fungicidal control of powdery mildew in field pea. Can J Plant Sci 76:933–935. https://doi.org/10.4141/cjps96-156

Wen L, Chen Y, Schnabel E, Crook A, Frugoli J (2019) Comparison of efficiency and time to regeneration of Agrobacterium-mediated transformation methods in *Medicago truncatula*. Plant Methods 15:1–10. https://doi.org/10.1186/s13007-019-0404-1

Wiermer M, Feys BJ, Parker JE (2005) Plant immunity: The EDS1 regulatory node. Curr Opin Plant Biol 8:383–389. https://doi.org/10.1016/j.pbi.2005.05.010

Wright E, Wang ZY (2015) *Medicago truncatula* transformation using cotyledonary explants. In: Agrobacterium protocols, vol 1223. Springer, New York, pp 35–41. https://doi.org/10.1007/978-1-4939-1695-5_3

Wyand RA, Brown JK (2003) Genetic and forma specialis diversity in *Blumeria graminis* of cereals and its implications for host-pathogen co-evolution. Mol Plant Pathol 4:187–198. https://doi.org/10.1046/j.1364-3703.2003.00167.x

Xiao S, Ellwood S, Calis O, Patrick E, Li T, Coleman M, Turner JG, Xiao S, Ellwood S, Callis O, Patrick E, Li T, Coleman M, Turner JG (2001) Broad-Spectrum mildew resistance in Arabidopsis mediated by RPW8. Science 291:118–120. https://doi.org/10.1126/science.291.5501.118

Yaege JR, Stuteville DL (2002) Reactions of accessions in the annual Medicago core germplasm collection to *Erysiphe pisi*. Plant Dis 86:312–315. https://doi.org/10.1094/PDIS.2002.86.3.312

Yamada K, Saijo Y, Nakagami H, Takano Y (2016) Regulation of sugar transporter activity for antibacterial defense in Arabidopsis. Science 354:1427–1430. https://doi.org/10.1126/science.aah5692

Yang S, Tang F, Caixeta ET, Zhu H (2013) Epigenetic regulation of a powdery mildew resistance gene in *Medicago truncatula*. Mol Plant 6:2000–2003. https://doi.org/10.1093/mp/sst106

Young ND, Debellé F, Oldroyd GED, Geurts R, Cannon SB, Udvardi MK, Benedito VA, Mayer KFX, Gouzy J, Schoof H, van de Peer Y, Proost S, Cook DR, Meyers BC, Spannagl M, Cheung F, de Mita S, Krishnakumar V, Gundlach H, Zhou S, Mudge J, Bharti AK, Murray JD, Naoumkina MA, Rosen B, Silverstein KAT, Tang H, Rombauts S, Zhao PX, Zhou P, Barbe V, Bardou P, Bechner M, Bellec A, Berger A, Bergès H, Bidwell S, Bisseling T, Choisne N, Couloux A, Denny R, Deshpande S, Dai X, Doyle JJ, Dudez AM, Farmer AD, Fouteau S, Franken C, Gibelin C, Gish J, Goldstein S, González AJ, Green PJ, Hallab A, Hartog M, Hua A, Humphray SJ, Jeong DH, Jing Y, Jöcker A, Kenton SM, Kim DJ, Klee K, Lai H, Lang C, Lin S, MacMil SL, Magdelenat G, Matthews L, McCorrison J, Monaghan EL, Mun JH, Najar FZ, Nicholson C, Noirot C, O'Bleness M, Paule CR, Poulain J, Prion F, Qin B, Qu C, Retzel EF, Riddle C, Sallet E, Samain S, Samson N, Sanders I, Saurat O, Scarpelli C, Schiex T, Segurens B, Severin AJ, Sherrier DJ, Shi R,

Sims S, Singer SR, Sinharoy S, Sterck L, Viollet A, Wang BB, Wang K, Wang M, Wang X, Warfsmann J, Weissenbach J, White DD, White JD, Wiley GB, Wincker P, Xing Y, Yang L, Yao Z, Ying F, Zhai J, Zhou L, Zuber A, Dénarié J, Dixon RA, May GD, Schwartz DC, Rogers J, Quétier F, Town CD, Roe BA (2011) The Medicago genome provides insight into the evolution of rhizobial symbioses. Nature 480:520–524. https://doi.org/10.1038/nature10625

Yuan M, Jiang Z, Bi G, Nomura K, Liu M, Wang Y, Cai B, Zhou JM, He SY, Xin XF (2021) Pattern-recognition receptors are required for NLR-mediated plant immunity. Nature 592:105–109. https://doi.org/10.1038/s41586-021-03316-6

Zhou P, Silverstein KAT, Ramaraj T, Guhlin J, Denny R, Liu J, Farmer AD, Steele KP, Stupar RM, Miller JR, Tiffin P, Mudge J, Young ND (2017) Exploring structural variation and gene family architecture with De Novo assemblies of 15 Medicago genomes. BMC Genomics 18:1–4. https://doi.org/10.1186/s12864-017-3654-1

Zhu F, Xi DH, Yuan S, Xu F, Zhang DW, Lin HH (2014) Salicylic acid and jasmonic acid are essential for systemic resistance against tobacco mosaic virus in *Nicotiana benthamiana*. Mol Plant-Microbe Interact 27:567–577. https://doi.org/10.1094/MPMI-11-13-0349-R

Transcriptional Networks in *Medicago truncatula*: Genomic and Functional Overview During Root Nodule Symbiosis

6

Akanksha Bhardwaj and Senjuti Sinharoy

Abstract

Root nodule symbiosis (RNS) is a mutualistic interaction between rhizobia and host legumes. Genetic and genomic resources generated by the Medicago community provide an ideal platform for understanding the symbiotic interaction between *Medicago truncatula* and its symbiotic partner *Sinorhizobium* (*Ensifer*) sp.. "Nod Factor," the chemical messenger secreted by rhizobia activates developmental cues to establish RNS. Root nodule development consists of a series of events starting from rhizobial recognition and infection at the root epidermal cells, transcriptomic changes in the symbiotically hallmarked root cells followed by the activation of several parallel signal transduction pathways. These coordinate the connectivity of the epidermal to inner root cell layers to initiate nodule primordia formation. Root nodule development is a finely tuned process where nodule organogenesis is synchronized with rhizobial infection, and the nodule number is controlled by a systemic signal. RNS is governed by the gene regulatory network driven by transcription factors. In this chapter, we present an overview of how transcription factor-guided transcriptional networks control cascade of events during nitrogen-fixing root nodule genesis.

6.1 Introduction

Plants require mineral nutrients for their growth and development. Nitrogen is one of the most important macronutrients. Being a component of nucleic acids, amino acids, cellular enzymes, chlorophyll, and several metabolites, nitrogen is crucial for the survival of living organisms. Plants mainly acquire nitrogen in the forms of nitrate (NO_3-) and ammonium (NH_4+) from soil (Crawford and Forde 2002). Plants have developed different mechanisms to cope up with nutrient limitations by trading their photoassimilates, through mutualistic interactions with arbuscular mycorrhizal (AM) fungi, nitrogen-fixing proteobacteria, cyanobacteria, and yeasts (De La Peña et al. 2018; Paul et al. 2020). These symbiotic relationships maintain the sustainability of the Earth ecosystem. Formation of the unique symbiotic organ "root nodule" is restricted to a monophyletic group of angiosperms, which include four orders, Fabales, Fagales, Cucurbitales, and Rosales. These four orders together constitute the nitrogen fixing clade (NFC). Legumes and the only nonlegume (Parasponia) interact with rhizobia (a group of α- and β-proteobacteria) and all other nonlegumes

A. Bhardwaj · S. Sinharoy (✉)
National Institute of Plant Genome Research,
New Delhi 110067, India
e-mail: ssinharoy@nipgr.ac.in

S. Sinharoy et al. (eds.), *The Medicago truncatula Genome*, Compendium of Plant Genomes,
https://doi.org/10.1007/978-3-030-90757-0_6

interact with Frankia (filamentous actinomycete bacteria) for the development of nodules. Nodule forming and non-nodule forming lineages are cluttered across the NFC (Soltis et al. 1995; Vessey et al. 2004; Raul et al. 2019). The most prevalent hypothesis indicates that RNS has a common evolutionary origin ~ 110 million years ago (Mya) in the base of NFC followed by multiple losses. Although, it is tricky to pinpoint the evolutionary origin of RNS (Werner et al. 2014; Griesmann et al. 2018). Two decades of research using model legumes (*Medicago truncatula* and *Lotus japonicus*) have revealed that lateral root development process has been co-opted during the development of root nodules.

Medicago truncatula (mentioned as Medicago in the rest of the text) has been used as a model legume for over twenty years, due to its short generation time, small diploid genome (2n = 2x = 16; 430 Mb), and self-pollinating capacity (Young et al. 2011). Medicago establishes a symbiotic relationship with *Sinorhizobium* (*Ensifer*) *meliloti* strains (Galibert et al. 2001) (genus name updated from Sinorhizobium to Ensifer (Young 2003)). Decades of international effort yielded the development of rich genetic resources for Medicago, which include ethyl methanesulfonate (EMS), fast neutron bombarded (FNB), γ-ray induced mutants, *RNAi* lines, and the most successful among all is the *Nicotiana tabacum retrotransposon* (*Tnt1*) induced insertional mutant population (Sagan et al. 1995; Frugier et al. 2000a; Penmetsa and Cook 2000; Tadege et al. 2008; Rogers et al. 2009; Pislariu et al. 2012). These resources accelerated the genetics, cell biology, and biochemical analysis using Medicago. Over 20 years of genetic research has advanced our knowledge about the molecular mechanism and evolution of RNS. In this chapter, we will summarize the development in the field of Medicago and Ensifer symbiosis. We will first describe molecular events that orchestrate the nodule developmental program and a brief overview of the signaling network. In the latter part of the chapter, we will focus on the transcription factors (TF) and the regulatory pathways controlled by the TF to establish the successful root nodule symbiosis (RNS).

6.2 Nodule—A *de novo* Root Organ

The RNS occurs via two coordinated processes: (a) infection of rhizobia in the epidermis and (b) *de novo* organogenesis in the pericycle/cortex (Crespi and Frugier 2009) (Fig. 6.1). *Medicago* and *Ensifer* communicate by exchanging chemical signals. *Ensifer* secretes Nodulation Factors (NFs) in response to plant secreted flavonoids. NF elicits a cascade of events in the host plant, which leads to the generation of a *de novo* symbiotic organ (Perret et al. 2000). Rhizobia co-opted the signal transduction pathway that is used by AM for the colonization to the host plant. The overlapping pathway that has been used for the AM and rhizobial colonization is called the common symbiosis signaling pathway (CSSP) (Oldroyd et al. 2011; Oldroyd 2013). Nodulation factors (NFs) are perceived by Medicago root hair cells. NF receptors *LysM-domain-containing receptor kinase 3* (*MtLYK3*) and *Nod factor perception* (*MtNFP*) bind to the NF and thereby activate CSSP. Activation of CSSP leads to root hair curling and pericyclic/cortical cell divisions (Arrighi et al. 2006; Smit et al. 2007). *Ensifer* traverses through the root hair cells, followed by the epidermal and cortical layers successively inside a tube-like structure composed of plant-derived cell walls and plasma membrane (composition from outside to inside) called Infection Threads (ITs) (Fig. 6.1). The IT is ramified in the cortex and delivers rhizobia to the dividing root cortical cells where endocytosis of the symbiont takes place. After endocytosis, bacteria are surrounded by a plant membrane-derived envelope and reside in plant cytoplasm as an organelle-like structure called "symbiosome." The bacteria after endocytosis are called "bacteroid." The "bacteroid" further differentiates into the nitrogen-fixing unit, via the repression of nitrogen assimilation genes and induction of the genes required for nitrogen fixation (Udvardi and Poole 2013; Hernandez et al. 2016).

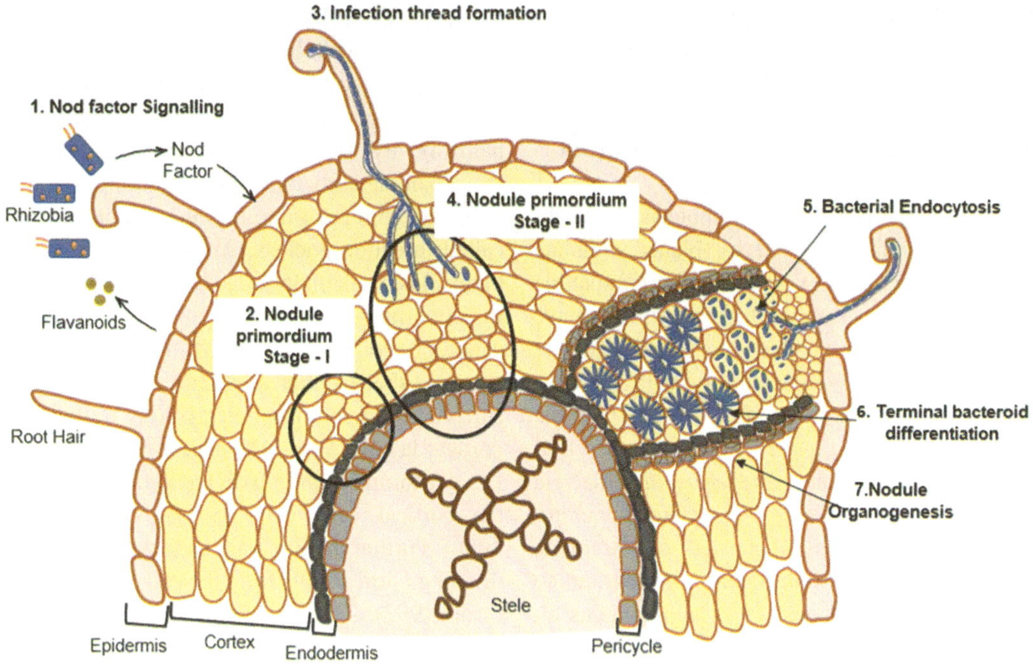

Fig. 6.1 Nodule–a de novo root organ development. Medicago root nodule development has multiple stages; **1. Nod factor signaling** starts with the secretion of flavonoids from Medicago root cells which are perceived by *Ensifer meliloti*. Rhizobia respond back by releasing Nod Factors (NFs). The NF signal perception triggers root hair curling and rhizobia entrapment. **2. Nodule primordium stage- I**, simultaneously with the establishment of rhizobial infection at epidermis, pericycle followed by inner cortical cell division started within 24 h post-inoculation, thus initiating nodule primordium formation. **3. Infection thread (IT) formation**, the rhizobia get entrapped into the curled root hair cells and make an entrance to the inner cortical layers via a tunnel-like structure IT. **4. Nodule primordium stage- II**, by 33–35 hpi, the mitotic division has started in the outer cortical cells that follow the division of the endodermal layer. **5. Bacterial endocytosis**, when IT reaches the nodule primordium, IT branches, and releases rhizobia into root cells. In these infected cells, rhizobia reside inside a host cell membrane-bound structure "symbiosome." **6. Terminal bacteroid differentiation**, rhizobia residing in the infected cells undergo differentiation via endoreduplication and become elongated or Y-shaped. **7. Nodule organogenesis**, dedifferentiation of root cortical, pericyclic, and endodermal cells lead to the formation of a novel organ-like structures "nodule." It contains a peripheral vascular bundle. Medicago develops indeterminate nodules having an active apical meristem and developmental gradient exists throughout the nodule. This gradient defines the nodule in different zones as described in Sect. 6.2

The CSSP activates calcium spiking in the root epidermis. This calcium spiking frequency and amplitude (called calcium signature) are different between AM symbiosis and RNS. The calcium signature is decoded by a *calcium calmodulin-dependent protein kinase* (*CCaMK*). The calcium spiking followed by activation of CCaMK triggers downstream transcription networks (Gleason et al. 2006; Charpentier and Oldroyd 2013) (Fig. 6.2a). The nodule organogenesis is under the control of the host molecular genetic network that is regulated by several transcription factors. The nodule-specific transcriptomic network determines the new organ identity and maintains a symbiotic cell environment.

Nodule development is a complex process consisting of several distinct microscopic and molecular events. Namely, (a) epidermal recognition of rhizobia and IT formation, (b) nodule organogenesis, (c) nodule number control via autoregulation of nodulation, and (d) accommodation of rhizobia and control of host immunity inside the infected cells.

6.2.1 Epidermal Recognition of Rhizobia and Infection Thread Formation

6.2.1.1 A Chemical Dialog and Mutual Recognition

This is a pioneering step of symbiosis establishment. Flavonoids secreted by plants stimulate nodulation factor (NF) exudation by compatible symbiont rhizobia strains. Different rhizobia have specific nod genes which ensure the structure of NF and hence the compatibility with the host (Barnett et al. 2001; Finan et al. 2001). The basic structure of NF has a β-1, 4-linked-N-acetyl-D-glucosamine backbone. *Ensifer meliloti* secrete NFs which are O-sulfated, O-acetylated lipochito-tetra-oligosaccharides. It consists of tetrameric glucosamine units with three acylated amino groups where the longest acyl chain contains bi-unsaturated 16 carbon atoms with sulfation at the reducing end of the chain (Lerouge et al. 1990). This NF is perceived by plasma membrane residing Medicago receptors complex MtNFP and MtLYK3. The signal perception by Nod Factor Receptors (NFRs) leads to calcium oscillations, followed by activation of nuclear signaling events and IT formation (Ané et al. 2002; Arrighi et al. 2006; Smit et al. 2007). The first well known symbiotic protein that gets activated in this signaling cascade is nuclear-localized *does not make infection 3* (MtDMI3/CCaMK). *DMI3-1* does not initiate IT or activate cortical cell divisions, resulting in non-nodulation phenotype. Despite the non-nodulating phenotype, *DMI3-1* maintains an intact calcium spiking response to the NF. Therefore, calcium spiking and decoding of the calcium signal by *DMI3* are both required for the activation of downstream components involved in IT formation and nodule organogenesis (Gleason et al. 2006). Though NF is the central chemical molecule secreted by rhizobia and drives the symbiosis, the legume host controls the level of the NF by using Nod Factor Hydrolases (MtNFH1). MtNFH1 degrades and cleaves the NF backbone and thereby controls rhizobial colonization and infection (Cai et al. 2018).

6.2.1.2 Rhizobia Infection Establishment

Successful decoding of the calcium signature by DMI3 and activation of downstream signaling pathway by transcriptional network leads to the rhizobial entry via root hair cells. The rhizobia get entrapped into the curled root hair cells called infection pockets. An invagination of the plant plasma membrane and cell wall from the infection pocket generates a tunnel-like structure called IT (Fournier et al. 2008). IT is a composite structure that facilitates bacteria arrival from the surface of the root hair microcolony to the inner cortical layers. After crossing the root hair cells, ITs get ramified and guide the symbiont toward the cortical cells (Monahan-Giovanelli et al. 2006). Further, rhizobia are endocytosed into the dividing cortical cells from the end of the IT (Rae et al. 1992). The *Interacting Protein of DMI3* (*MtIPD3*) is a unique coil-coil domain-containing transcription factor. IPD3 is involved in IT formation and the release of bacteria. DMI3 phosphorylates and activates IPD3 (Horváth et al. 2011; Ovchinnikova et al. 2011). *IPD3-like* (*IPD3L*) gene in Medicago shows redundant functions similar to *MtIPD3* (Jin et al. 2018). DELLA1, 2, and 3 stabilize DMI3-IPD3 complex formation and phosphorylation of IPD3 by DMI3 (Fonouni-Farde et al. 2016). In a parallel signaling pathway two GRAS domain-containing transcription factors, *Nodulation Signaling Pathway* (*MtNSP*) *1* and *2* guide IT formation. NSP1 and NSP2 form heterodimers and bind to the *Nodulation Responsive Element* (*NRE1*) motif thereby activating NF induced genes (Kaló et al. 2005; Smit et al. 2005; Hirsch et al. 2009). DMI3-IPD3 and NSP1-NSP2 complexes upregulate the downstream transcription factor *Nodule INception* (*NIN*) (Singh et al. 2014; Jin et al. 2016). NIN is a master regulator of nodulation and it regulates almost every aspect of nodule development that includes IT formation, nodule organogenesis, symbiosome development, and autoregulation of nodulation (AON). Noteworthy, DMI3, IPD3, and NSP2 belong to the CSSP. Hence, specific decoding of calcium signature and activation of *NIN* by the

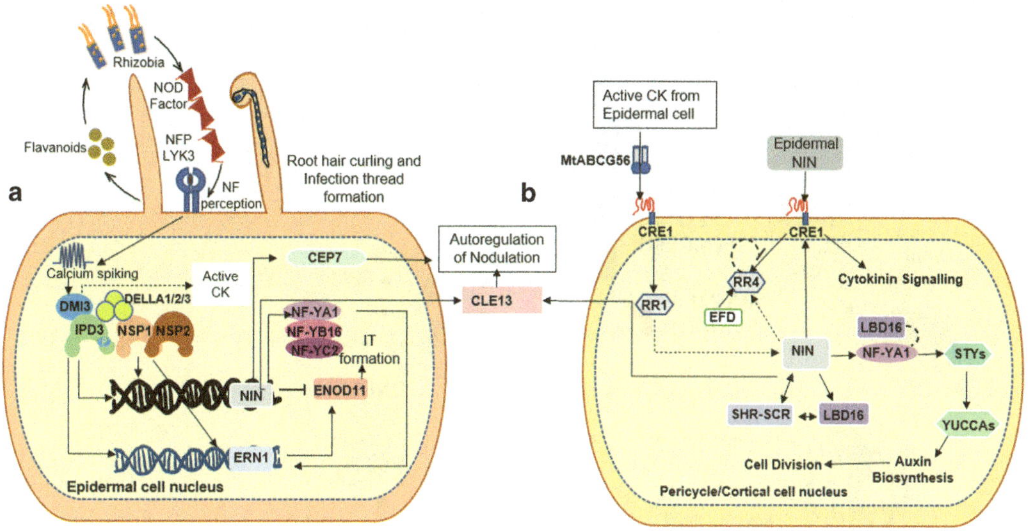

Fig. 6.2 Schematic representation of TF networks in root epidermal cells and cortical cells during nodule development. a. Transcription factors and signaling events in the epidermal and root hair cells during rhizobial infection, root hair curling, IT formation, and propagation. Epidermal root hair cells secrete flavonoids that act as an attractant to rhizobia. Rhizobia exude NF. These NFs are perceived by nod factor receptors NFP and LYK3 (Arrighi et al. 2006; Smit et al. 2007). NF perception triggers calcium spiking in the root hair and epidermal cell nuclei. NF-activated calcium signature is decoded by DMI3, which phosphorylates the IPD3 (TF) (Gleason et al. 2006). The NF signaling and DMI3 regulate CK biosynthesis genes and generate active cytokinin in the epidermal cells. Two GRAS-domain-containing TFs, MtNSP1 and NSP2, form heterodimers (Kaló et al. 2005; Smit et al. 2005). DELLA1/2/3 interacts with MtNSP2 and MtIPD3 and thus forms a five-protein complex (Fonouni-Farde et al. 2016). DMI3-IPD3 and NSP1-NSP2 complexes upregulate *NIN* expression. NIN directly activates the *NF-YA1* gene (Laporte et al. 2014). MtNF-YA1, MtNF-YB16, and MtNF-YC2 form a heterotrimeric complex (Baudin et al. 2015). ERN1 TF gets activated by IPD3, NSP1/NSP2, and NF-YA1/A2. ERN1 activates *ENOD11* expression. NIN represses *ENOD11* expression (Cerri et al. 2012; Liu et al. 2019a). NIN also activates *CEP7* and *CLE13* peptide genes which participate in AON (Laffont et al. 2020). **b**. Transcription factors activated signaling events during pericycle or cortical cell division to form nodule primordia. Active CK is transported from the epidermis via ABCG-family transporters mainly MtABCG56 to pericycle/cortical cell (Jarzyniak et al. 2021). CK activates *NIN* expression probably by type-B response regulator RR1 (Liu et al. 2019c). NIN and CRE1 show feedback regulation as epidermal NIN positively regulates *CRE1* expression and pericycle or cortical NIN activates *CRE1* expression (Vernié et al. 2015). CRE1 mediated cytokinin signaling triggers type A-response regulator RR4. EFD also activates *RR4*. RR4 inhibits CK signaling. NIN positively regulates *RR4* expression (Vernié et al. 2008). NIN directly activates *NF-YA1* and *LBD16* (Laporte et al. 2014; Schiessl et al. 2019) expression. NIN regulates *LBD16* expression, and it depends on the SHR-SCR transcription factor module (Dong et al. 2021). NF-YA1 activates *STY* and *YUCCA*s. YUCCAs promote local auxin biosynthesis and auxin accumulation stimulates cell divisions (Schiessl et al. 2019). Bold arrows indicate direct activation and dotted arrows indicate possible direct/indirect activation.

above-mentioned transcriptional network leads to the activation of the nodulation-specific pathway (Liu et al. 2019a). NIN directly activates the *Nuclear Factor Y* gene (*NF-YA1*) (Laporte et al. 2014). MtNF-YA1, MtNF-YB16, and MtNF-YC2 form heterotrimeric complexes, and control IT formation and nodule organogenesis (Baudin et al. 2015). All three TFs, NIN, ERN1, and NF-YA1, were involved in IT initiation and propagation (discussed in Sect. 6.3.1) (Liu et al. 2019a) (Fig. 6.2a). Additionally, genes like *MtENOD11* and *MtENOD12*, integral membrane proteins *FLOTILLINs* (*MtFLOT2*, *MtFLOT4*), and *Symbiotic Remorin* (*MtSYMREM1*) are crucial for the IT progression and bacterial release. A lipid raft is generated at the root hair cell tip, containing NF-receptor, MtFLOT2, MtFLOT4, and MtSYMREM1, is involved in IT initiation

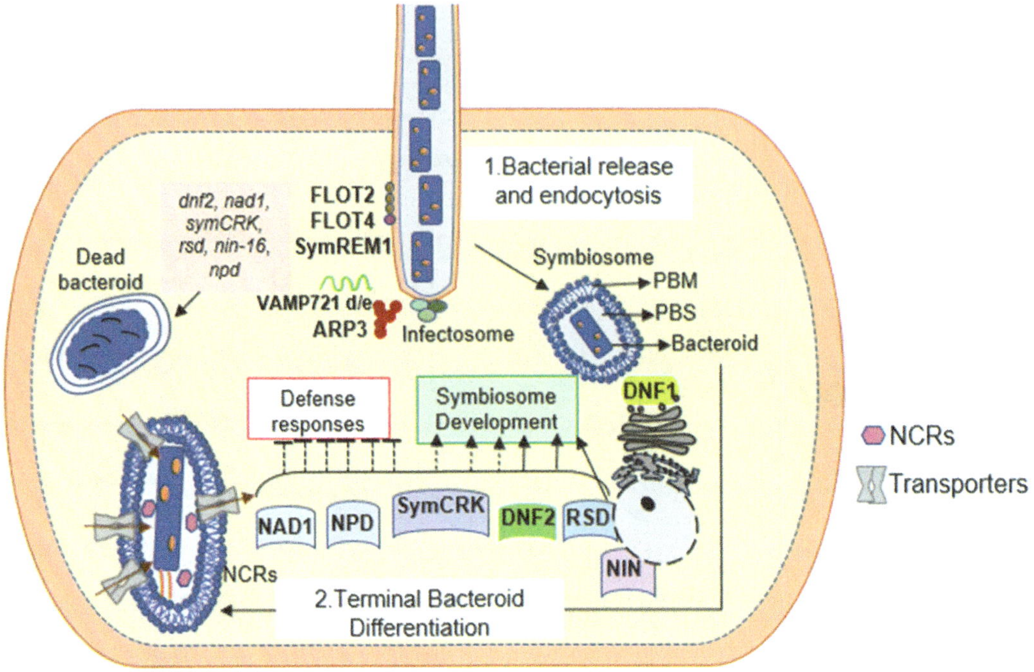

Fig. 6.3 Bacterial endocytosis and bacteroid differentiation in infected cells of nodule. Rhizobia get entry to dividing inner root cortical cells by IT. **1.** Bacterial release and endocytosis; IT growth is directed by "infectosome" protein complexes and integral membrane proteins FLOT2, FLOT4, and SymREM1 are also necessary for IT progression and bacterial release into root cortical cells (Liu et al. 2019b). For IT growth and bacterial release cytoskeleton remodeling is necessary, ARP3 facilitates this rearrangement (Gavrin et al. 2015b). VAMP721 d/e proteins mediate vesicle delivery to the plasma membrane thus promote endocytosis of rhizobia (Xie et al. 2012). After being released in the plant cell cytoplasm, rhizobia get surrounded by a plant-derived membrane called peribacteroid membrane (PBM) and are known as bacteroid. Peribacteroid space is the space between PBM and bacteroid. This organelle-like structure where bacteroids reside is known as symbiosome. **2.** Terminal bacteroid differentiation; bacteroids undergo terminal differentiation regulated by NCR peptides. *DNF1* gene codes for a signal peptidase complex localized to ER-Golgi and are required for secretory protein delivery toward symbiosome. Bacteroids get elongated due to endoreduplication, and endoreduplication is controlled by DNF1 derived NCR peptides (Wang et al. 2010). Nuclear-localized transcription factor RSD and NIN also regulate bacteroid differentiation (Sinharoy et al. 2013; Liu et al. 2021). Additionally, three genes, *SymCRK*, *NAD1*, and *NPD1*, might be participating in bacteroid differentiation. The connection between these genes is not known. *DNF2* gene is also required for bacteroid persistence in symbiosome and thought to inhibit defense-triggered responses. The single mutant of six genes *dnf2*, *rsd-1*, *nin-16*, *npd*, *symCRK*, and *nad1* shows early senescence with disrupted differentiation. These six genes thus possibly inhibit plant defense responses against microbial growth in the wild type nodule. Dotted lines represent possible relations and solid lines/arrows represent known functions (Bourcy et al. 2013; Sinharoy et al. 2013; Berrabah et al. 2014b; Wang et al. 2016; Pislariu et al. 2019; Liu et al. 2021)

and propagation (Andriankaja et al. 2007; Haney and Long 2010; Lefebvre et al. 2010). A protein complex known as "infectosome," which contains MtVPY (VAPYRIN), MtLIN (LUMPY INFECTION), and MtEXO70 H4 (EXOCYST) and ensures the directionality of IT (Liu et al. 2019b) (Fig. 6.3).

6.2.2 Nodule Organogenesis—From the Developmental Perspective

A detailed and precise description of the dedifferentiation of different root cell layers that leads to the development of nodules has been described in an outstanding nodule ontology analysis entitled fate map study (Xiao et al. 2014). Medicago fate map helps us to understand the nodule developmental dynamics. Dedifferentiation of root cortical, pericyclic, and endodermal cells leads to the formation of a novel organ-like structure "nodule." Legume root nodule is an organ "*sui generis*" as it contains peripheral vascular bundle as opposed to the central root-like vasculature (Hirsch et al. 1997). Noteworthy, nonlegumes form nodules with a central vasculature like lateral root (Soyano et al. 2021). Medicago develops indeterminate type nodules with an active apical meristem. Meristem keeps on adding cells throughout the lifespan of the nodule, and hence, a special developmental gradient exists in a nitrogen-fixing mature nodule. The developmental gradient is otherwise absent in the determinate nodules (example model plant *Lotus japonicus*). A mature Medicago nodule has distinct development zones starting from the apical meristem (ZI- furthest from the root), followed by infection zone (ZII), i.e., a region of infection, interzone (IZ), which is a region between infection and nitrogen fixation zone, and a region that can be distinctly visualized by the accumulation of starch granules, nitrogen fixation zone (ZIII). The above-mentioned zones symbolize various stages of bacterial progression ranging from free-living form (present inside the IT) to nitrogen-fixing bacteroids. During this process, the rhizobia gradually progress through different metabolic phases and ultimately perform nitrogen fixation (Vasse 1990). A senescence zone exists only in a mature Medicago nodule, present at the base of the nodule (Guerra et al. 2010) (see Chap. 7 Fig. 7.1).

Nodule development in Medicago starts with the anticlinal mitotic division of the pericyclic cells and is followed by the division of inner cortical cells within 24 h post-inoculation (hpi). By 33–35 hpi, mitotic division has started in the outer cortical cells followed by division of the endodermal layers. Around 8 cell layers of inner cortical cells get infected with rhizobia and are the sole contributor in the infection zone of young nodule primordia. The rhizobial infection triggers several rounds of endoreduplication in the host cells. Later both infected and uninfected cells, and peripheral nodule tissues were added by the nodule meristem. In a young nodule, the pericycle and endodermal cell layers give rise to the nodule vascular bundle and the outer cortical cell divisions generate nodule meristem (Xiao et al. 2014). Medicago apical meristem can be divided into two parts, a) nodule central meristem (NCM) is the middle part of the meristem which adds infected and uninfected cells to the central infection zone and b) nodule vascular meristems (NVM) are present at the tip of peripheral vasculatures. Multiple NVMs surround the NCM. Synchronized division of NCM and NVMs gives rise to the formation of a typical cylindrical nodule structure. Nodule morphogenesis is promoted by a plant genetic program, where cell division and differentiation are regulated by *MtPLETHORA* (*MtPLTs*), *cyclins,* and TF KNOTTED-like homeobox (KNOX). *MtPLT1* and *MtPLT2* express in NVMs, whereas *MtPLT3* and *MtPLT4* express in the NCM. These four *PLTs* play an important role in NCM and NVMs formations and their maintenance (Franssen et al. 2015).

6.2.3 Symbiosome and Bacteroid

IT grows in a polarized fashion toward dividing cortical cells, and symbiotic bacteria are released from the end of each IT branch. Root nodules provide a comfortable home for the symbiont. Each infected cell accommodates thousands of symbionts. The endocytosis of rhizobia generates an organelle-like structure "symbiosome" and the symbiont inside the symbiosome is called "bacteroid." Symbiosome consists of bacteroids, peribacteroid space (PBS), peribacteroid membrane (PBM), and symbiosome membrane (SM) (Fig. 6.3). SM is derived from the plant plasma membrane and throughout the nodule development, exocytotic vesicles add membrane materials to the SM. A functional symbiosome (nitrogen-fixing) needs differentiation of both symbiotic partners (De Bruijn 2019). In Medicago nodules, bacteroids get elongated, undergo endoreduplication, and lose their ability to regrow in a free-living condition. Inside the nodule, bacteroids get adjusted to a symbiotic lifestyle, which includes the establishment of leghemoglobin-driven hypoxic conditions. Low oxygen concentration acts as an activation signal for the genes that encode the nitrogenase subunit enzymes (*nif* and *fix*) and protects the oxygen-labile nitrogenase complex from permanent oxidation (De La Peña et al. 2018). SM acts as an interface to facilitate the transport of plant-derived photosynthates, ammonia assimilates (amino acids), phosphate, sulfates, metal ions (copper, iron, molybdenum, zinc, etc.), and carbon source, etc. PM contains H^+-ATPases which create a pH gradient and this membrane potential drives metabolic exchange and provides the reducing environment to bacteroids (Clarke et al. 2014).

6.2.3.1 Bacterial Endocytosis and Differentiation

The first major molecular event that leads to the rhizobium entry into the plant cell is "endocytosis." Local cell wall degradation takes place at the tip of the IT. This cell wall-free IT is called an "infection droplet." Infection droplet formation allows close contact of the IT membrane and the newly divided cortical cell membrane. This interaction facilitates endocytosis of the rhizobia (Catalano et al. 2007). Cellular exocytotic vesicle-associated membrane proteins (VAMPs), namely MtVAMP721d and MtVAMP721e, localize at the site of the rhizobial release point. *RNAi*-mediated silencing of *MtVAMP721d* and *MtVAMP721e* blocks endocytosis of rhizobia. Medicago nodulation pectate lyase (NPL) is required for the degradation of the cell wall inside the IT and the propagation of IT. Further, MtNPL and other pectin-degrading enzymes are involved in the formation of cell wall-free infection droplets in Medicago (Xie et al. 2012). How MtNPL or other pectin-degrading enzymes are targeted to the IT and cell contact points are not clear. The hint that *MtVAMP721d* and *MtVAMP721e* could be involved in delivering pectin-containing enzymes came from a study in a heterologous system (Gavrin et al. 2016). Further, proper regulation of the host cell exocytotic pathway is necessary for nitrogen-fixing symbiosome development (Roy et al. 2020). The vesicle movement toward the symbiosome is controlled via the cytoskeleton network. Live-cell imaging using actin-related protein ARP3:GFP fusion revealed the change of actin cytoskeleton dynamics during endocytosis of rhizobia (Gavrin et al.). In a separate study, it has been depicted that repression of *MtVAMP721a* during nodule development has a positive impact on symbiosome development (Sinharoy et al. 2013). Apart from the cellular exocytosis and endocytosis pathways, other proteins present at the tip of the IT during infection droplet formation plays a pivotal role in rhizobial endocytosis. Molecular scaffold proteins SYMREM1, FLOT2, FLOT4, and symbiosis receptor kinase (SymRK) are involved in the endocytosis of rhizobia (Haney and Long 2010; Lefebvre et al. 2010; Saha et al. 2016).

6.2.3.2 Terminal Differentiation of Bacteroid (TBD)

Immediately, after endocytosis, symbionts inside the IT do not fix nitrogen and resemble free-living bacteria as per their size, shape, and metabolic status. Eventually, the bacteroids divide and further differentiate into nitrogen-fixing and metabolically distinct forms. Terminally, differentiated bacteroid is a hallmark of the inverted repeat lacking clade (IRLC) family. After the bacteroids stop dividing, they change their size, shape, volume, DNA content, membrane porosity, and become enlarged (2 μm to 8 μm). Ultimately, they take an elongated or Y-shaped structure. The morphological changes in the bacteroid shape occur mainly due to the endoreduplication of the bacterial genome. These elongated bacteroids also lose viability, i.e., the capacity of regrowth in a free-living form. The change in the rhizobial shape is called terminal bacteroid differentiation (TBD). Medicago and other legumes showing TBD has single rhizobium inside the symbiosomes, whereas legumes that do not show TDB has multiple bacteroids inside the symbiosomes (Mergaert et al. 2003; Montiel et al. 2016; Mandal et al. 2020).

Medicago-derived *nodule-specific-cysteine-rich peptides* are the crucial factors behind TBD. Medicago genome contains 789 *NCR* genes encoding NCR peptides. NCR peptides are small cysteine-rich peptides that possess an N-terminal signal peptide. Plant secretory pathways target NCR peptides to symbiosomes. The delivery of proteins to the symbiosomes occurs through the endoplasmic reticulum (ER). Endopeptidase complexes residing in the ER cleave off the signal peptides from NCRs and this causes the delivery of NCRs toward symbiosomes. NCRs promote endoreduplication of bacteroids by promoting DNA replication and modifying or inhibiting bacterial cytokinesis (Haag and Mergaert 2019). NCR035 and NCR247 show antimicrobial properties and generate porosity in the *Ensifer meliloti* membrane. Even though *NCR*s are absent from the *Lotus japonicus* genome, ectopic expression of *MtNCR035* promotes TBD (Van De Velde et al. 2010; Farkas et al. 2014). Two more NCRs, *NCR169* and *NCR211*, were identified by a forward genetic screen establishing the specific role of a few *NCRs* among this big gene family. Hence, despite huge diversity, few NCRs retained specific functions (Horvath et al. 2015; Kim et al. 2015).

6.2.4 Accommodation of Rhizobia and Control of Host Immunity Inside the Infected Cells

How many rhizobia are present in a single infected cell inside the nitrogen fixation zone? It is difficult to specifically come up with a numerical value, but this number is in the range of thousands. The infected cells need to squeeze their cytoplasm and all other organelles in much lesser space to accommodate a large number of rhizobia. In Medicago, rhizobia express the flagellar regulon in the nitrogen fixation zone, indicating that the pathogen-associated molecular patterns (PAMPs) trigger immunity (PTI) is supposed to be induced inside the infected cells (Roux et al. 2014). Until now, there is no experimental characterization of any pattern recognition receptor (PRR) having a role in rhizobial PAMP recognition. A co-inoculation study was conducted in the *Medicago-Ensifer* system with *Ralstonia solanacearum*, which provides evidence that suppressed immune responses occur during the rhizobial infection. These pathogen-infected nodules can develop weak defense reactions against pathogenic *R. solanacearum*. Additionally, the nodule can confine pathogens preventing them from spreading to nearby nonsymbiotic plant cells (Benezech et al. 2020). In Medicago, five major genes have been linked with suppression of the defense response a noncanonical phosphatidylinositol-specific phospholipase C (*DNF2*), *symbiotic cysteine-rich receptor-like kinase* (*SymCRK*), *nodules with activated defense 1* (*NAD1*), *regulator of symbiosome differentiation* (*RSD*), and *nodule-specific PLAT domain protein* (*NPD1-5*) (Bourcy et al. 2013; Sinharoy et al. 2013; Berrabah et al. 2014a; Wang et al. 2016; Pislariu et al. 2019). Recently, *NIN* has also been linked to control defense responses inside an infected cell. A *nod* + and *fix- nin* mutant showing a reduction in the *NIN* expression level (*nin-16*) has been characterized, and *nin-16* nodules have a normal infection zone, with proper bacterial endocytosis but an abolished fixation zone with profound defense responses. Additionally, *nin-16* mutants also have a defect in symbiosome formation. Transcriptomic studies using *nin-16* nodules showed reduced expression of *SymCRK, NAD1* and *RSD* genes, thus establishing a connection in the signaling pathway among these genes (Liu et al. 2021).

In the wildtype context, defense response has never been observed inside the nodule but in *symcrk*, *dnf2*, *npd1*, *rsd*, *nad1*, and *nin-16* mutants strong defense responses lead to arrest of symbiotic cell growth-promoting early senescence (Bourcy et al. 2013; Sinharoy et al. 2013; Berrabah et al. 2014b; Wang et al. 2016; Pislariu et al. 2019; Liu et al. 2021). The precise role of DNF2 and SymCRK is unknown although their structural similarities with PRRs suggest their role in PTI. *DNF2, SymCRK, RSD,* and *NAD1* all directly or indirectly influence TBD (Berrabah et al. 2014a; Alunni and Gourion 2016). At the tissue level, *RSD* and *NIN* expression reached the maximum at infection zone then dropped significantly, whereas *SymCRK* and *NAD1* expression was observed from infection zone and are maximum at fixation zone. What is the connection among these genes at the molecular level is unanswered. Transcriptome analysis and more precisely zone-specific profiling of mRNAs associated with ribosomes (translatome) will develop our insights about symbiosome development and the molecular mechanism behind active suppression of defense responses.

6.2.5 Autoregulation of Nodulation (AON)

Symbiotic nitrogen fixation by legumes is an adaptive strategy to survive only in the nitrogen deficit environment. Nodule development is an energy-demanding process and demands ~25% of plant photosynthate. Hence, the host plant keeps tight control over nodule number. To maintain proper energy and metabolic status for healthy plant growth, the above-ground shoot and below-ground root communicate to control the nodule number which termed as autoregulation of nodulation (AON). AON-control starts as early as 24 hpi along with the initiation of the root cortical cell division (Lagunas et al. 2019). CLAVATA3/embryo surrounding region-related (MtCLE12/MtCLE13) are two peptide hormones that systemically control nodule numbers (Mortier et al. 2010). Root-derived CLE peptides are perceived by shoot- localized leucine-rich receptor kinase called *super numeric nodule* (*SUNN*). *Root-determined nodulation* (*MtRDN1*) encodes a hydroxyproline-O-arabinosyl transferase. *MtRDN1* adds L-arabinose moiety to the MtCLE12 peptide specifically, thereby creating diversity between these two peptides. The biological implication of hydroxyproline-O-arabinosylation of CLE12 is not known (Yoro et al. 2019). Noteworthy, a high concentration of nitrogen in the soil negatively impacts nodulation and it is also an integral part of AON (Reid et al. 2011). Further, nitrogen-mediated responses to lateral root development and AON-mediated nodule number control are linked processes. C-terminally encoded peptide (MtCEP) is the second peptide hormone family that controls nodule number but in a positive way. Exogenous application of MtCEP1 peptides increases nodule number and represses lateral root development (Imin et al. 2013). In the CEP family, only two peptides, MtCEP7/13, are expressed in response to the rhizobial inoculation in a NIN and cytokinin response 1 (MtCRE1) receptor-dependent manner. It is presumed that CEPs are perceived by compact root architecture 2 (MtCRA2) receptor-like-kinase, a protein that controls both lateral root initiation and nodule development (Laffont et al. 2020). Noteworthy, symbiotic nitrogen fixation is reduced immediately if nitrogen concentration increases in soil. The reduction of nitrogen fixation efficiency at the later stage is also connected to the AON pathway. Medicago NIN-like protein (NLP) mainly NLP1 translocates to the nucleus in presence of high nitrogen concentration. NLP1 heterodimerizes with NIN through the PB1 domain. NLP1 and NIN heterodimer suppress NIN-mediated activation of nodulation-related genes in an unknown mechanism (Lin et al. 2018). Further, NLP1 activates MtCLE35 and MtCLE35 suppresses nodule number in an AON-dependent manner (Luo et al. 2021). Thus, a master circuitry driven by NIN and NLPs through two antagonistic peptide hormones (CEPs and CLEs) drives the AON-control by finely tuning lateral root and nodule development. Recently, two GATA family TFs, HANABA TARANU (HAN1 and HAN2), have been linked with nodule number control. *Tnt1* double mutant, *mthan1* and *mthan2*, generates the double number of nodules compared to the wildtype. *HAN1* and *HAN2* show inducibility upon nitrate treatment and directly or indirectly activate *NCR* genes (Xu et al. 2021). Investigations are needed to understand how HAN1 and HAN2 are connected with the AON pathway.

6.3 Transcription Factor and their Targets Drive the Development of Root Nodule Symbiosis

In the last two decades, transcriptomic studies, functional genomics, cell biology, and biochemical characterization using Medicago root nodule at different developmental stages as well as using different mutants have generated an overall understanding of nodule development (elaborated in Chap. 7). In the following section, we will elucidate our understanding of the transcriptomic networks that guide (1) the epidermal rhizobial infection and (2) pericyclic/cortical nodule organogenesis. A list of TF that has been implicated during nodule development has been given in Table 6.1.

6.3.1 Transcriptomic Network in Nod Factor Signaling, Rhizobial Infection in the Epidermis

The NF signaling cascade initiates legume-rhizobium symbiosis, which leads to the dynamic reprogramming of plant cells. After NF signal perception, CSSP generates calcium oscillation followed by decoding of the calcium signature by CCaMK/DMI3. IPD3, which acts downstream to the calcium signaling, gets phosphorylated by CCaMK and leads to the initiation of chains of transcriptional activation (see Sect. 6.2.1.2). NSP1 and NSP2 are two GRAS family TFs, which initiate a parallel signaling pathway along with the CCaMK/IPD3 module. NSP1 maintains constitutive expression and localizes to the nuclei of epidermal and cortical cell layers and root hair cells (Smit et al. 2005). NSP1 forms a heterodimer with NSP2 (which lacks a DNA binding domain). NSP1-NSP2 heterodimer controls downstream gene expression. Like most GRAS family proteins, NSP2 also has leucine-rich regions at the N-terminus that act as a nuclear localization signal. NSP2 shows constitutive expression in shoot and root, but also can be induced after the addition of NF (Kaló et al. 2005). NSP2 is post-transcriptionally regulated by *miR171h* (Devers 2019). NSP1-NSP2 heterodimer activates *NIN* expression. *IPD3* also maintains a steady expression in the root but shows strong inducibility downstream of NF signaling. IPD3/DMI3 and NSP1/NSP2 complexes bind to the promoters of the early NF inducible genes. NSP1 binds to the NRE-box and IPD3/DMI3 complex binds to the CYCLOPS responsive *cis element* contains a palindromic sequence (*CYC-box*). DELLAs, which are classically known as the negative regulators of gibberellin signaling, interact with MtNSP2 and MtIPD3 and thus form a five-protein complex (Gallego-Bartolomé et al. 2012; Roy et al. 2020) (Fig. 6.2a). Hence, early nodulin gene expression downstream of the NF signaling pathway can either be activated by NSP1/NSP2 heterodimer or IPD3/DMI3 complex or the five-protein complex mentioned earlier. When and which protein complex activates a particular gene expression during NF signal transduction is not clear. The coexistence of *NRE-box* and *CYC-box* in the promoters of *ERN1*, *NIN*, and cell wall-associated protein (ENOD11) suggests that these genes might be controlled by the five-protein complex (Hernandez et al. 2016; Liu et al. 2019a).

TFs, which show nodule exclusive expressions, are NIN, NF-YA1, and ethylene responsive factor required for nodulation 1 (ERN1). NIN belongs to the RWP-RK transcription factor family and is a major regulator of infection and organogenesis. NIN binds to the promoter of *NF-YA1* and activates its expression. NIN and ERN1 signaling pathways are interconnected at earlier stages of rhizobial infection. *ERN1* gets activated by IPD3, NSP1/NSP2, and NF-YA1/A2. Again, NIN suppresses the expression of *ERN1* target *ENOD11* and thereby controls nodule number (Cerri et al. 2012; Liu et al. 2019a). NIN-Like proteins (NLP) in Arabidopsis are involved in nitrogen sensing. The N-terminal part of NLPs gets phosphorylated and targeted to the nucleus mainly under low-nitrogen conditions. Though NIN is a member of the NLP family, it is constitutively localized to the nucleus. Further, *nin-1* mutant cannot be complemented by NLPs, which implies that NLPs neofunctionalization leads to the evolution of NIN, a master regulator of nodule development (Liu and Bisseling 2020). Single-cell transcriptomics of wildtype, *nin-1*, *ern1*, and *nf-ya1* help in understanding how these TFs control IT propagation. Over 5000 genes get down-regulated in *nin-1* mutant root hair cells, establishing NIN as a central hub in the infection event (also see Chap. 7). NIN, on one hand, represses defense-related genes and on the other hand activates cell wall modifying, gibberellin biosynthesis, and mineral assimilation-related genes. *NF-YA1* and a direct target of NIN, together with NIN control several aspects of nodule development. Whereas ERN1 controls only infection events. Some of the other important direct targets of NIN whose mutation individually cause a defect in IT formation are rhizobium polar growth (*RPG*), *cystathionine-β-synthase-like1* (*CBS1*), *IPD3L*, and *nodulation pectate lyase* (*NPL*) (Liu et al. 2019a).

6.3.2 Role of TFs in Nodule Inception and Organogenesis: Recruitment of Lateral Root Developmental Program for Nodule Organogenesis

Like epidermal infection, NIN also regulates nodule organogenesis in collaboration with MtNF-YA1. The complete absence of NIN in non-nodulating plants emphasizes its specific recruitment in nodulation. Orthologs of NIN in nonlegume nodulating plants from order Fagales have a role in nodule formation (Liu and Bisseling 2020). A conserved cytokinin responsive element presents at ~15 to 20 kb upstream to the *MtNIN* start codon, designated as the CE (for *cytokinin response elements*-containing) region. This *CE region* has several cytokinin response regulatory elements (10 putative elements in Medicago) and controls *NIN* expression in the root pericycle and cortex. The remote regulatory *CE-element* is present in both model legumes (Lotus and Medicago). The bioinformatic prediction in many other legumes establishes the evolutionary conservation of CK mediated activation of *NIN* in the pericycle (Liu et al. 2019c). Active CK is synthesized in the Medicago root epidermis after rhizobial infection and gets transported via ABCG family transporters to the root pericycle and cortex. Thus, CK functions as a mobile messenger downstream of NF signaling that establishes connectivity between epidermis and pericycle (Jarzyniak et al. 2021). Pericyclic and cortical NIN activate cortical CK signaling by directly binding to *MtCRE1* (Vernié et al. 2015), suggesting that NF-driven recruitment of NIN inside the CK signaling pathway is one of the evolutionary adaptations that lead to the initiation of nodule organogenesis. NIN also activates *NF-YA1* in the pericycle/cortex (Laporte et al. 2014). NF-YA1 activates *SHORT INTERNODES/STYLISH* (*STY*) group of transcription factors and STY activates flavin monooxygenases (*YUCCA*). Expression of auxin biosynthesis gene (*YUCCA*s) in the inner root cell layers generates a local auxin peak (Shrestha et al. 2020). In a nutshell, interconnected CK and NIN signaling followed by NIN and NF-YA1 medicated activation of local auxin biosynthesis and auxin-mediated cell division leads to the formation of nodule primordium (Soyano et al. 2013).

NF-YA proteins heterodimerize with other NF-Y subunits and participate in multiple steps of nodule development (Baudin et al. 2015). The NIN and NF-YA1 node is a connecting link of the transcriptional network co-optimized for nodule organogenesis from lateral root development. NF-YA1 form complexes with asymmetric leaves 2-like (LjASL18)/ lateral organ boundaries domain (MtLBD16) to promote nodule growth. ASL18/LBD16 is a major regulator of lateral root development, and it works downstream of the auxin signaling pathway to control lateral root formation (Schiessl et al. 2019; Soyano et al. 2021). In legumes, *ASL18/LBD16* gets activated by NIN which leads to its recruitment to the nodule development. Multiple NIN binding sites are present in the single intron of *ASL18/LBD16*. As NIN works under the CK signaling pathway, hence, NIN-mediated activation of *ASL18/LBD16* also functions under the CK signaling pathway.

The intronic NIN binding site is conserved in Medicago and other legumes. It is hypothesized that along with neofunctionalization of NIN in legumes, recruitment of NIN binding site in *LBD16* intron is another major predisposition event that leads to the evolution of nodule forming trait (Soyano et al. 2019).

Two GRAS family transcription factors, SHORT-ROOT (SHR) and SCARECROW (SCR), are major regulators of lateral root development. SHR acts upstream to SCR and maintains cortical and epidermal cellular patterning (radial root patterning). The SHR-SCR transcriptional module gets recruited in nodule development. The use of *pAtSCR* and *pMtSCR* constructs in Medicago shows the *pAtSCR* expression is restricted in the root endodermis similar to Arabidopsis. Conversely, *pMtSCR* shows expression in legume cortical cells. Further, reduction in nodule number in *mtscr* plants establishes a role of SCR in cortical nodule organogenesis. Hence, diversification in the promoter element leading to the broader expression of *MtSCR* gene is a prerequisite for nodule development. *MtSHR* expression is restricted to the stele in Medicago, like Arabidopsis. Converse to Arabidopsis, MtSHR accumulates in Medicago root cortex and epidermis. Cortex-specific expression of dominant repression fusion version of MtSHR1 leads to the reduction in nodule number, establishing the role of epidermal and cortical MtSHR in nodule development. Rhizobial inoculation does not affect *MtSHR1/2* transcript levels but stabilizes or increases SHR proteins in epidermal and cortical cells in NIN- NSP1-NSP2 dependent manner. Further, ectopic expression of *SHR* triggers cortical cell division in NIN-NSP1-NSP2 independent manner. Noteworthy, induction of MtLBD16 downstream of NF signaling is dependent on SHR-SCR module and induction of SCR depends on MtLBD16 (Dong et al. 2021). Taken together, SHR-SCR and LBD16, major lateral root developmental circuits, get recruited downstream of the NIN and NSP1-NSP2 signaling pathway during nodule primordium formation.

Leguminous nodules have peripheral vascular bundles originate from dividing endodermal and pericyclic cells. Homeotic genes of the NOOT-BOP-COCH-LIKE (NBCL) family encode MtNOOT1/2 co-transcriptional regulators. *MtNOOT1/2* expresses in a partially overlapping manner in the nodule peripheral vascular bundle, pericycle, and endodermal cells. MtNOOT1/2 inhibits ectopic meristem formation from nodule meristem thereby maintaining nodule identity (Couzigou et al. 2012; Magne et al. 2018).

The next level of regulation during nodule development comes through chromatin remodeling factors. Histone deacetylases (MtHDT1/2/3) regulate *3-hydroxy-3-methylglutaryl coenzyme a reductase* (*MtHMGR1*). *MtHMGR1* interacts with *SymRK/DMI2* and promotes nodule primordia formation. *MtHMGR1* is positively regulated by HDTs. *MtHDTs* show expression in roots as well as in the nodule meristem and invasion zone. *MtHDT*-RNAi plants have decreased nodule primordia. Noteworthy, *MtHDT*-RNAi plants show root meristem defective phenotypes like Arabidopsis, suggesting shared genetic modules between lateral root and nodule development (Kevei et al. 2007; Kondorosi et al. 2007; Li 2021) (Fig. 6.2b).

6.4 Control of Symbiosome Development by Transcription Factors: A Largely Unexplored Area

How TFs control the symbiosome development is still obscure. SM, the major interface between plant and symbiont, is loaded with plant transporters. Is there a master TF that activates all these transporters or are there different TFs that control different transporters? The answer to this question is still is a black box. The first transcription factor indirectly linked to bacteroid differentiation is an ethylene response factor required for nodule differentiation (EFD), which forms fix- nodules. EFD controls both early and late nodule development. The *efd* plant produces three-fold more nodules than its wild type counterpart, establishing EFD as a negative regulator of nodule number. EFD plays a pivotal role in the distal invasion zone. The *efd* nodule contains numeroua branched ITs indicating a delay in endocytosis of rhizobium. Further, EFD activates *RR4*, a type-A cytokinin response regulator which negatively controls CK signaling. It is presumed that EFD- mediated suppression of CK signaling is necessary for the differentiation of the distal invasion zone cells, which are leaving meristem (Vernié et al. 2008). Thus, a reduced amount of CK in the distal invasion zone cells is a prerequisite for endocytosis and further differentiation of bacteroids.

The second TF that shows a defect in symbiosome development phenotype is RSD. RSD is a member of the cystine-2/ histidine-2 (C_2H_2) family of TF, which also contains the EAR type repressor domain. RSD lacks any activation domain but contains a dominant repression domain (Sinharoy et al. 2013). RSD belongs to the family of C_2H_2 TF, two of the TFs in this family have been reported as dual regulators (activator and repressor both). In *rsd* nodules, the defect started from the distal invasion zone like *efd* and ultimately generates fix-phenotype. Unlike *efd*, *rsd* nodules do not have any IT or release defect rather *rsd* nodules contain ~ 1.5-fold more symbiosome in each cell at the distal invasion zone (Bharadwaj et al., unpublished data). RSD represses vesicular associated membrane protein-encoding gene *MtVAMP721a* by directly binding to its promoter. In Arabidopsis, VAMP721/722 delivers defense-related substrates through focal secretion at the site of fungal attack (Kwon et al. 2008). In legumes, similar focal secretion is necessary for the endocytosis of rhizobia (see sect. 6.3.2.1). In this respect, RSD probably helps in symbiosome development by restricting the default cellular vesicular transport pathway. Transcriptome study indicated that RSD has many targets apart from MtVAMP721a (Sinharoy et al. 2013). Apart from the defect in symbiosome development, *rsd* nodules also show a strong defense response (see sect. 6.2.4). The third TF related to symbiosome development is NIN (see sect. 2.4). How these three TFs and their networks are interconnected, and how these TFs suppress defense responses while supporting an amenable environment for functional nitrogen fixation is an open question that needs to be addressed.

Table 6.1 Functionally characterized transcription factors at different nodulation stages

Nodulation stage	Gene symbol	Gene ID Mt5.0	TF family	Mutagen	References
Nod factor signaling	*MtIPD3*	MtrunA17Chr5g0409341	C2C2-CO-like	Fast Neutron mutagenesis	Horváth et al. (2011)
Nod factor signaling pathway	*MtNSP1*	MtrunA17Chr8g0344101	GRAS	EMS	Smit et al. (2005)
Nod factor signaling pathway	*MtNSP2*	MtrunA17Chr3g0114841	GRAS	EMS	Kaló et al. (2005)
Nodule primordia, downstream to CK signaling	*MtbHLH476*	MtrunA17Chr5g0400621	bHLH	RNAi	Ariel et al. (2012)
Rhizobial preinfection, root hair curling	*MtARF16a*	MtrunA17Chr1g0199681	ARF	*Tnt*1	Breakspear et al. (2014)
Symbiosome development	*MtARP3*	MtrunA17Chr8g0381261	ARF	RNAi	Gavrin et al. (2015b)
Nodule organogenesis, CK signaling	*MtKNOX3*	MtrunA17Chr1g0149651	Homobox-WOX	RNAi	Azarakhsh et al. (2015)

(continued)

Table 6.1 (continued)

Nodulation stage	Gene symbol	Gene ID Mt5.0	TF family	Mutagen	References
Maintenance of nodule meristem and identity	*MtNOOT1*	MtrunA17Chr7g0256321	ABTB	*Tnt*1	Couzigou et al. (2012)
Nodule identity regulation	*MtNOOT2*	MtrunA17Chr1g0171771	ABTB	*Tnt*1	Magne et al. (2018)
Nodule development and nodule number	*MtNIN*	MtrunA17Chr5g0448621	Nin-like	FNB, *Tnt*1	Marsh et al. (2007)
Nodule meristem initiation	*MtNF-YA1*	MtrunA17Chr1g0177091	Hap2	RNAi, EMS	Combier et al. (2006)
Infection thread formation, nodule organogenesis	*MtERN1*	MtrunA17Chr7g0253424	AP2-EREBP	Fast Neutron mutagenesis	Andriankaja et al. (2007)
Nod factor signaling	*MtERN2*	MtrunA17Chr6g0461271	AP2-EREBP	*Tnt*1, TILLING	Andriankaja et al. (2007)
Nodule organogenesis	*MtLBD16*	MtrunA17Chr7g0260971	AS2-LOB	Tnt1	Soyano et al. (2019)
Nodule organogenesis	*MtSHR1*	MtrunA17Chr5g0401131	GRAS	*Tnt*1	Dong et al. (2021)
Nodule organogenesis	*MtSHR2*	MtrunA17Chr4g0054071	GRAS	*Tnt*1	Dong et al. (2021)
Nodule organogenesis	*MtSCR*	MtrunA17Chr7g0245601	GRAS	*Tnt*1	Dong et al. (2021)
Nitrate response on nodule formation and infection	*MtNLP1*	MtrunA17Chr3g0143921	Nin-like	*Tnt1*, RNAi	Lin et al. (2018)
Nodule differentiation and nodule number	*MtEFD*	MtrunA17Chr4g0002631	AP2-EREBP	Deletion TILLING	Vernié et al. (2008)
Nodule organogenesis	*MtRR4*	MtrunA17Chr5g0414931	RR-A type	Deletion TILLING	Vernié et al. (2008)
Symbiosome development	*MtRSD*	MtrunA17Chr7g0239441	C2H2	*Tnt*1	Sinharoy et al. (2013)
Rhizobial invasion and nodule cell differentiation	*MtZPT2-1*	MtrunA17Chr3g0134711	C2H2	Antisensing transgenic lines	Frugier et al. (2000b)
Nodule vasculature patterning	*MtbHLH1*	MtrunA17Chr3g0132471	bHLH	CRES-T	Godiard et al. (2011)
Nodule senescence	*MtbHLH2*	MtrunA17Chr5g0411341	bHLH	*Tnt*1, TALEN	Deng et al. (2019)
Nodule senescence	*MtNAC969*	MtrunA17Chr4g0042561	NAM	RNAi	De Zélicourt et al. (2012)
Nodule number control	*MtHAN1*	MtrunA17Chr5g0404131	C2C2-GATA	*Tnt*1	Xu et al. (2021)
Nodule number control	*MtHAN2*	MtrunA17Chr8g0373231	C2C2-GATA	*Tnt*1	Xu et al. (2021)

Concluding remarks

Functional genomic studies with an amalgamation of genetic, transcriptomic, and cell biological approaches have enlightened our understanding of NF signaling, epidermal rhizobial infection, and cortical nodule organogenesis. The transcriptional regulation of symbiosome development and symbiotic cell differentiation is poorly understood. The genesis of the symbiosome, its smooth transition to the functional nitrogen-fixing stage and further nodule senescence, are all finely-tuned processes like other nodule developmental programs. The connecting links that join the transitions of bacteroid development remain largely unknown. Recently, NIN has been identified as a factor that might play a central role in connecting the transcriptional networks during symbiosome formation. A systematic investigation is needed to answer many unanswered questions that has been highlighted in the chapter.

Acknowledgements We thank NIPGR for core grant and DBT (Department of Biotechnology)-eLibrary Consortium (DeLCON), India for providing access to e-resources. Akanksha Bhardwaj was supported by CSIR (09/803(0145)/2018-EMR-I).

References

Alunni B, Gourion B (2016) Terminal bacteroid differentiation in the legume-rhizobium symbiosis: nodule-specific cysteine-rich peptides and beyond. New Phytol 211:411–417. https://doi.org/10.1111/nph.14025

Andriankaja A, Boisson-Dernier A, Frances L et al (2007) AP2-ERF transcription factors mediate nod factor-dependent Mt ENOD11 activation in root hairs via a novel cis-regulatory motif. Plant Cell 19:2866–2885. https://doi.org/10.1105/tpc.107.052944

Ané J-M, Kiss GB, Riely BK et al (2002) Medicago truncatula DMI1 required for bacterial and fungal symbioses in legumes. Annu Rev Biochem 269:30. https://doi.org/10.1126/science.1093038

Ariel F, Brault-Hernandez M, Laffont C et al (2012) Two direct targets of cytokinin signaling regulate symbiotic nodulation in medicago truncatula. Plant Cell 24:3838–3852. https://doi.org/10.1105/tpc.112.103267

Arrighi J-FX, Barre A, Ben Amor B et al (2006) The medicago truncatula lysine motif-receptor-like kinase gene family Includes NFP and new nodule-expressed genes 1[W]. Plant Physiol 142:265–279. https://doi.org/10.1104/pp.106.084657

Azarakhsh M, Kirienko AN, Zhukov VA et al (2015) KNOTTED1-LIKE HOMEOBOX 3: a new regulator of symbiotic nodule development. J Exp Bot 66:7181–7195. https://doi.org/10.1093/jxb/erv414

Baudin M, Laloum T, Lepage A et al (2015) A phylogenetically conserved group of nuclear factor-Y transcription factors interact to control nodulation in legumes. Plant Physiol 169:2761–2773. https://doi.org/10.1104/pp.15.01144

Benezech C, Berrabah F, Jardinaud MF et al (2020) Medicago-sinorhizobium-ralstonia co-infection reveals legume nodules as pathogen confined infection sites developing weak defenses. Curr Biol 30:351-358.e4. https://doi.org/10.1016/j.cub.2019.11.066

Barnett MJ et al (2001) Nucleotide sequence and predicted functions of the entire Sinorhizobium meliloti pSymA megaplasmid. Proc Natl Acad Sci USA 98:9883–9888

Berrabah F, Bourcy M, Cayrel A et al (2014a) Growth conditions determine the DNF2 requirement for symbiosis. PLoS ONE 9:91866. https://doi.org/10.1371/journal.pone.0091866

Berrabah F, Bourcy M, Eschstruth A et al (2014b) A nonRD receptor-like kinase prevents nodule early senescence and defense-like reactions during symbiosis. New Phytol 203:1305–1314. https://doi.org/10.1111/nph.12881

Bourcy M, Brocard L, Pislariu CI et al (2013) Medicago truncatula DNF2 is a PI-PLC-XD-containing protein required for bacteroid persistence and prevention of nodule early senescence and defense-like reactions. New Phytol 197:1250–1261. https://doi.org/10.1111/nph.12091

Breakspear A, Liu C, Roy S et al (2014) The root hair "infectome" of medicago truncatula uncovers changes in cell cycle genes and reveals a requirement for auxin signaling in rhizobial infectionw. Plant Cell 26:4680–4701. https://doi.org/10.1105/tpc.114.133496

Cai J, Zhang LY, Liu W et al (2018) Role of the nod factor hydrolase MtNFH1 in regulating nod factor levels during rhizobial infection and in mature nodules of medicago truncatula. Plant Cell 30:397–414. https://doi.org/10.1105/tpc.17.00420

Catalano CM, Kirk, Czymmek J, et al (2007) Medicago truncatula syntaxin SYP132 deWnes the symbiosome membrane and infection droplet membrane in root nodules. Planta 225:541–550. https://doi.org/10.1007/s00425-006-0369-y

Cerri MR, Frances L, Laloum T et al (2012) Medicago truncatula ERN transcription factors: regulatory interplay with NSP1/NSP2 GRAS factors and expression dynamics throughout rhizobial infection. Plant Physiol 160:2155–2172. https://doi.org/10.1104/pp.112.203190

Charpentier M, Oldroyd GED (2013) Nuclear calcium signaling in plants. Plant Physiol 163:496–503. https://doi.org/10.1104/pp.113.220863

Clarke VC, Loughlin PC, Day DA et al (2014). Transport Processes of the Legume Symbiosome Membrane. https://doi.org/10.3389/fpls.2014.00699

Combier JP, Frugier F, De Billy F et al (2006) MtHAP2-1 is a key transcriptional regulator of symbiotic nodule development regulated by microRNA169 in Medicago truncatula. Genes Dev 20:3084–3088. https://doi.org/10.1101/gad.402806

Couzigou JM, Zhukov V, Mondy S et al (2012) NODULE ROOT and COCHLEATA maintain nodule development and are legume orthologs of arabidopsis BLADE-ON-PETIOLE genesw oa. Plant Cell 24:4498–4510. https://doi.org/10.1105/tpc.112.103747

Crawford NM, Forde BG (2002) Molecular and developmental biology of inorganic nitrogen nutrition. Arab B 1:e0011. https://doi.org/10.1199/tab.0011

Crespi M, Frugier F (2009) Erratum: De novo organ formation from differentiated cells: root nodule organogenesis (Sci. Signal. (2009) 2 (er1)). Sci Signal 2:1–9. https://doi.org/10.1126/scisignal.253er1

De Bruijn FJ (2019) Rhizobial release, symbiosomes, and bacteroid formation: introduction. Model Legum Medicago truncatula 589–590. https://doi.org/10.1002/9781119409144.ch73

De La Peña TC, Fedorova E, Pueyo JJ, Mercedes Lucas M (2018) The symbiosome: legume and rhizobia co-evolution toward a nitrogen-fixing organelle? Front Plant Sci 8:1–26. https://doi.org/10.3389/fpls.2017.02229

De Zélicourt A, Diet A, Marion J et al (2012) Dual involvement of a medicago truncatula NAC transcription factor in root abiotic stress response and symbiotic nodule senescence. Plant J 70:220–230. https://doi.org/10.1111/j.1365-313X.2011.04859.x

Deng J, Zhu F, Liu J et al (2019) Transcription factor bHLH2 represses CYSTEINE PROTEASE77 to negatively regulate nodule senescence. Plant Physiol 181:1683–1703. https://doi.org/10.1104/pp.19.00574

Devers EA (2019) Mir171h restricts root symbioses and shows, like its target NSP2, a complex transcriptional regulation in Medicago truncatula. Model Legum Medicago truncatula 975–986. https://doi.org/10.1002/9781119409144.ch124

Dong W, Zhu Y, Chang H, et al (2021) An SHR-SCR module specifies legume cortical cell fate to enable nodulation. 586 | Nat | 589. https://doi.org/10.1038/s41586-020-3016-z

Farkas A, Maróti G, Dürgo H et al (2014) Medicago truncatula symbiotic peptide NCR247 contributes to bacteroid differentiation through multiple mechanisms. Proc Natl Acad Sci U S A 111:5183–5188. https://doi.org/10.1073/pnas.1404169111

Finan TM et al (2001) The complete sequence of the 1,683-kb pSymB megaplasmid from the N2-fixing endosymbiont Sinorhizobium meliloti. Proc Natl Acad Sci USA 98:9889–9894

Fonouni-Farde C, Tan S, Baudin M et al (2016) DELLA-mediated gibberellin signalling regulates Nod factor signalling and rhizobial infection. Nat Commun 7. https://doi.org/10.1038/ncomms12636

Fournier J, Timmers ACJ, Sieberer BJ et al (2008) Mechanism of infection thread elongation in root hairs of Medicago truncatula and dynamic interplay with associated rhizobial colonization. Plant Physiol 148:1985–1995. https://doi.org/10.1104/pp.108.125674

Franssen HJ, Xiao TT, Kulikova O et al (2015) Root developmental programs shape the Medicago truncatula nodule meristem. Dev 142:2941–2950. https://doi.org/10.1242/dev.120774

Frugier F, Poirier S, Atrice B et al (2000a) A Krü ppel-like zinc finger protein is involved in nitrogen-fixing root nodule organogenesis

Frugier F, Poirier S, Satiat-Jeunemaître B et al (2000b) A Kruppel-like zinc finger protein is involved in nitrogen-fixing root nodule organogenesis. Genes Dev 14:475–482. https://doi.org/10.1101/gad.14.4.475

Galibert F, Finan TM, Long SR, et al (2001) The composite genome of the legume symbiont Sinorhizobium meliloti. Science (80-) 293:668–672. https://doi.org/10.1126/science.1060966

Gallego-Bartolomé J, Minguet EG, Grau-Enguix F et al (2012) Molecular mechanism for the interaction between gibberellin and brassinosteroid signaling pathways in Arabidopsis. Proc Natl Acad Sci 109:13446–13451. https://doi.org/10.1073/PNAS.1119992109

Gavrin A, Chiasson D, Ovchinnikova E et al (2016) VAMP721a and VAMP721d are important for pectin dynamics and release of bacteria in soybean nodules. New Phytol 210:1011–1021. https://doi.org/10.1111/nph.13837

Gavrin A, Jansen V, Ivanov S et al (2015) ARP2/3-mediated actin nucleation associated with symbiosome membrane is essential for the development of symbiosomes in infected cells of Medicago truncatula root nodules. Mol Plant-Microbe Interact 28:605–614. https://doi.org/10.1094/MPMI-12-14-0402-R

Gleason C, Chaudhuri S, Yang T et al (2006). Nodulation independent of rhizobia induced by a calcium-activated kinase lacking autoinhibition. https://doi.org/10.1038/nature04812

Godiard L, Lepage A, Moreau S et al (2011) MtbHLH1, a bHLH transcription factor involved in medicago truncatula nodule vascular patterning and nodule to plant metabolic exchanges. New Phytol 191:391–404. https://doi.org/10.1111/j.1469-8137.2011.03718.x

Griesmann M, Chang Y, Liu X, et al (2018) Phylogenomics reveals multiple losses of nitrogen-fixing root nodule symbiosis. Science (80-) 361. https://doi.org/10.1126/science.aat1743

Guerra JCP, Coussens G, de Keyser A et al (2010) Comparison of developmental and stress-induced nodule senescence in Medicago truncatula. Plant Physiol 152:1574–1584. https://doi.org/10.1104/pp.109.151399

Haag AF, Mergaert P (2019) Terminal bacteroid differentiation in the Medicago–rhizobium interaction–a tug of war between plant and bacteria. Model Legum Medicago truncatula 600–616. https://doi.org/10.1002/9781119409144.ch75

Haney CH, Long SR (2010) Plant flotillins are required for infection by nitrogen-fixing bacteria. 107. https://doi.org/10.1073/pnas.0910081107

Hernandez G, Mathesius U, Libault M, et al (2016) A Snapshot of Functional Genetic Studies in Medicago truncatula. Front Plant Sci | www.frontiersin.org 1:1175. https://doi.org/10.3389/fpls.2016.01175

Hirsch AM, Larue TA, Doyle J (1997) Is the legume nodule a modified root or stem or an organ sui generis ? CRC Crit Rev Plant Sci 16:361–392. https://doi.org/10.1080/07352689709701954

Hirsch S, Kim J, Muñoz A et al (2009) GRAS proteins form a DNA binding complex to induce gene expression during nodulation signaling in medicago truncatulaw. Plant Cell 21:545–557. https://doi.org/10.1105/tpc.108.064501

Horvath B, Domonkos A, Kereszt A et al (2015) Loss of the nodule-specific cysteine rich peptide, NCR169, abolishes symbiotic nitrogen fixation in the Medicago truncatula dnf7 mutant. Proc Natl Acad Sci U S A 112:15232–15237. https://doi.org/10.1073/pnas.1500777112

Horváth B, Yeun LH, Domonkos Á et al (2011) Medicago truncatula IPD3 is a member of the common symbiotic signaling pathway required for rhizobial and mycorrhizal symbioses. / 1345 MPMI 24:1345–1358. https://doi.org/10.1094/MPMI

Imin N, Mohd-Radzman NA, Ogilvie HA, Djordjevic MA (2013) The peptide-encoding CEP1 gene modulates lateral root and nodule numbers in Medicago truncatula. J Exp Bot 64:5395–5409. https://doi.org/10.1093/jxb/ert369

Jarzyniak K, Banasiak J, Jamruszka T et al (2021) Early stages of legume–rhizobia symbiosis are controlled by ABCG-mediated transport of active cytokinins. Nat Plants 7:428–436. https://doi.org/10.1038/s41477-021-00873-6

Jin Y, Chen Z, Yang J et al (2018) IPD3 and IPD3L function redundantly in rhizobial and mycorrhizal symbioses. Front Plant Sci 9:1–12. https://doi.org/10.3389/fpls.2018.00267

Jin Y, Liu H, Luo D et al (2016) DELLA proteins are common components of symbiotic rhizobial and mycorrhizal signalling pathways. Nat Commun 7. https://doi.org/10.1038/ncomms12433

Kaló P, Gleason C, Edwards A, et al (2005) Nodulation signaling in legumes requires NSP2, a member of the GRAS family of transcriptional regulators. Science (80-) 308:1786–1789. https://doi.org/10.1126/science.1110951

Kevei Z, Lougnon G, Mergaert P et al (2007) 3-Hydroxy-3-methylglutaryl coenzyme A reductase1 interacts with NORK and is crucial for nodulation in Medicago truncatula. Plant Cell 19:3974–3989. https://doi.org/10.1105/tpc.107.053975

Kim M, Chen Y, Xi J et al (2015) An antimicrobial peptide essential for bacterial survival in the nitrogen-fixing symbiosis. Proc Natl Acad Sci U S A 112:15238–15243. https://doi.org/10.1073/pnas.1500123112

Kondorosi A, Endre G, Kondorosi E (2007) 3-Hydroxy-3-Methylglutaryl Coenzyme a reductase1 interacts with NORK and is crucial for nodulation in Medicago truncatula. 19:3974–3989. https://doi.org/10.1105/tpc.107.053975

Kwon C, Neu C, Pajonk S et al (2008) Co-option of a default secretory pathway for plant immune responses. Nature 451:835–840. https://doi.org/10.1038/nature06545

Laffont C, Ivanovici A, Gautrat P et al (2020) The NIN transcription factor coordinates CEP and CLE signaling peptides that regulate nodulation antagonistically. Nat Commun 11:1–13. https://doi.org/10.1038/s41467-020-16968-1

Lagunas B, Achom M, Bonyadi-Pour R et al (2019) Regulation of resource partitioning coordinates nitrogen and rhizobia responses and autoregulation of nodulation in medicago truncatula. Mol Plant 12:833–846. https://doi.org/10.1016/J.MOLP.2019.03.014

Laporte P, Lepage A, Fournier J et al (2014) The CCAAT box-binding transcription factor NF-YA1 controls rhizobial infection. J Exp Bot 65:481–494. https://doi.org/10.1093/jxb/ert392

Lefebvre B, Timmers T, Mbengue M et al (2010) A remorin protein interacts with symbiotic receptors and regulates bacterial infection. Proc Natl Acad Sci U S A 107:2343–2348. https://doi.org/10.1073/pnas.0913320107

Lerouge P, Roche P, Fauchert C et al (1990) Symbiotic host-specificity of Rhizobium me/i/oti is determined by a sulphated and acylated glucosamine oligosaccharide signal LETTERS TO NATURE

Li H (2021) Plant-specific histone deacetylases are essential for early and late stages of Medicago nodule development. 1591–1605. https://doi.org/10.1093/plphys/kiab140

Lin J, shun, Li X, Luo ZL et al (2018) NIN interacts with NLPs to mediate nitrate inhibition of nodulation in Medicago truncatula. Nat Plants 4:942–952. https://doi.org/10.1038/s41477-018-0261-3

Liu CW, Breakspear A, Guan D et al (2019) NIN acts as a network hub controlling a growth module required for rhizobial infection. Plant Physiol 179:1704–1722. https://doi.org/10.1104/pp.18.01572

Liu CW, Breakspear A, Stacey N et al (2019) A protein complex required for polar growth of rhizobial infection threads. Nat Commun 10:1–17. https://doi.org/10.1038/s41467-019-10029-y

Liu J, Bisseling T (2020) Evolution of nin and NIN-like genes in relation to nodule symbiosis. Genes (basel) 11:1–15. https://doi.org/10.3390/genes11070777

Liu J, Rasing M, Zeng T et al (2021) NIN is essential for development of symbiosomes, suppression of defence and premature senescence in Medicago truncatula nodules. New Phytol 230:290–303. https://doi.org/10.1111/nph.17215

Liu J, Rutten L, Limpens E et al (2019) A remote cis-regulatory region is required for nin expression in the pericycle to initiate nodule primordium formation in medicago truncatula. Plant Cell 31:68–83. https://doi.org/10.1105/tpc.18.00478

Luo Z, Lin J shun, Zhu Y et al (2021) NLP1 reciprocally regulates nitrate inhibition of nodulation through SUNN-CRA2 signaling in Medicago truncatula. Plant Commun 2:100183. https://doi.org/10.1016/j.xplc.2021.100183

Magne K, Couzigou JM, Schiessl K et al (2018) MtNODULE ROOT1 and MtNODULE ROOT2 are essential for indeterminate nodule identity. Plant Physiol 178:295–316. https://doi.org/10.1104/pp.18.00610

Mandal D, Srivastava D, Sinharoy S (2020) Optimization of hairy root transformation for the functional genomics in chickpea: a platform for nodule developmental studies BT - Legume genomics: methods and protocols. In: Jain M, Garg R (eds). Springer US, New York, NY, pp 335–348

Marsh JF, Rakocevic A, Mitra RM et al (2007) Medicago truncatula NIN is essential for rhizobial-independent nodule organogenesis induced by autoactive calcium/calmodulin-dependent protein kinase. Plant Physiol 144:324–335. https://doi.org/10.1104/pp.106.093021

Mergaert P, Nikovics K, Kelemen Z et al (2003) A novel family in Medicago truncatula consisting of more than 300 nodule-specific genes coding for small, secreted polypeptides with conserved cysteine motifs. Plant Physiol 132:161–173. https://doi.org/10.1104/pp.102.018192

Monahan-Giovanelli H, Pinedo CA, Gage DJ (2006) Architecture of infection thread networks in developing root nodules induced by the symbiotic bacterium Sinorhizobium meliloti on Medicago truncatula. Plant Physiol 140:661–670. https://doi.org/10.1104/pp.105.072876

Montiel J, Szucs A, Boboescu IZ et al (2016) Terminal bacteroid differentiation is associated with variable morphological changes in legume species belonging to the inverted repeat-lacking clade. Mol Plant-Microbe Interact 29:210–219. https://doi.org/10.1094/MPMI-09-15-0213-R

Mortier V, den Herder G, Whitford R et al (2010) CLE peptides control medicago truncatula nodulation locally and systemically. Plant Physiol 153:222–237. https://doi.org/10.1104/pp.110.153718

Oldroyd GED (2013) Speak, friend, and enter: Signalling systems that promote beneficial symbiotic associations in plants. Nat Rev Microbiol 11:252–263. https://doi.org/10.1038/nrmicro2990

Oldroyd GED, Murray JD, Poole PS, Downie JA (2011). The rules of engagement in the legume-rhizobial symbiosis. https://doi.org/10.1146/annurev-genet-110410-132549

Ovchinnikova E, Journet E-P, Chabaud M et al (2011) IPD3 Controls the Formation of Nitrogen-Fixing Symbiosomes in Pea and Medicago Spp. / 1333 MPMI 24:1333–1344. https://doi.org/10.1094/MPMI

Paul K, Saha C, Nag M et al (2020) A tripartite interaction among the basidiomycete rhodotorula mucilaginosa, N2-fixing endobacteria, and rice improves plant nitrogen nutrition. Plant Cell 32:486–507. https://doi.org/10.1105/tpc.19.00385

Penmetsa RV, Cook DR (2000) Production and characterization of diverse developmental mutants of medicago truncatula 1

Perret X, Staehelin C, Broughton WJ (2000) Molecular basis of symbiotic promiscuity

Pislariu CI, Murray JD, Wen JQ et al (2012) A Medicago truncatula tobacco retrotransposon insertion mutant collection with defects in nodule development and symbiotic nitrogen fixation. Plant Physiol 159:1686–1699. https://doi.org/10.1104/pp.112.197061

Pislariu CI, Sinharoy S, Torres-Jerez I et al (2019) The nodule-specific PLAT domain protein NPD1 is required for nitrogen-fixing symbiosis. Plant Physiol 180:1480–1497. https://doi.org/10.1104/pp.18.01613

Rae AL, Bonfante-Fasolo P, Brewin NJ (1992) Structure and growth of infection threads in the legume symbiosis with Rhizobium leguminosarum. Plant J 2:385–395. https://doi.org/10.1111/j.1365-313X.1992.00385.x

Raul B, Kryvoruchko I, Benedito VA et al (2019) Root nodule development in model versus non-canonical plants BT - plant biotechnology: progress in genomic era. In: Gaur RK (ed) Khurana SMP. Springer Singapore, Singapore, pp 397–428

Reid DE, Ferguson BJ, Hayashi S et al (2011) Molecular mechanisms controlling legume autoregulation of nodulation. Ann Bot 108:789–795. https://doi.org/10.1093/aob/mcr205

Rogers C, Wen J, Chen R, Oldroyd G (2009) Deletion-Based reverse genetics in Medicago truncatula. Plant Physiol 151:1077–1086. https://doi.org/10.1104/pp.109.142919

Roux B, Rodde N, Jardinaud MF et al (2014) An integrated analysis of plant and bacterial gene expression in symbiotic root nodules using laser-capture microdissection coupled to RNA sequencing. Plant J 77:817–837. https://doi.org/10.1111/tpj.12442

Roy S, Liu W, Nandety RS et al (2020) Celebrating 20 years of genetic discoveries in legume nodulation and symbiotic Nitrogen fixation. Plant Cell 32:15–41. https://doi.org/10.1105/tpc.19.00279

Sagan M, Morandi D, Tarenghi E, Duc G (1995) Selection of nodulation and mycorrhizal mutants in the model plant Medicago truncatula (Gaertn.) after γ-ray mutagenesis. Plant Sci 111:63–71. https://doi.org/10.1016/0168-9452(95)04229-N

Saha S, Paul A, Herring L et al (2016) Gatekeeper tyrosine phosphorylation of SYMRK is essential for synchronizing the epidermal and cortical responses in root nodule symbiosis. Plant Physiol 171:71–81. https://doi.org/10.1104/pp.15.01962

Schiessl K, Lilley JLS, Lee T et al (2019) NODULE INCEPTION recruits the lateral root developmental program for symbiotic nodule organogenesis in medicago truncatula. Curr Biol 29:3657-3668.e5. https://doi.org/10.1016/j.cub.2019.09.005

Shrestha A, Zhong S, Therrien J et al (2020) Lotus japonicus Nuclear Factor YA1, a nodule emergence stage-specific regulator of auxin signalling. New Phytol. https://doi.org/10.1111/nph.16950

Singh S, Katzer K, Lambert J et al (2014) CYCLOPS, A DNA-binding transcriptional activator, orchestrates symbiotic root nodule development. Cell Host Microbe 15:139–152. https://doi.org/10.1016/j.chom.2014.01.011

Sinharoy S, Torres-Jerez I, Bandyopadhyay K et al (2013) The C2H2 transcription factor REGULATOR OF SYMBIOSOME DIFFERENTIATION represses transcription of the secretory pathway gene VAMP721a and promotes symbiosome development in Medicago truncatula. Plant Cell 25:3584–3601. https://doi.org/10.1105/tpc.113.114017

Smit P, Limpens E, Geurts R et al (2007) Medicago LYK3, an entry receptor in rhizobial nodulation factor signaling. Plant Physiol 145:183–191. https://doi.org/10.1104/pp.107.100495

Smit P, Raedts J, Portyanko V et al (2005) NSP1 of the GRAS protein family is essential for rhizobial nod factor-induced transcription. Science (80-) 308:1789–1791. https://doi.org/10.1126/science.1111025

Soltis DE, Soltis PS, Morgant DR et al (1995) Chloroplast gene sequence data suggest a single origin of the predisposition for symbiotic nitrogen fiLxation in angiosperms

Soyano T, Kouchi H, Hirota A, Hayashi M (2013). NODULE INCEPTION directly targets NF-Y subunit genes to regulate essential processes of root nodule development in Lotus Japonicus. https://doi.org/10.1371/journal.pgen.1003352

Soyano T, Liu M, Kawaguchi M, Hayashi M (2021) Leguminous nodule symbiosis involves recruitment of factors contributing to lateral root development. Curr Opin Plant Biol 59:102000. https://doi.org/10.1016/j.pbi.2020.102000

Soyano T, Shimoda Y, Kawaguchi M, Hayashi M (2019) A shared gene drives lateral root development and root nodule symbiosis pathways in Lotus. Science (80-) 366:1021–1023. https://doi.org/10.1126/science.aax2153

Tadege M, Wen J, He J et al (2008) Large-scale insertional mutagenesis using the Tnt1 retrotransposon in the model legume Medicago truncatula. Plant J 54:335–347. https://doi.org/10.1111/j.1365-313X.2008.03418.x

Udvardi M, Poole PS (2013) Transport and metabolism in legume-rhizobia symbioses. Annu Rev Plant Biol 64:781–805. https://doi.org/10.1146/annurev-arplant-050312-120235

Van De Velde W, Zehirov G, Szatmari A et al (2010) Plant peptides govern terminal differentiation of bacteria in symbiosis. Science (80-) 327:1122–1126. https://doi.org/10.1126/science.1184057

Vasse J, de Billy F, Camut S, Truchet G (1990) Correlation between ultrastructural differentiation of bacteroids and nitrogen fixation in alfalfa nodules. J Bacteriol 172:4295–4306

Vernié T, Kim J, Frances L et al (2015) The NIN transcription factor coordinates diverse nodulation programs in different tissues of the medicago truncatula root. Plant Cell 27:3410–3424. https://doi.org/10.1105/tpc.15.00461

Vernié T, Moreau S, De Billy F et al (2008) EFD is an ERF transcription factor involved in the control of nodule number and differentiation in Medicago truncatula. Plant Cell 20:2696–2713. https://doi.org/10.1105/tpc.108.059857

Vessey JK, Pawlowski K, Bergman B (2004) Root-based N 2-fixing symbioses: Legumes, actinorhizal plants, Parasponia sp. and cycads

Wang C, Yu H, Luo L et al (2016) NODULES WITH ACTIVATED DEFENSE 1 is required for maintenance of rhizobial endosymbiosis in Medicago truncatula. New Phytol 212:176–191. https://doi.org/10.1111/nph.14017

Wang D, Griffitts J, Starker C et al (2010) A nodule-specific protein secretory pathway required for nitrogen-fixing symbiosis. Science (80-) 327:1126–1129. https://doi.org/10.1126/science.1184096

Werner GDA, Cornwell WK, Sprent JI, et al (2014) ARTICLE A single evolutionary innovation drives the deep evolution of symbiotic N 2-fixation in angiosperms. https://doi.org/10.1038/ncomms5087

Xiao TT, Schilderink S, Moling S et al (2014) Fate map of Medicago truncatula root nodules. Dev 141:3517–3528. https://doi.org/10.1242/dev.110775

Xie F, Murray JD, Kim J, et al (2012) | 633-638 PLANT BIOLOGY Downloaded at Natl Inst of Plant Genome Research on. PNAS 109. https://doi.org/10.1073/pnas.1113992109

Xu Y, Wang H, Lu Z et al (2021) Developmental Analysis of the GATA Factor HANABA TARANU mutants in medicago truncatula reveals their roles in nodule formation. Front Plant Sci 12:1–14. https://doi.org/10.3389/fpls.2021.616776

Yoro E, Nishida H, Ogawa-Ohnishi M et al (2019) PLENTY, a hydroxyproline O-arabinosyltransferase, negatively regulates root nodule symbiosis in Lotus japonicus. J Exp Bot 70:507–517. https://doi.org/10.1093/jxb/ery364

Young JM (2003) The genus name Ensifer Casida 1982 takes priority over Sinorhizobium Chen et al. 1988, and Sinorhizobium morelense Wang et al. 2002 is a later synonym of Ensifer adhaerens Casida 1982. Is the combination "Sinorhizobium adhaerens" (Casida 1982) Willems et. Int J Syst Evol Microbiol 53:2107–2110. https://doi.org/10.1099/ijs.0.02665-0

Young ND, Debellé F, D Oldroyd GE et al (2011) The medicago genome provides insight into the evolution of rhizobial symbioses. https://doi.org/10.1038/nature10625

Understanding of Root Nodule Development at Level of System Biology as Obtained by High Throughput Transcriptomic Approach

7

Akanksha Bhardwaj and Senjuti Sinharoy

Abstract

Nodule development starts with dedifferentiation of the already differentiated root tissues. Dedifferentiation involves a complex but finely tuned, coherent host, and symbiont crosstalk that ultimately leads to the development of an extraordinary new organ called "root nodule." This developmental process is regulated by innumerable interconnected transcriptional networks. Medicago generates a cylindrical-shaped indeterminant nodule where a developmental gradient persists from the nodule tip to the base. Both plant and symbiont undergo a huge transcriptomic change as they adjusted to the mutualistic lifestyle. Transcriptome analysis on nodule developmental time series, mutant nodules, laser-capture microdissection (LCM) of separate nodule zones, and single-cell transcriptomics of root hair cells allowed researchers to understand the complex developmental circuit during nodule development at a spatio-temporal resolution. In this chapter, we will focus on the advancement in the transcriptomics study that leads to the understanding of Medicago nodule development in finer details.

A. Bhardwaj · S. Sinharoy (✉)
National Institute of Plant Genome Research,
New Delhi 110067, India
e-mail: ssinharoy@nipgr.ac.in

7.1 Introduction

Medicago truncatula generates the indeterminate type of nodules that retain a persistent apical meristem. Mutualistic interaction of Medicago with its symbiotic partner *Ensifer meliloti* leads to the establishment of nitrogen-fixing root nodules. Nodule development starts in Medicago with the recognition of the rhizobial Nod Factor (NF) in the plant epidermis. Almost instantly with the application of the NF, calcium oscillation has been observed in the root hair cell nuclei. This calcium oscillation leads to the activation of the downstream signaling pathway (see Chap. 6). Within 24-h post-inoculation (hpi), transcriptomic changes have been observed in the root epidermis and cortex (Jardinaud et al. 2016). Root nodule development (RNS) is a complex process. It is characterized by several cellular, morphological, and physiological events. The precise transcriptomic analysis leads to the understanding of the nodule development in finer details. The transcriptomics era started with the EST sequencing and microarray analysis in 1998 and geared up with the advancement of the high throughput next-generation sequencing approaches and release of the Medicago genome sequences (Covitz et al. 1998; Benedito et al. 2008; Pecrix et al. 2018). In this chapter, we will highlight the transcriptomic studies and how these studies generated a system biology level understanding of the Medicago nodule development.

S. Sinharoy et al. (eds.), *The Medicago truncatula Genome*, Compendium of Plant Genomes,
https://doi.org/10.1007/978-3-030-90757-0_7

7.2 EST Sequencing and Microarray – First Generation of Nodule Transcriptomics

EST sequencing was the first high throughput approach for the identification of genes and facilitate functional genomic studies during RNS. As root hairs are the primary site of rhizobial recognition and entry, the first EST sequencing was conducted to detect root hair cells-enriched cDNA. In this study, root hair cells were mechanically harvested from the whole root. This study is the first example of EST sequencing as well as single-cell transcriptomics in Medicago. Sequencing of the root hair-enriched cDNA library produced ~890 ESTs (Covitz et al. 1998). EST sequencing sets a platform for the development of other genomic tools such as DNA microarray.

The next major upliftment in this area was the design of custom-made oligonucleotides probe-based chips in collaboration with Affymetrix (http://www.affymetrix.com). This Affymetrix GeneChip contains ~10,000 plant probe sets and the probes from the complete *Ensifer meliloti* genome. The first use of Affymetrix GeneChip identified differential expression of 584 Medicago and 1288 *Ensifer* genes (Barnett et al. 2004). This study was followed by another microarray-based transcriptome profiling of wild type and three "*defective in nitrogen fixation*" (*dnf*) series mutants (*dnf1*, *dnf2* and *dnf7*) to understand the global transcriptional changes between wild type and these mutants. Noteworthy, *dnf* mutants generate non-nitrogen fixing nodules (Starker et al. 2006).

This Affymetrix GeneChip array was further used to develop a centralized platform with a web server called "The *Medicago truncatula* gene expression atlas" (MtGEA) described in Chap. 1. Parallel to the development of Affymetrix microarray, a second oligonucleotide probe-based microarray was designed with ~6000 probes identified by a separate EST sequencing study from uninfected roots, mycorrhizal roots, and young root nodules named as root interaction transcriptome (Mt6k-RIT) (Küster et al. 2004). This array was further upgraded in collaboration with Samuel Roberts Noble Foundation to 16086–70 mer oligonucleotides-based Medicago Genome Oligo Set 1.0 (Mt16kOLI) microarrays and subsequently upgraded to Mt16kOLI1Plus by adding 384 new genes, primarily transcription factors (Tellström et al. 2007; Moreau et al. 2011). The transcriptomic resources have seen continuous development between 1998–2010. In between this period, few microarray chips were also designed independently for root nodule developmental studies (see Table 7.1). During this period, several microarray-based studies enriched our knowledge about RNS. As mentioned earlier, many of these microarray chips contain plant and bacterial genes, in this section, we are only highlighting the results obtained from the plant part. The bacterial gene expression was summarized in Sect. 7.5.

A path-breaking study conducted by Eva Kondorosi's group (Maunoury et al. 2010) on wild type and different non-functional (fix−) nodules developed by plant and bacterial mutant's demarcated two waves of transcriptional switches. The first wave of transcriptional programming is required during nodule organogenesis. During this phase, mainly, transient activation of cell cycle and protein synthesis takeplace. The second wave of transcriptional programming is required during bacteroids differentiation. At this stage, mainly, the genes belonging to the secretory pathway, transmembrane, secretory proteins, or peptides were induced. The same studies compared the transcriptome of wild type and mutant symbiotic partners and found that different symbiotic mutants are showing three different transcriptomic signatures- (i) plants or bacterial mutants that are devoid of infection or contain only ITs show root-like transcriptomic signature, (ii) mutants where plant cells were differentiated and infected but the bacteroids did not differentiate passed the first transcriptome switch but not the second one, (iii) mutants nodules where both plant cell and bacteroids were fully differentiated

Table 7.1 Microarray-based transcriptomics approaches conducted to study *Medicago truncatula-Ensifer meliloti* symbiosis transcriptional reprogramming

Microarray developed	Technology	Probes	Year	References
Affymetrix GeneChip	Photolithographic masking oligonucleotide microarray	Complete *Ensifer meliloti* genome, 10 k from plant	2004, 2015	Barnett et al. (2004), Lang and Long (2015)
Affymetrix GeneChip	Photolithographic masking oligonucleotide microarray	9935 from plant	2004, 2006	Mitra et al. (2004), Starker et al. (2006)
Affymetrix GeneChip	*Mediacgo* Genome array (MtGEA v1) and functional genomics annotation to genes (MtGEA v2)	50,900 from plant	2008, 2009	Benedito et al. (2008), He et al. (2009)
Unigene microarray	cDNA microarray	2366 from plant	2010	Maunoury et al. (2010)
Mt6k-RIT	cDNA microarray (accession number A-MEXP-80; http://www.ebi.ac.uk/arrayexpress)	6359 from plant	2004	Küster et al. (2004)
Mt16kOLI1Plus, Mt16KPlus	Oligonucleotide microarray (accession number A-MEXP-138; http://www.ebi.ac.uk/arrayexpress)	16,470 from plant	2007,2011	Tellström et al. (2007), Moreau et al. (2011)

but are non-functional passed both transcriptome switches similar to the wild type nodule. These transcriptomic studies postulated that nodule development goes through different developmental phases. Interestingly, few transcription factors are involved in almost every developmental phase and they govern different transcriptional hubs (see Chap. 6).

7.3 Medicago Genome Versions and Next Generation Sequencing-Based Transcriptomes

Publication of Medicago whole-genome sequence provided a platform for a better and comprehensive understanding of nodule development. To date, four *M. truncatula* genome sequences and corresponding gene annotations have been released. The effort of the International Medicago Genome Annotation Group (IMGAG) led to the first release of the Medicago draft genome (Mt3.5.1) (Young et al. 2011). This sequencing was conducted based on BAC assembly and Illumina shotgun sequence, producing a genome sequence with ~94% of coverage. IMGAG further published an improved version of the Medicago genome (Mt4.0). This version was generated by de novo whole-genome shotgun assembly using Illumina and 454 reads. The scaffolds generated by the de novo whole-genome shotgun method was anchored onto the previously generated pseudomolecules (Tang et al. 2014). In a parallel timeframe, a laser dissection RNA-Seq experiment was conducted on nodule tissue (see Sect. 7.4.2). While conducting this study, the group developed a genome version which was similar to Mt4.0 termed as *Mt20120830-LIPM*. This version was created by the Laboratory of Plant-Microbe Interactions (LIPM) Toulouse, INRA, France, based on Mt3.5.1 release (Young et al. 2011) and combining other available *M. truncatula* genome sequencing data (Roux et al. 2014). The latest and most advanced version of the *M. truncatula* genome is published by LIPM, and this version was created based on PacBio sequencing. The long PacBio reads led to a substantially improved *M. truncatula* genome (Pecrix et al. 2018). The most interesting feature of Mt5.0 is the annotation of the long non-coding RNAs (lncRNAs).

Genome sequencing revealed many interesting features such as the Medicago genome consisting of more synteny blocks compared to other legumes. Synteny blocks are conserved regions within two sets of chromosomes that depict the ancestry. The phylogenomic analyzes suggest that the Medicago genome has undergone whole-genome duplication (WGD) ~58 million years ago (MYA). RNA sequencing revealed 963 WGD-derived gene pairs in the *M. truncatula* genome. Among these duplicated pairs, many got recruited in root nodule symbiosis. For example, nod factor receptor *Nod Factor Perception* (*NFP*) and a transcription factor *ERF required for nodulation 1* (*ERN1*) both have paralogues that participate in mycorrhizal signaling as well. These phylogenomic analyzes based on transcriptomic data and genome sequences suggest sub- and/or neo-functionalization of Medicago genes for nodulation (Young et al. 2011).

7.4 High-Resolution Transcriptomics Studies and Understanding of the Root Nodule Symbiosis

Transcriptomics approaches with multicellular tissues and organs like nodules or whole roots possessing heterogeneous cells lead to dilution of the transcriptome. Transcriptome dilutions restrict the identification of both low-abundant transcripts and moderate-to-high abundant transcripts when its expression is restricted to a single-cell layer. Homeotic regulators, transcription factors, and chromatin regulators are the control hubs and usually express transiently, hence transcriptome coupled with cell type-specific expression or based on laser microdissection (LCM) has the potential to provide much accuracy.

7.4.1 Isolated Root Hair Cell-Based Transcriptomics

Root hair cells are the first contact and entry point of rhizobia. Rhizobia recognition leads to activation of a chain of events in the root hair cells, a. curling, b. entrapment of rhizobia and micro-colony formation, and c. infection thread (IT) formation. Mechanical dissection and isolation of the root hair cells are not complicated to achieve. The first root hair-specific single-cell transcriptomics were attempted in 1998 using a microarray chip (Covitz et al. 1998). More recently after the publication of the Medicago genome sequence, microarray and NGS-based root hair transcriptomics were conducted in wild type and transcription factor mutants (Breakspear et al. 2014). The root hairs (single-cell outgrowth from the epidermis) were mechanically isolated before and after infection with *Ensifer meliloti*, and in some cases, the same strain was defective in nod factor synthesis. Profiling of root hair cells after *Ensifer* inoculation gave a comprehensive understanding of genetic events during early stages of rhizobial infection (Nod factor perception particularly), termed as "Infectome." The major finding from this transcriptomic analysis is that phytohormones, primarily auxin, and cytokinin (CK) regulates the transcriptional reprogramming of epidermal cells. Auxin promotes rhizobial infection. An *Auxin Response Factor* (*ARF16a*) was found to be essential for IT progression. *Lonely guy* genes (*LOG*) get induced during rhizobial inoculation which generate active CK and this further trigger CK signaling pathway by response regulators and two-component system (TCS) (Breakspear et al. 2014). ABCG family of the transporter (*ABCG56*) gets activated in the root hair cells

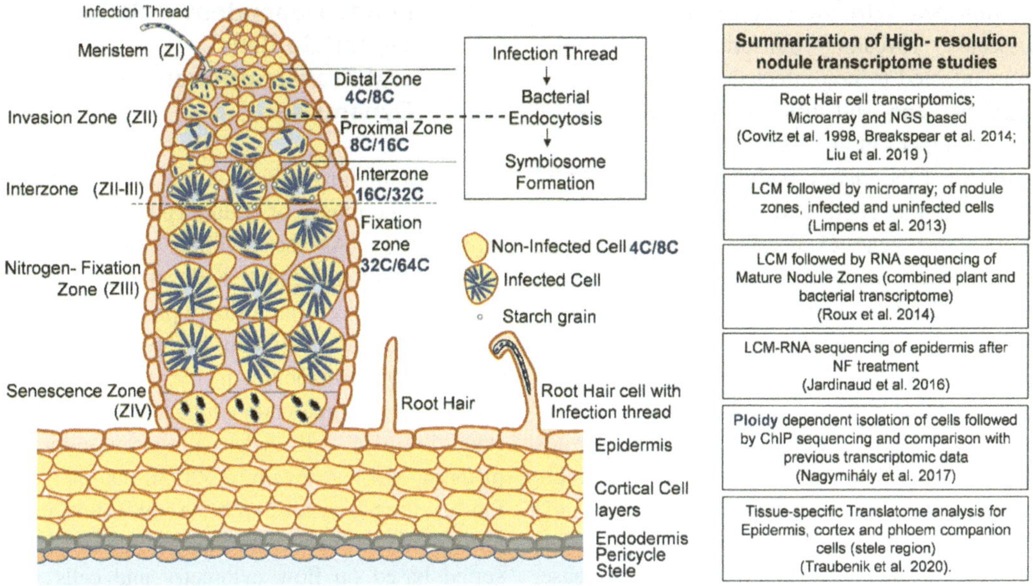

Fig. 7.1 Diagrammatic representation of the mature indeterminate Medicago nodule with the summarization of the single-cell or tissue-specific transcriptome and translatome studies

and this transporter plays a pivotal role in active CK transport during symbiosis (Jarzyniak et al. 2021) (see Chap. 6, Fig. 6.2).

Three transcription factors (TFs) activated immediately after rhizobial infection are *Nodule INception* (*NIN*), *Nuclear Factor Y1* (*NF-YA1*), and *ERF required for Nodulation 1* (*ERN1*). Among them, NIN is the master regulator of infection and organogenesis. NIN directly activates *NF-YA1* expression. Further, NIN and ERN1 signaling pathways are intertwined (see Chap. 6). Root hair cell transcriptomic analysis had been conducted on *nin-1, ern1* and *nf-ya1* mutants after inoculating with wild type and non-NF-producing rhizobium. This study highlighted that among 1124 genes upregulated after rhizobium inoculation in root hair cells, 43% get down-regulated in *nin-1* root hair cells, whereas only 9.1% are down-regulated in the *ern1*. Hence, established NIN as a central regulator of root hair transcriptome. This study also identified the putative direct common targets of NIN, NF-YA1, and ERN1 (Liu et al. 2019). Thus, these transcriptomic studies drew a broader image of rhizobial infection through root hair cells.

7.4.2 Laser-Capture Microdissection (LCM) coupled with Microarray and RNA-Sequencing

The second set of high-resolution transcriptome data came from laser-capture microdissection (LCM) of different nodule zones coupled with microarray/RNA sequencing. In one attempt nodule zones, infected and uninfected cells were isolated using LCM (Fig. 7.1). A special development gradient exists from the top to the bottom across the cylindrical indeterminate nodule of Medicago. The nitrogen fixation zone of Medicago nodules contains highly endoreduplicated, bigger, rhizobium-containing cells, and less endoreduplicated, smaller cells that are devoid of rhizobia. Nodule zones and cells present in each zone are at different developmental phases. Hence, it is expected that different nodule zones should have different transcriptomic signatures (Limpens et al. 2013). The first zone-specific transcriptomic analysis was conducted using Affymetrix Medicago GeneChips containing 50, 900 probe sets. This analysis presented cell type-specific gene enrichment where infected and non-infected cells were separated and further dissected, revealing transcriptional dynamics from infected to non-infected cells, meristem to infection zone, and distal to proximal infection zone (Limpens et al. 2013).

LCM of the mature nodule zone followed by RNA-Seq was conducted (Roux et al. 2014) to characterize gene expression across Medicago indeterminate nodule zones (Vasse et al. 1990). They used five nodule zones representing (i) the apical meristem region with small cells (ZI); (ii) distal invasion zone, where endocytosis of rhizobia takes place (ZIIa); (iii) proximal invasion zone, where differentiation of bacteroids takes place (ZIIb); (iv) interzone II–III, where major starch accumulation takes place; and (v) nitrogen fixation zone ZIII (Fig. 7.1). This sequencing data were compiled in a database, SYMbiMICS (https://iant.toulouse.inra.fr/symbimics/), which is a vital platform for exploring gene expression and understanding its role during Medicago nodule development (Roux et al. 2014).

Another LCM-RNA sequencing analysis was conducted after 4 and 24 h post Nod Factors (NFs) treatment followed by LCM of root regions responsive to NF coupled to RNA sequencing (Jardinaud et al. 2016). This study identified around 1,070 genes being regulated by NF. Further, approximately 300 genes showed 10-folds upregulation after NF treatment. Among them, 44 genes were symbiosis specific. These 44 genes include *NCRs* that are common to mycorrhizal and nodulated roots. This approach identified two important receptor kinases, *SYMBIOSIS LEUCINE-RICH RECEPTOR KINASE (SYMRK)* and *LYSINE MOTIF RECEPTOR KINASE (LYK3)*. These two were highly upregulated in the epidermis. Additionally, receptor kinase *LYK10* is also expressed in epidermal cells. Hormone signaling-related genes such as CK and strigolactone pathway genes and flavonoids biosynthesis genes show NF-induced expression. This study identified strong but transient activation of genes after NF treatment in the root epidermis. In summary, epidermal regulation of NF responsive genes is highly complex (Jardinaud et al. 2016).

7.4.3 Ploidy Dependent Isolation of Nodule Cells and Transcriptional Dynamics

The third high-resolution study was conducted based on the ploidy of nodule cells (Nagymihály et al. 2017). As the dividing cells leave the meristem, they start endoreduplication. In the distal invasion zone, cells are usually present in 4C and get infected with rhizobium. In the invasion zone, the bacteroid (rhizobium inside the plant-bound membrane) also gets endoreduplicated. During this stage, the plant cell nucleus keeps on endoreduplicating and reaches up to 32C in the interzone. Further, the ploidy level of these cells is 64C in the nitrogen fixation zone. Infected nodule cells were sorted based on flow cytometry and cells with different ploidy (4C, 8C, 16C, and 32C) were isolated. Chromatin immunoprecipitation (ChIP) was conducted on the DNA that was isolated from these cells against H3K27me3 and/or anti-H3K9ac histone marks. The ChIP data were further compared with the LCM nodule zone-specific RNA-Seq-based transcriptomics data. This study revealed that methylation and histone modification mediate epigenetic control of nodule-specific genes like *NCRs* and *ENOD12*. The increasing ploidy levels alter chromatin accessibility thus gene expression is affected. Symbiotic cells undergo successive rounds of endoreduplication, parallel to the terminal differentiation of endosymbionts. Expression of those genes responsible for terminal bacteroid differentiation (TBD) such as *NCRs* (see Chap. 6) were found to be epigenetically regulated (Nagymihály et al. 2017). Hence, this study opened the avenue of research in the area of chromatin dynamics during nodule development, a largely unknown area.

7.4.4 Translating Ribosome Affinity Purification

Another technique for examining cell-specific expression is translating ribosome affinity purification (TRAP). In this technique, ribosomal proteins tagged with an epitope (usually FLAG) are expressed in plant tissues. Immunopurification of polysomes (i.e., transcripts bound to two or more ribosomes) followed by RNA-Seq estimates translational control of gene expression (translatome). The first Medicago nodule translatome was conducted by Traubenik's group (Traubenik et al. 2020). To understand the translational control of gene expression, *p35S::FLAG-RPL18* was transformed into Medicago followed by both transcriptome and translatome analysis. Comparison between transcriptome and translatome highlighted ~ 65% of the differentially upregulated genes identified by transcriptome, which did not show enrichment in nodule translatome. Underlining, transcriptional and translational responses are partially uncoupled during nodule development. In the same study, TRAP has been combined with Medicago epidermal cell-specific *EXPANSIN7* (*MtEXP7*), Arabidopsis cortex specific *CORTEX SPECIFIC TRANSCRIPT (AtCO2*), and phloem companion cell-specific *SUCROSE TRANSPORTER 2 (AtSUC2)* promoters to understand tissue-specific gene expression during Medicago nodule development. These tissue-specific ribosome-pulldown experiments have been designed specially to understand the early recognition and nodule primordium initiation. The analysis highlighted the high accumulation of *DMI2*-mRNA in the polysomes during rhizobial infection, suggesting a large amount of DMI2 protein is getting synthesized during infection. Some transcripts show cell-specific abundance such as *ERN1*, *NF-YA1*, and *NPL* show more enrichment in epidermal cells compared to cortical cells (Reynoso et al. 2013; Pan et al. 2018; Traubenik et al. 2020). Cell type-specific translatome analysis holds high potential and has proven to be a method of choice in animal models or Arabidopsis as TRAP can be performed on intact tissues as opposed to LCM or FACS mediated tissue isolation. Hence, a high amount of mRNA can be obtained without cumbersome methodology or techniques that can change gene expression. Cell/tissue-specific TRAP is going to be a method of choice to answer many unanswered questions in nodule biology.

Medicago nodule possess a) meristematic zone (ZI) containing mitotically active cells; b) invasion zone (ZII) where bacterial endocytosis takes place and is subdivided into two subzones vis distal (ploidy level 4C/8C) and proximal (ploidy level 4C/8C) invasion zone; c) Interzone (ploidy level 16C/32C), a zone distinctly visualized due to accumulation of starch; d) nitrogen fixation zone (ZIII) containing a mixture of infected cells (ploidy level 32C/64C) and uninfected cells (ploidy level 4C/8C) where enzymatic fixation of di-nitrogen takes place; and e) senescence zone (ZIV) presents only in older nodules. Based on the nodule developmental gradient, several tissue-specific transcriptomic studies and ploidy-based ChIP sequencing have been conducted. Infected root hair cell transcriptomic study was conducted in wild type and three transcription factor mutants (Breakspear et al. 2014; Liu et al. 2019), and epidermal cells after NF treatment were dissected by LCM (Jardinaud et al. 2016) to understand the early stages of plant-rhizobia recognition. A translatome study was conducted using different tissue-specific promoters *EXPANSION7*, *CORTEX SPECIFIC TRANSCRIPT*, and *SUCROSE TRANSPORTER 2*, respectively, for epidermis, cortex, and phloem-companion cells (Fig. 7.1) (Traubenik et al. 2020).

7.5 Understanding from the Bacterial Part

In root nodule symbiosis, symbiotic bacteria play the solo part in the enzymatic fixation of the gaseous di-nitrogen. Host cell physioxia plays a pivotal role in the onset of nitrogen fixation. In a mature indeterminate nodule, leghemoglobin accumulates only in the nitrogen fixation zone. Accumulation of leghemoglobin leads to the hypoxic condition in the nodule. Hypoxia induces bacterial nitrogen fixation operon (nif and fix genes), thus promoting enzymatic fixation of nitrogen. Like the plant, sequencing of the symbiont genome accelerated the SNF research. *Ensifer meliloti* whole-genome consists of a chromosome (3.65 Mb) and two megaplasmids (pSymA and pSymB). This genome sequence has been assembled in 2001 (Galibert et al. 2001). The bacterial genome sequence leads to the incorporation of *S. meliloti* specific probe sets in the Affymetrix-based Medicago GeneChip. In the gene chip, a total of 9,935 probe sets are present that correspond to the Ensifer genome. Hence, whenever a microarray has been attempted using Affymetrix GeneChip, consolidated analysis of plant and/or bacterial gene expression can be obtained depending on the sample preparation. Bacteroid transcriptome analysis highlighted that transcription factors that activate nod operon (nodD), NF biosynthesis genes, transporter genes, such as sitA, exopolysaccharide abundance regulatory genes syrM and syrA, adenylate cyclases genes (cyaE, cyaF1, cyaF6), genes responsible for nitrogen fixation (nifA, nifH nifN, fixK1, fixN1, fixL etc.), and phenylacetic acid catabolism genes paaABCD, were highly induced inside nodules. Additionally, nodule transcriptomic studies revealed that ~33% of symbiosis-associated genes were located on the pSymA plasmid (Barnett et al. 2004).

Nodule transcriptome study was conducted using bacterial mutants defective in different stages of nodule development keeping the plant background as wild type to understand how plant gene expression is coupled with the bacterial gene expression. Bacterial *exoY* mutant that is defective in succinoglycan production cannot initiate successful IT formation (Cheng and Walker 1998) and produces small nodule bumps. Transcriptomics of these nodules resemble plant roots. Further, inoculation of *bacA* mutant forms small white nodules where bacterial endocytosis is normal, but the bacteroids are small and undifferentiated. The transcriptome signature of this mutant resembles partly with root but shows the first wave of nodule-specific gene expression (see Sect. 7.2). Further, *E. meliloti* mutants (such as *nifH*, *nifA*, *fixG*, *fixJ*,and *fixK*) where nodule organogenesis per se gets completed but cannot start the fixation of atmospheric di-nitrogen in spite of showing signature of complete nodule organogenesis. This suggests enzymatic fixation of nitrogen in the root nodule happens after the completion of the nodule development (Maunoury et al. 2010).

The advancement in the understanding of bacterial gene expression came from the LCM study that has already been described above (Sect. 7.4.2). The LCM coupled with RNA-Seq analysis was conducted for both plant and bacterial genes (Roux et al. 2014). In the LCM analysis, 7799 bacterial genes were detected to be expressed inside the nodule. These genes were categorized in 13 different clusters as per their expression patterns. The integrated analysis showed fine coordination between plant and bacterial gene expression profiles, for example, bacterial genes for cell division (ftsK, minCDE), DNA replication genes (repC1, repC2), and genes controlling cell cycles (ctrA, divJ) were highly expressed in proximal and distal invasion zone where host and bacterial cells were continuing their development, whereas these genes show negligible expression in the nitrogen fixation zone. The LCM analysis also raised many unanswered questions. The master regulator of the flagellar operon (visN) shows expression from the interzone, and its expression is the highest in the nitrogen fixation zone, indicating the whole operon is functional in this zone. In the invasion and nitrogen fixation zone, the rhizobium is present inside the plant-derived membrane envelope (Roux et al. 2014). What is the function of flagella at this stage? Why does not the pathogen-associated molecular patterns (PAMPs) trigger immunity (PTI) during this stage (See Chap. 6)? There are definitely more questions than answers.

7.6 Transcriptomic Studies Unfold Many Unanswered Questions—The Future Direction of Nitrogen Fixation Research

Dedifferentiation of already differentiated root cells was the first evolutionary innovation that nodule-forming plants must have learned. Hence, it can be visualized easily that chromatin remodeling and change in the epigenetic mark were the first few steps that would have happened inside the plant nucleus concomitantly with the perception of the NF. TFs might have played a vital role just after the opening/remodeling of chromatin. Nodule-specific upregulation of TFs would have happened either (a) under NF signaling or (b) under the regulation of hormones (Breakspear et al. 2014; Nagymihály et al. 2017). Further, the most recent version of the Medicago genome along with the LCM mediated transcriptome analysis highlighted that small and long non-coding RNAs (lncRNAs) are expressed specifically to different nodule zones (Pecrix et al. 2018). lncRNA is antisense non-coding transcripts. lncRNA-mediated gene regulation has been broadly studied in mammalian cells. They can play multiple roles such as a enhancer of transcription by interacting with the chromatin (recruits chromatin remodeling complex to activate transcription); (b) decoy to sequester transcription factors (sequester proteins with regulatory function); (c) as guide RNA (carries protein molecules to their target genes); and (d) modulator of transcriptional expression (acts as a scaffold to recruit the molecular complex to target genes) (Batley et al. 2020). Only a limited number of long non-coding RNAs have been functionally characterized in plants, especially in legumes. The recent release of the Medicago genome has annotated thousands of long non-coding RNAs showing positive or negative correlations with neighboring mRNAs and being induced in the nodule. Such profiles have been seen for many vital genes that have already been genetically characterized for their roles during nodule development such as *NSP2, IPD3, NIN, SymCRK, ERN2, DME, DNF1,* and *NCRs* (for the role of these genes see Chap. 6). The positively correlated lncRNA-mRNA pairs are *DME, NSP2, NIN, IPD3, RSD, SYMREM1,* and *SymCRK.* Interestingl,y sense and antisense transcripts of *EFD* show different expression profiles in nodule zones (Pecrix et al. 2018). Sense transcript of *EFD* shows maximum expression in early nodule stages and at distal and proximal invasion zone, whereas the antisense transcript expresses maximum in the interzone and fixation zone in mature nodules. It hints that *lncRNAs* might regulate these symbiosis-related genes' expression. In Medicago, *ENOD40* (*Early Nodulin 40*) was the first characterized lncRNA which binds to RNA-Binding protein (RBP) and promotes cytoplasmic relocalization of this protein during nodule formation. Thus, *lncRNA*-mediated regulation of nodule development needs further attention (Campalans et al. 2004; Mergaert et al. 2020).

7.7 Conclusion

Gradual advances from EST sequencing to next-generation sequencing built up a platform for the Medicago genome and the nodule-specific gene expression profile. Further, transcriptomics coupled with functional genomics studies have drawn a skeleton structure of the molecular mechanism of RNS. Over two decades of research also highlighted the complexity of the molecular interaction that guides this marvelous organ development. Recent advances in the next generation sequencing (NGS) technology combined with cell/tissue-specific transcriptomic, ChIP, and translatome analysis highlight the hidden complexity that guides nodule development. These new techniques hold enormous potential and the growing information that came out from these analyzes will help to create a superior understanding of RNS.

Acknowledgements We thank NIPGR for core grant and Department of Biotechnology (DBT)-eLibrary Consortium (DeLCON), India for providing access to e-resources. Akanksha Bhardwaj was supported by CSIR (09/803(0145)/2018-EMR-I).

References

Barnett MJ, Toman CJ, Fisher RF, Long SR (2004) A dual-genome Symbiosis Chip for coordinate study of signal exchange and development in a prokaryote-host interaction. Proc Natl Acad Sci U S A 101:16636–16641. https://doi.org/10.1073/pnas.0407269101

Batley J, Willmann MR, Budak H, et al (2020) Long Non-coding RNA in Plants in the Era of Reference Sequences. https://doi.org/10.3389/fpls.2020.00276

Benedito VA, Torres-Jerez I, Murray JD, et al (2008) A gene expression atlas of the model legume Medicago truncatula. Plant J 55:504–513. https://doi.org/10.1111/j.1365-313X.2008.03519.x

Breakspear A, Liu C, Roy S, et al (2014) The root hair "infectome" of medicago truncatula uncovers changes in cell cycle genes and reveals a requirement for auxin signaling in rhizobial infectionw. Plant Cell 26:4680–4701. https://doi.org/10.1105/tpc.114.133496

Campalans A, Kondorosi A, Crespi M (2004) Enod40, a short open reading frame-containing mRNA, induces cytoplasmic localization of a nuclear RNA binding protein in Medicago truncatula. Plant Cell 16:1047–1059. https://doi.org/10.1105/tpc.019406

Cheng HP, Walker GC (1998) 1998-Succinoglycan Is Required for Initiation and Elongation of Infection Threads.pdf. J Bacteriol 180:5183–5191

Covitz PA, Smith LS, Long SR (1998) Expressed sequence tags from a root-hair-enriched Medicago truncatula cDNA library. Plant Physiol 117:1325–1332. https://doi.org/10.1104/pp.117.4.1325

Galibert F, Finan TM, Long SR, et al (2001) The composite genome of the legume symbiont Sinorhizobium meliloti. Science (80) 293:668–672. https://doi.org/10.1126/science.1060966

He J, Benedito VA, Wang M, et al (2009) The Medicago truncatula gene expression atlas web server. BMC Bioinformatics 10. https://doi.org/10.1186/1471-2105-10-441

Jardinaud MF, Boivin S, Rodde N, et al (2016) A laser dissection-RNAseq analysis highlights the activation of cytokinin pathways by nod factors in the Medicago truncatula root epidermis. Plant Physiol 171:2256–2276. https://doi.org/10.1104/pp.16.00711

Jarzyniak K, Banasiak J, Jamruszka T, et al (2021) Early stages of legume–rhizobia symbiosis are controlled by ABCG-mediated transport of active cytokinins. Nat Plants 7:428–436. https://doi.org/10.1038/s41477-021-00873-6

Küster H, Hohnjec N, Krajinski F, et al (2004) Construction and validation of cDNA-based Mt6k-RIT macro- and microarrays to explore root endosymbioses in the model legume Medicago truncatula. J Biotechnol 108:95–113. https://doi.org/10.1016/j.jbiotec.2003.11.011

Lang C, Long SR (2015) Transcriptomic analysis of Sinorhizobium meliloti and Medicago truncatula symbiosis using nitrogen fixation-deficient nodules. Mol Plant-Microbe Interact 28:856–868. https://doi.org/10.1094/MPMI-12-14-0407-R

Limpens E, Moling S, Hooiveld G, et al (2013) Cell- and Tissue-Specific Transcriptome Analyses of Medicago truncatula Root Nodules. PLoS One 8:. https://doi.org/10.1371/journal.pone.0064377

Liu CW, Breakspear A, Guan D, et al (2019) NIN acts as a network hub controlling a growth module required for rhizobial infection. Plant Physiol 179:1704–1722. https://doi.org/10.1104/pp.18.01572

Maunoury N, Redondo-Nieto M, Bourcy M, et al (2010) Differentiation of symbiotic cells and endosymbionts in Medicago truncatula nodulation are coupled to two transcriptome-switches. PLoS One 5:. https://doi.org/10.1371/journal.pone.0009519

Mergaert P, Kereszt A, Kondorosi E (2020) Gene Expression in Nitrogen-Fixing Symbiotic Nodule Cells in Medicago truncatula and Other Nodulating Plants. Plant Cell 32:42–68. https://doi.org/10.1105/tpc.19.00494

Mitra RM, Shaw SL, Long SR (2004) Six nonnodulating plant mutants defective for Nod factor-induced transcriptional changes associated with the legume-rhizobia symbiosis. Proc Natl Acad Sci U S A 101:10217–10222. https://doi.org/10.1073/pnas.0402186101

Moreau S, Verdenaud M, Ott T, et al (2011) Transcription reprogramming during root nodule development in Medicago truncatula. PLoS One 6:. https://doi.org/10.1371/JOURNAL.PONE.0016463

Nagymihály M, Veluchamy A, Györgypál Z, et al (2017) Ploidy-dependent changes in the epigenome of symbiotic cells correlate with specific patterns of gene expression. Proc Natl Acad Sci U S A 114:4543–4548. https://doi.org/10.1073/pnas.1704211114

Pan H, Stonoha-Arther C, Wang D (2018) Medicago plants control nodulation by regulating proteolysis of the receptor-like kinase DMI2. Plant Physiol 177:792–802. https://doi.org/10.1104/pp.17.01542

Pecrix Y, Staton SE, Sallet E, et al (2018) Whole-genome landscape of Medicago truncatula symbiotic genes. Nat Plants 4:1017–1025. https://doi.org/10.1038/s41477-018-0286-7

Reynoso MA, Blanco FA, Bailey-Serres J, et al (2013) Selective recruitment of mRNAs and miRNAs to polyribosomes in response to rhizobia infection in Medicago truncatula. Plant J 73:289–301. https://doi.org/10.1111/tpj.12033

Roux B, Rodde N, Jardinaud MF, et al (2014) An integrated analysis of plant and bacterial gene expression in symbiotic root nodules using laser-capture microdissection coupled to RNA sequencing. Plant J 77:817–837. https://doi.org/10.1111/tpj.12442

Starker CG, Parra-Colmenares AL, Smith L, et al (2006) Nitrogen fixation mutants of Medicago truncatula fail to support plant and bacterial symbiotic gene expression. Plant Physiol 140:671–680. https://doi.org/10.1104/pp.105.072132

Tang H, Krishnakumar V, Bidwell S, et al (2014) An improved genome release (version Mt4.0) for the model legume Medicago truncatula. BMC Genomics 15:1–14. https://doi.org/10.1186/1471-2164-15-312

Tellström V, Usadel B, Thimm O, et al (2007) The lipopolysaccharide of Sinorhizobium meliloti suppresses defense-associated gene expression in cell cultures of the host plant Medicago truncatula. Plant Physiol 143:825–837. https://doi.org/10.1104/pp.106.090985

Traubenik S, Reynoso MA, Hobecker K, et al (2020) Reprogramming of root cells during nitrogen-fixing symbiosis involves dynamic polysome association of coding and noncoding RNAs. Plant Cell 32:352–373. https://doi.org/10.1105/tpc.19.00647

Vasse J, de Billy F, Camut S, Truchet G (1990) Correlation between ultrastructural differentiation of bacteroids and nitrogen fixation in alfalfa nodules. J Bacteriol 172:4295–4306

Young ND, Debellé F, Oldroyd GED, et al (2011) The Medicago genome provides insight into the evolution of rhizobial symbioses. Nature 480:520–524. https://doi.org/10.1038/nature10625

Whole Genome Sequencing Identifies a *Medicago truncatula Tnt1* Insertion Mutant in the *VTL8* Gene that is Essential for Symbiotic Nitrogen Fixation

8

Jingya Cai, Vijaykumar Veerappan, Taylor Troiani, Kirankumar S. Mysore, Jiangqi Wen, and Rebecca Dickstein

Abstract

Legumes carry out symbiotic nitrogen (N) fixation by forming N-fixing nodules with rhizobia during N deficiency. In an effort to discover new legume genes essential to nodule function using the *Medicago truncatula–Sinorhizobium meliloti* model legume–rhizobia model system, a forward genetics approach was employed using lines of *M. truncatula* mutagenized with the tobacco-retrotransposon *Tnt1*, from which N-fixation defective (Fix$^-$) plants were recovered. The *Tnt1* mutagenized lines are in the R108 genetic background. One of these lines, NF11322, has a *Tnt1* insertion interrupting a novel nodule-specific gene annotated as an iron transporter. Phylogeny analysis of the encoded protein, now known as *M. truncatula* Vacuolar iron Transporter-Like 8, MtVTL8, shows that it is a homolog of the *Saccharomyces cerevisiae* vacuolar iron/manganese transporter CCC1 protein, *Arabidopsis thaliana* vacuolar iron transporter VIT1 protein, and *Lotus japonicus* SEN1 protein, required during nodule formation in *L. japonicus*. The mutation in *MtVTL8* co-segregates with the Fix$^-$ phenotype and mutants show distinctive similarities with a deletion mutant in the A17 genetic background that includes the *MtVTL8* gene. This evidence demonstrates that the defect in NF11322 in *MtVTL8* is the causative mutation in this mutant line. *MtVTL8* encodes an iron transporter, implicated directly in transporting iron to the internalized rhizobia within nodule infected cells and essential for symbiotic nitrogen fixation.

8.1 Introduction

Random mutagenesis is a powerful genetic tool, and includes transposon insertion, treatment with chemical mutagens such as ethyl methanesulfonate and fast neutron bombardment. The transposable element from *Nicotiana tabacum* type 1, *Tnt1*, is a 5.3 kb retrotransposon that has found use in mutagenizing *Medicago truncatula* (d'Erfurth et al. 2003; Tadege et al. 2008; Pislariu et al. 2012; Sun et al. 2019). It transposes via reversely transcribing an RNA copy and then integrating into the plant genome (Cui et al. 2013). In *M. truncatula*, *Tnt1* is only active during tissue culture. *Tnt1* was originally

J. Cai · V. Veerappan · T. Troiani · R. Dickstein (✉)
Department of Biological Sciences and BioDiscovery Institute, University of North Texas, Denton, TX 76203, USA
e-mail: beccad@unt.edu

V. Veerappan
Department of Biology, Eastern Connecticut State University, Willimantic, CT 06226, USA

K. S. Mysore · J. Wen
Oklahoma State University, Ardmore, OK 73401, USA

S. Sinharoy et al. (eds.), *The Medicago truncatula Genome*, Compendium of Plant Genomes,
https://doi.org/10.1007/978-3-030-90757-0_8

transformed into *M. truncuatula* via *Agrobacterium tumefaciens* and once present can transpose again via somatic embryogenesis, obviating the need for re-transformation (d'Erfurth et al. 2003; Tadege et al. 2008). In *M. truncatula*, scientists have used leaf explants from R108 containing five *Tnt1* insertions as starting material for additional rounds of mutagenesis (Tadege et al. 2008; Sun et al. 2019). *Tnt1* is stable during seed to seed propagation, so *Tnt1* is an efficient tool for large scale insertion mutagenesis. In *M. truncatula*, *Tnt1* insertion correlates with gene methylation and inserts preferentially in exons (Tadege et al. 2008; Sun et al. 2019). It has been used to create a robust library of mutants (Pislariu et al. 2012). Individual leaf explants were grown into independent mutant lines with *Tnt1* insertions in the genome. On average, *Tnt1* inserts at 80 different places for each line during tissue culture, and it is reported that 21,741 independent lines have been generated indicating around 1.7 million insertions (Sun et al. 2019).

Forward genetic screens were carried out at the Noble Research Institute for over a decade to identify R1 progeny from *Tnt1* mutagenized *M. truncatula* lines for defective phenotypes including symbiotic nitrogen fixation (SNF) nodulation phenotypes. Scientists described six basic defective nodule morphologies: non-nodule (Nod^-), ineffective nodule (Nod^+/Fix^-), partially ineffective nodule ($Nod^+/Fix^{+/-}$), defective nodule formation, and nitrogen fixation ($Nod^{+/-}/Fix^-$), delayed nodule formation and effective nitrogen fixation ($dNod^{+/-}$, Fix^+) and supernodulation ($Nod^{++}/Fix^{+/-}$) (Pislariu et al. 2012). Once the mutant plants are isolated, genomic DNA is extracted and subjected to various sequencing strategies to uncover the *Tnt1* insertion sites, including whole genome sequencing and *Tnt1*-capture sequencing (Sun et al. 2019). The *Tnt1* insertion sites can then be used to design gene-specific primers, which when used together with *Tnt1*-specific primers within polymerase chain reactions provide genotyping for each *Tnt1* insertion site. When these are used in a population that is genetically segregating for the phenotype of interest, for example, a BC_1F_2 back-crossed population, the underlying causative mutation can be genetically mapped (Veerappan et al. 2016). Clues to potential causative mutations in the *Tnt1* mutants can also be provided by correlation of mutation position in specific genes with genes that are solely or highly expressed in nodules. The latter can be determined via the publicly available Affymetrix-based *M. truncatula* gene expression atlas (Benedito et al. 2008) and/or the RNAseq-based Symbimics nodule-expression site (Roux et al. 2014). Through this process, the putative gene that causes the mutant phenotype can be identified (Veerappan et al. 2016).

Iron is an important cofactor in many enzymes and proteins, such as the photosystem complex involved in photosynthesis, cytochrome C reductase involved in electron transport chain, cytochrome P450 involved in stress and leghemoglobin, and rhizobial nitrogenase complex involved in SNF (Van Loon 2009; Kato et al. 2019). In cultivated soils, iron has a concentration of 20 to 40 g/kg, but it mainly exists as an insoluble ferric (Fe III) hydroxide complex in soil, which is unavailable for plants (Colombo et al. 2014). Plants including dicots and non-grass monocots mobilize iron by releasing protons to acidify the rhizosphere, reduce Fe III into Fe II using reductase enzymes at the root surface, and transport it into the cytoplasm of root cells (Connolly et al. 2003; Van Loon 2009; Kato et al. 2019). The absorbed iron is then translocated via the vasculature to other parts of the plant. In plant cells, iron mainly exists in the reduced ferrous (Fe^{2+}) form and is stored in vacuoles for later usage. Iron levels are strictly regulated as excessive iron causes cell damage by producing radicals. Many transporters are involved in iron translocation (Brear et al. 2013; Tejada-Jiménez et al. 2015; Kryvoruchko et al. 2018).

During legume SNF, iron plays an important role as it is the cofactor of leghemoglobin involved in buffering oxygen level in nodules and is a component of the nitrogenase complex responsible for bacteroid nitrogen fixation. Iron is thought to be released from the xylem, transported to the infected nodule cells, which can then be transported into the cytosol (González-

Guerrero et al. 2016). There are two proposed routes for iron entering into infected cells: an apoplastic route and a symplastic route. For the symplastic route, membrane importers located on pericycle cells are involved, such as those in the yellow stripe-like (YSL) transporter family, natural resistance-associated macrophage protein (NRAMP) transporter family and Zrt-, Irt-like protein (ZIP) transporter family (Brear et al. 2013; Tejada-Jiménez et al. 2015; Kryvoruchko et al. 2018). These transporter families are believed to be involved in importing iron across the plasma membrane of the cells inside pericycle cells into the infected cells (Brear et al. 2013) (Brear et al. 2013). Less is known about the transporters involved in the mobilizing iron into the bacteroid (Walton et al. 2020).

Here, we report an iron transporter gene, *M. truncatula* Vacuolar iron Transporter-Like 8, *MtVTL8,* identified via forward genetics in the *M. truncatula* R108 *Tnt1* genetic background. When mutated, it results in nitrogen fixation defective (Fix^-) nodules.

8.2 Materials and Methods

8.2.1 Mutant Propagation and Growth

M. truncatula Tnt1 line NF11322 was back crossed to wild type R108 using the keel petal incision technique and crosses confirmed (Veerappan et al. 2014). BC_1F_2 seeds were scarified using concentrated sulfuric acid (H_2SO_4) for 4 min with shaking, washed five times with sterilized water, surface sterilized with 6% commercial bleach for 90 s with shaking, and then washed for five times with sterilized water. The treated seeds were imbibed by submerging them in sterilized water with rotary shaking at 90 rpm at 25 °C in the dark. The seeds were decanted and fresh sterile water was added every 30 min for 4 h. Seeds were then vernalized at 4 °C in dark for 3 days and germinated on 1% water agar petri dishes with plates inverted in the dark at 25 °C for two days. Seedlings were transferred to a modified aeroponic system described in the *Medicago* handbook (Barker et al. 2006) with Lullien's medium for aeroponic growth (Lullien et al. 1987). Seedlings were grown for 5 days in Lullien's medium and then 5 days in Lullien's medium without NH_4NO_3. Then the plants were inoculated with *Sinorhizobium meliloti 41* (*Rm41)* containing a *hemA:lacZ* reporter. Control plants were grown in complete Lullien's medium without inoculation. The plants were grown at 20 °C with a 16 h light/8 h dark cycle. Plants were collected 15 days post inoculation (dpi) or at the indicated time.

8.2.2 DNA Manipulations

Whole genome sequencing was carried out and analyzed as previously described (Veerappan et al. 2016). For genotyping, gDNA was extracted using the Dellaporta method (Dellaporta et al. 1983). PCR was used to genotype back crossed NF11322 BC_1F_2 plants. For this, the Green GoTaq® DNA polymerase kit (Promega Inc, Madison, WI, Catalog#M3001) was utilized, using the manufacturer's recommendations. Gene-specific primers for MtV TL8 were Medtr4g094335-1F, AAACAGACCA ATCACACATTCA, and Medtr4g094335-1R, AAAGGAACCAAACCACCAATAG which yield a 567 bp product. Tnt1-F1-TCCTTGTTG GATTGGTAGCCAACTTTGTTG was used with Medtr4g094335-1R to yield a 725 bp fragment to confirm the *Tnt1* insertion. The PCR program was 96 °C for 5 min followed by 29 cycles of 96 °C for 30 s, 60 °C for 30 s, 72 °C for 80 s, and 72 °C for 10 min.

8.2.3 Gene Expression

The Symbimics website (https://iant.toulouse.inra.fr/symbimics/) was used to search for nodule-specific and nodule-enhanced expressions (Pecrix et al. 2018).

8.2.4 Phylogenetic Analysis

Phylogenetic analysis was carried out using Maximum Parsimony. Mega 7 software was employed (Kumar et al. 2016). The GenBank accession numbers for the protein sequences in the phylogeny are ScCCC1, NP_013321.1; AtVIT1, NP_178286.1; AtVTL1, NP_173538.2; OsVIT1, XP_015636127.1; OsVIT2, KAB811 0390.1; LjSEN1, BAL46698.1; GmN-21, NP_0 01236825.2; TgVIT1, BAH59029.1; EgVIT1, XP_010066557.1; MtVTL4, XP_013457227.1; MtVTL8, XP_013457231.1; Mt5g068580, XP_0 24640109.1; Mt1g099010, XP_003592125.1; Mt2g008110, XP_003593128.2; Mt4g094328, KEH31259.1; Mt4g094330, KEH31259.1; Mt4g 094332, XP_013457230.1; Mt6g034975, XP_0 13451809.1; Mt7g076320, XP_024625458.2; Mt8g012900, XP_013444147.1; Mt8g105790, XP_003630980.1; Mt8g105810, XP_0036309 81.2; Mt0119s0020, XP_013442645.1; and Mt0119s0010, XP_013442645.1.

8.2.5 Microscopy

Nodules were grown aeroponically as described above and harvested at 15 dpi. They were fixed, embedded in Epon Araldite resin and sectioned with glass knives as described previously (Veereshlingam et al. 2004). Staining was with toluidine blue.

8.2.6 Results and Discussion

Forward genetic screening of the *Tnt1* insertion mutants in the R108 genetic background (Tadege et al. 2008; Moll et al. 2017) is a good strategy to identify novel genes involved in symbiotic nitrogen fixation (SNF). This is because the nitrogen fixation defective (Fix^{-}) genes are mutated and tagged by the *Tnt1* element at the same time. The disadvantage of this strategy is that the *Tnt1* mutants contain many *Tnt1* inserts per line. Once genes are identified, molecular tools can be used to gain insight into their functions. One *M. truncatula* Fix^{-} mutant line is NF11322, identified in a forward genetic screen. We used whole genome sequencing as a way to identify all the *Tnt1* insertion sites in this line, similar to methods reported previously (Veerappan et al. 2016), finding a total of 59 high-confidence and 18 low-confidence putative *Tnt1* sites. One of these is in the vacuolar iron transporter-like gene now known as *M. truncatula* Vacuolar iron Transporter-Like 8, *MtVTL8,* (Walton et al. 2020), Medtr4g094335 in the Mt4.0 genome annotation (Tang et al. 2014). NF11322 has a *Tnt1* insertion at 196 bp downstream of the ATG of this nodule-specific gene (Fig. 8.1). A fast neutron mutagenized deletion allele in the A17 genetic background (Young et al. 2011; Tang et al. 2014) was previously described that contains a large deletion spanning the region containing *VTL8*, called 13U. Most of the nodulation defects in this line are ascribed to the deletion in *VTL8* (Walton et al. 2020). We call our new allele *mtvtl8-2*.

To confirm that *mtvtl8-2* is the causative mutation in the NF11322 mutant *M. truncatula* line, co-segregation analysis was employed. Mutants were back crossed to wild-type R108. Their F_2 progeny shows a phenotype segregation ratio of approximately three Fix^{+} to one Fix^{-} in 210 F_2 progeny scored (Table 8.1). We carried out genotyping for the *MtVTL8* and *mtvtl8-2* alleles was on this population. The results show an approximate genotype segregation ratio of one (*MtVTL8/MtVTL8*) to two (*MtVTL8/ mtvtl8-2*) to one (*mtvtl8-2/ mtvtl8-2*) (Table 8.1). All the plants that show Fix^{-} nodule morphology have the *mtvtl8-2* mutation in both *MtVTL8* alleles.

Fig. 8.1 *Tnt1* insertion site in *MtVTL8* in *M. truncutula* NF11322. *Tnt1* is found 196 bp downstream of the Medtr4g094335 gene's ATG. Black bars represent the exon and white areas show the 5' and 3' untranslated regions

Table 8.1 Co-segregation of Fix$^-$ phenotype with *Tnt1* insertion in *MtVTL8*. F_2 plants from a Fix$^-$ NF11322 plant from were back-crossed into R108 were tested for nodule phenotype, wild type like (WTL) or Fix$^-$. The same plants were genotyped for their *MtVTL8* alleles, either homozygous wild type (*VTL8/VTL8*), heterozygous (*VTL8/vtl8*), or homozygous for the *Tnt1* insertion (*vtl8/vtl8*) *P < 0.05 is calculated using chi-square (χ^2) test at two-degrees of freedom

Line	Phenotype	Gene	Genotype (*VTL8/VTL8*: *VTL8/vtl8*: *vtl8/vtl8*)	χ^2
R108*NF11322 BC_1F_2	159 WTL:51Fix$^-$	*VTL8*	59:100:51	1.086*

This is consistent with the *mtvtl8-2* allele causing the nodulation defect in line NF11322.

Phylogeny analysis of the encoded protein and similar genes from *M. truncatula* show that MtVTL8 is a homolog of the *Saccharomyces cerevisiae* vacuolar iron/manganese transporter CCC1 protein, *Arabidopsis thaliana* vacuolar iron transporter VIT1 protein. It is the *M. truncatula* ortholog of the *Lotus japonicus* SEN1 protein required during nodule formation in *L. japonicus* (Fig. 8.2). In the *ljsen1* mutant, rhizobia are endocytosed normally into symbiosomes, but no nitrogen fixation takes place (Suganuma et al. 2003; Hakoyama et al. 2012). There are fifteen *VTL* homologs in *M. truncatula* and only two of them (*MtVTL4* and *MtVTL8*) are nodule specific (Walton et al. 2020). Interestingly, in the A17 U13 mutant, both nodule-

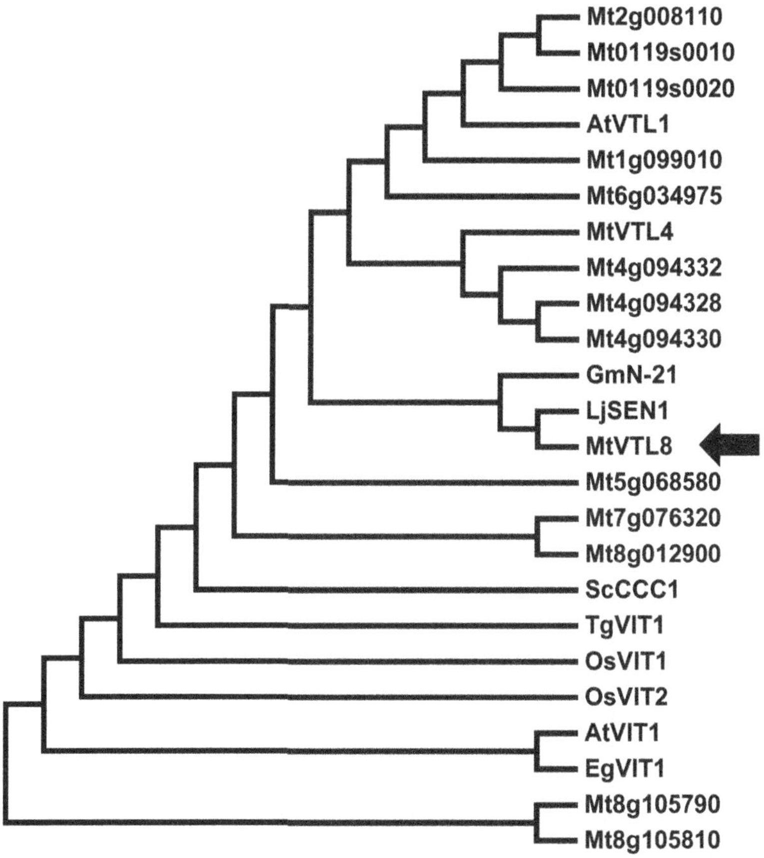

Fig. 8.2 Phylogenic tree shows that MtVTL8's closest homolog is *L. japonicus* SEN1. The arrow points to MtVTL8. The tree was made using maximum parsimony with amino acid sequences using MEGA 7 software. The accession numbers for each protein are in the Materials and Methods

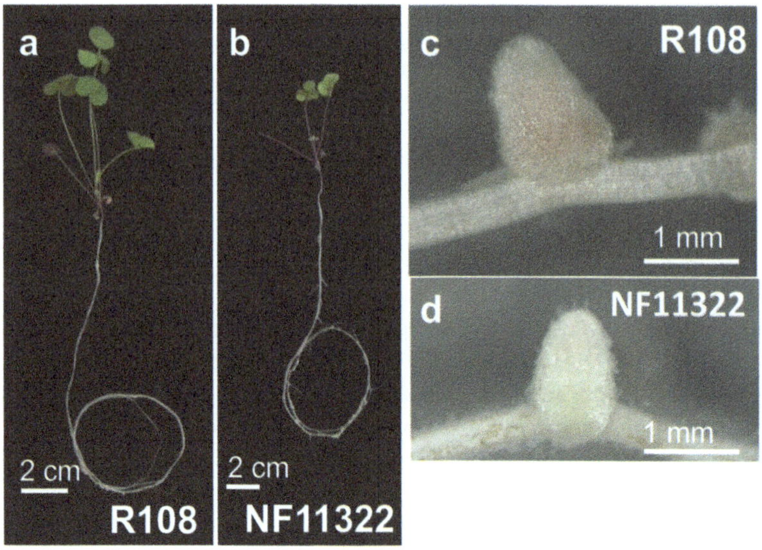

Fig. 8.3 Phenotype of NF11322 plants and nodules. Whole plant images of **a**, wild-type (R108) and **b**, NF11322. The NF11322 plant is smaller than R108 with pale green leaves indicating a N deficiency. Images of nodules: **c**, R108 nodule, **d**, NF11322 nodule both 15 dpi after inoculation with *S. meliloti* strain Rm41. The R108 nodules are pink, indicative of leghemoglobin, expressed late in nodule development, while the NF11322 nodules are white, indicating that they are Fix^{-}

specific *VTL* genes are deleted (Walton et al. 2020).

The *mtvtl8-2* plants were grown in symbiotic conditions with *Sinorhizobium meliloti* strain *Rm41*. At 15 days post inoculation, *mtvtl8-2* plants were found to be smaller than wild type R108 (Fig. 8.3a, b). The leaves were pale and chlorotic, as compared to the deep green leaves found in R108, indicative of a defect in symbiotic nitrogen fixation. The *mtvtl8-2* allele forms small white nodules when inoculated with *S. meliloti,* an SNF deficiency symptom, at 15 dpi with *Rm41* (Fig. 8.3c, d). Taken together, this data indicates that the *mtvtl8-2* mutant cannot fix nitrogen.

Growth of the *mtvtl8-2* plants were examined in a time course (Fig. 8.4). Plant dry weight from the plants grown under symbiotic conditions were checked at 10 days post inoculation (dpi), 15 dpi, 20 dpi and 25 dpi. *Mtvtl8-2* plants are significantly smaller compared with R108 at 15 dpi, 20 dpi, and 25 dpi (Fig. 8.4a). These data show that under symbiotic conditions, *mtvtl8-2* plants have defective growth compared with R108, and this may be caused by the defective nodules in *mtvtl8-2* plants. As a control to determine whether the defective growth of *mtvtl8-2* plants was related to N status, *mtvtl8-2* plants and R108 controls were grown under full ammonium nitrate. Plant dry weights were collected at 20 days, 25 days, 31 days, and 35 days, corresponding to 10, 15, 21, and 25 dpi, respectively, in symbiotic conditions. In these full N conditions, plant weights of *mtvtl8-2* plants were comparable to those of R108 controls, with the exception of one time point (Fig. 8.4b). These data suggest that the *MtVTL8* gene is required under symbiotic, but not non-symbiotic conditions. Additionally, we counted the number of nodules per plant in *mtvtl8-2* plants vs. wild type in symbiotic conditions. We found that the *mtvtl8-2* plants had comparable numbers of nodules at almost all data points checked (Fig. 8.4c). These data show that the *MtVTL8* gene does not participate in autoregulation of nodule number.

MtVTL8 does not contain any introns. Its coding sequence (CDS) is 708 bp, which encodes a 235 amino acid protein that is predicted to be a membrane transporter, consistent

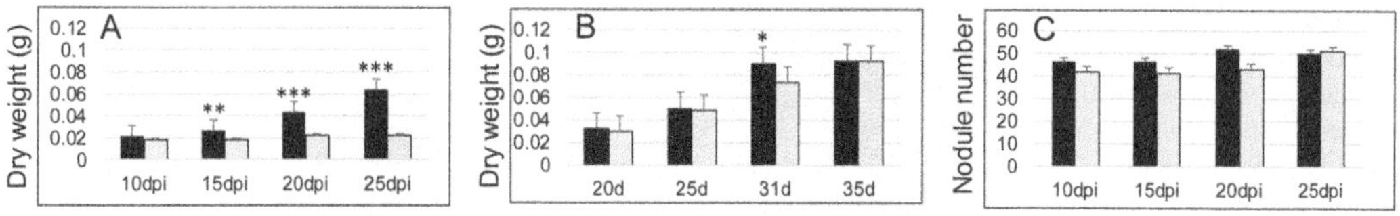

Fig. 8.4 Time course of plant growth in symbiotic vs. non-symbiotic conditions. Wild-type R108 (black bars) and *mtvtl8* mutant (gray bars) plants were grown in aeroponic chambers in symbiotic conditions for 5 days on media containing 5 mM NH_4NO_3, N starved for 5 days and then inoculated with *S. meliloti* Rm41 **a**, **c**. As control, plants were grown continuously in non-symbiotic conditions with 5 mM NH_4NO_3 **b**. Thus, with these conditions, 10 days post-inoculation (10 dpi) **a**, **c** is the same time point as 20 days growth **b**. Dry weight of wild type R108 and NF11322 at different dpi **a** and days of growth on full N **b** were recorded. The number of nodules per plant were evaluated **c** for inoculated plants. N = 10. Error bars represent standard error. Student's two-tail t test was used to evaluate differences, * = $p < 0.1$, ** = $p < 0.01$, *** = $p < 0.001$

with its annotation as an iron transporter. *VTL8* was not included on the Affymetrix chip that was used for gene expression data in the *M. truncatula* gene expression atlas (MtGEA) (Benedito et al. 2008), and thus its expression throughout *M. truncatula* cannot be assessed. However, data from the RNA-seq-based Symbimics database (Roux et al. 2014) shows it to be highly expressed in nodules, although this database lacks data from many other tissues of *M. truncatula*. Based on the Symbimics database (https://iant.toulouse.inra.fr/symbimics/) in the *M. truncatula* genotype A17, *MtVTL8* is mainly expressed in nodule and highly expressed in nodule interzone (IZ) (Fig. 8.5).

The *mtvtl8-2* nodules at 15 dpi were examined in more detail by light microscopy after toluidine blue staining (Fig. 8.6). The *mtvtl8-2* nodules were found to be smaller than those of wild type (compare Fig. 8.6b-a). The *mtvtl8-2* nodule meristem and zone II appear similar to wild type, with rhizobia released normally into symbiosomes (Fig. 6c, d). The bacteroids expand apparently normally in zone II, but when they reach the IZ to the nitrogen fixing zone III, they suffer a collapse (Fig. 6e, f). This pattern of nodule failure is consistent with the *MtVTL8* gene being expressed at the highest level in the IZ and necessary for the rhizobial bacteroids in nodules to develop into nitrogen fixation competency.

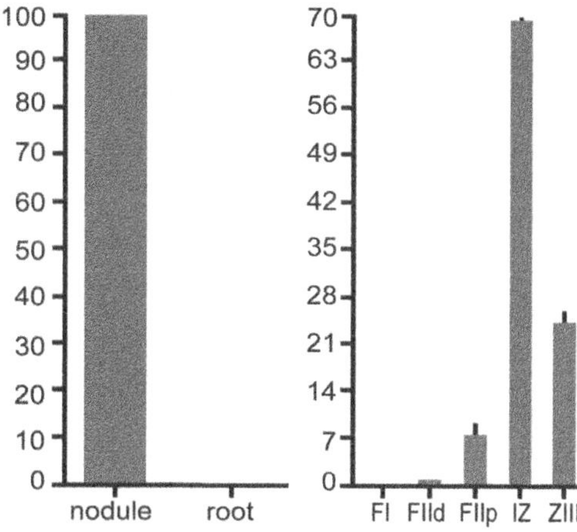

Fig. 8.5 *MtVTL8* is expressed in nodules, with highest expression in the interzone (IZ). Left, *MtVTL8* expression in nodules vs roots. Right, *MtVTL8* expression in nodule zones. FI, meristem; FIId, and FIIp, distal side and proximal sides, respectively, of the infection zone; IZ, interzone; ZIII, N-fixing zone. Data comes from the Symbimics database (https://iant.toulouse.inra.fr/symbimics/)

Together these data indicate that the *MtVTL8* gene is required for *M. truncatula* growth under symbiotic conditions and is consistent with the finding that it is a symbiosome membrane iron transporter (Walton et al. 2020). The *mtvtl8-2* allele has similar characteristics with the 13U allele (Walton et al. 2020), even though the two mutants are in different genetic backgrounds, R108 and A17, respectively.

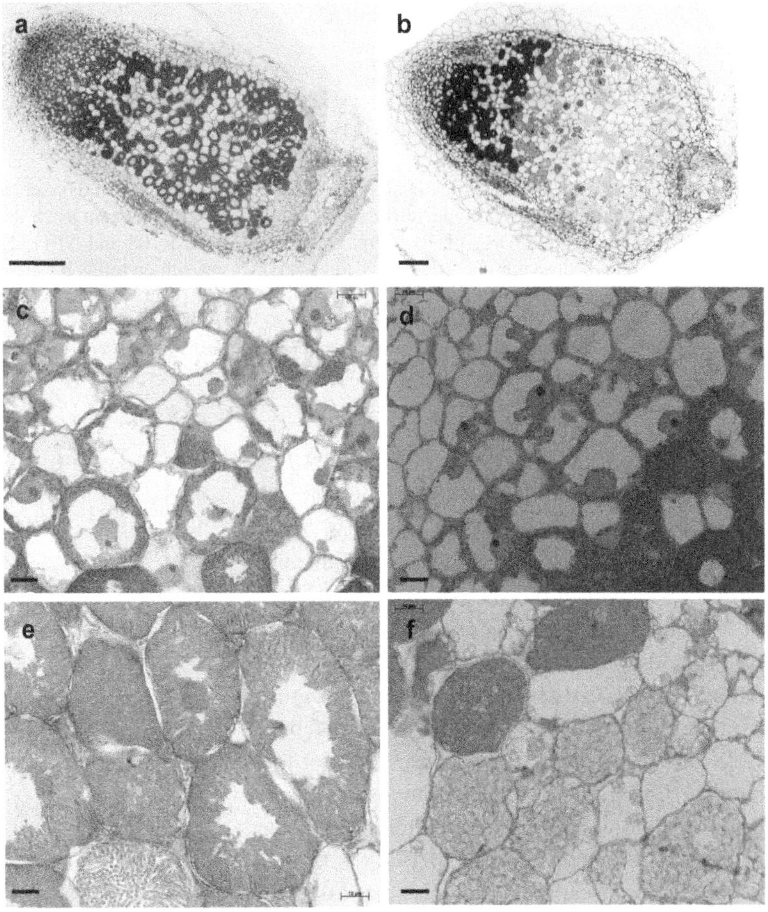

Fig. 8.6 NF11322 nodule phenotype. **a**, **b**, Light microscopy whole nodule images of wild-type R108 and NF11322 15 dpi nodules stained with toluidine blue. A, R108; B, NF11322. **c**, **d**, Higher magnification of zone II. **c**, R108; **d**, NF11322. The NF11322 cells appear very similar to those of R108 in this nodule zone. **e**, **f**, Higher magnification of zone III cells. E, R108; F, NF11322. The NF11322 cells contain rhizobial bacteroids that fail at this stage. The bacteroids lose their elongated morphology, compared to R108. The central vacuole found in R108 zone III cells is no longer in evidence in NF11322 cells. Scale bars: A = 200 μm; B = 100 μm; C–F = 10 μm

In conclusion, using a forward genetics approach, we uncovered a novel Fix$^-$ mutant in the *M. truncatula* R108 background, in a *Tnt1* mutagenized population, line NF11322. Using whole genome sequencing, we identified all the *Tnt1* insertions in the line. The causative mutation was determined using genetic co-segregation together with expression analysis to identify likely insertions. It is in *MtVTL8*, and shows similarities to the 13U deletion mutant with a defect in both *MtVTL8* and *MtVTL4*, previously described (Walton et al. 2020). *MtVTL8* encodes an iron transporter localized to the symbiosome membrane.

Acknowledgements We thank Janine Sherrier, her lab and the University of Delaware BioImaging Center for helpful advice with fixing, embedding, sectioning and staining nodules. This research was partially supported by the United States National Science Foundation grants NSF IOS-1127155, NSF IOS-1733470, and NSF IOS-2139351.

References

Barker DG, Pfaff T, Moreau D, Groves E, Ruffel S, Lepetit M, Whitehand S, Maillet F, Nair RM, Journet E-P (2006) Growing M. truncatula: choice of substrates and growth conditions The Medicago truncatula handbook. ISBN 0–9754303–1–9. http://www.noble.org/MedicagoHandbook/

Benedito VA, Torres-Jerez I, Murray JD, Andriankaja A, Allen S, Kakar K, Wandrey M, Verdier J, Zuber H, Ott T, Moreau S, Niebel A, Frickey T, Weiller G, He J, Dai X, Zhao PX, Tang Y, Udvardi MK (2008) A gene expression atlas of the model legume Medicago truncatula. Plant J 55:504–513

Brear EM, Day DA, C SPM (2013) Iron: an essential micronutrient for the legume-rhizobium symbiosis. Front Plant Sci 13:359

Colombo C, Palumbo G, He J-Z, Pinton R, Cesco S (2014) Review on iron availability in soil: interaction of Fe minerals, plants, and microbes. J Soils Sediments 14:538–548

Connolly EL, Campbell NH, Grotz N, Prichard CL, Guerinot ML (2003) Overexpression of the FRO2 ferric chelate reductase confers tolerance to growth on low iron and uncovers posttranscriptional control. Plant Physiol 133:1102–1110

Cui Y, Barampuram S, Stacey MG, Hancock CN, Findley S, Mathieu M, Zhang Z, Parrott WA, Stacey G (2013) Tnt1 retrotransposon mutagenesis: A tool for soybean functional genomics. Plant Physiol 161:36–47

d'Erfurth I, Cosson V, Eschstruth A, Lucas H, Kondorosi A, Ratet P (2003) Efficient transposition of the Tnt1 tobacco retrotransposon in the model legume Medicago truncatula. Plant J 34(1):95–106

Dellaporta SL, Wood J, Hicks JB (1983) A plant DNA minipreparation: version II. Plant Mol Biol Rep 1 (4):19–21

González-Guerrero M, Escudero V, Saéz Á, Tejada-Jiménez M (2016) Transition metal transport in plants and associated endosymbionts: arbuscular mycorrhizal fungi and rhizobia. Front Plant Sci 7:1088

Hakoyama T, Niimi K, Yamamoto T, Isobe S, Sato S, Nakamura Y, Tabata S, Kumagai H, Umehara Y, Brossuleit K, Petersen TR, Sandal N, Stougaard J, Udvardi MK, Tamaoki M, Kawaguchi M, Kouchi H, Suganuma N (2012) The integral membrane protein SEN1 is required for symbiotic nitrogen fixation in Lotus japonicus nodules. Plant Cell Physiol 53:225–236. https://doi.org/10.1093/pcp/pcr167

Kato T, Kumazaki K, Wada M, Taniguchi R, Nakane T, Yamashita K, Hirata K, Ishitani R, Ito K, Nishizawa T, Nureki O (2019) Crystal structure of plant vacuolar iron transporter VIT1. Nature Plants 5:308–315

Kryvoruchko IS, Routray P, Sinharoy S, Torres-Jerez I, Tejada-Jiménez M, Finney LA, Nakashima J, Pislariu CI, Benedito VA, González-Guerrero M, Roberts DM, Udvardi MK (2018) An iron-activated citrate transporter, MtMATE67, is required for symbiotic nitrogen fixation. Plant Physiol 176:2315–2329

Kumar S, Stecher G, Tamura K (2016) MEGA7: Molecular evolutionary genetics analysis version 7.0 for bigger datasets. Mol Biol Evol 33:1870–1874

Lullien V, Barker DG, de Lajudie P, Huguet T (1987) Plant gene expression in effective and ineffective root nodules of alfalfa (Medicago sativa). Plant Mol Biol 9:469–478

Moll KM, Zhou P, Ramaraj T, Fajardo D, Devitt NP, Sadowsky MJ, Stupar RM, Tiffin P, Miller JR, Young ND, Silverstein KAT, Mudge J (2017) Strategies for optimizing BioNano and Dovetail explored through a second reference quality assembly for the legume model, Medicago Truncatula. BMC Genomics 18:578

Pecrix Y, Staton S, Sallet E, Lelandais-Brière C, Moreau S, Carrère S, Blein T, Jardinaud MF, Latrasse D, Zouine M, Zahm M, Kreplak J, Mayjonade B, Satgé C, Perez M, Cauet S, Marande W, Chantry-Darmon C, Lopez-Roques C, Bouchez O, Bérard A, Debellé F, Muños S, Bendahmane A, Bergès H, Niebel A, Buitink J, Frugier F, Benhamed M, Crespi M, Gouzy J, Gamas P (2018) Whole-genome landscape of Medicago truncatula symbiotic genes. Nat Plants 4:1017–1025

Pislariu CI, Murray JD, Wen J, Cosson V, Muni RRD, Wang M, Benedito VA, Andriankaja A, Cheng X, Jerez IT, Mondy S, Zhang S, Taylor ME, Tadege M, Ratet P, Mysore KS, Chen R, Udvardi MK (2012) A Medicago truncatula tobacco retrotransposon insertion mutant collection with defects in nodule development and symbiotic nitrogen fixation. Plant Physiol 159 (4):1686–1699

Roux B, Rodde N, Jardinaud M-F, Timmers T, Sauviac L, Cottret L, Carrère S, Sallet E, Courcelle E, Moreau S, Debellé F, Capela D, de Carvalho-Niebel F, Gouzy J, Bruand C, Gamas P (2014) An integrated analysis of plant and bacterial gene expression in symbiotic root nodules using laser-capture microdissection coupled to RNA sequencing. Plant J 77:817–837

Suganuma N, Nakamura Y, Yamamoto M, Ohta T, Koiwa H, Akao S, Kawaguchi M (2003) The Lotus japonicus Sen1 gene controls rhizobial differentiation into nitrogen-fixing bacteroids in nodules. Mol Genet Genomics 269:312–320

Sun L, Gill US, Nandety RS, Kwon S, Mehta P, Dickstein R, Udvardi MK, Mysore KS, Wen J (2019) Genome-wide analysis of flanking sequences reveals that Tnt1 insertion is positively correlated with gene methylation in Medicago truncatula. Plant J 98:1016–1119

Tadege M, Wen J, He J, Tu H, Kwak Y, Eschstruth A, Cayrel A, Endre G, Zhao PX, Chabaud M, Ratet P, Mysore KS (2008) Large-scale insertional mutagenesis using the Tnt1 retrotransposon in the model legume Medicago truncatula. Plant J 54:335–347

Tang H, Krishnakumar V, Bidwell S, Rosen B, Chan A, Zhou S, Gentzbittel L, Childs KL, Yandell M, Gundlach H, Mayer KF, Schwartz DC, Town CD (2014) An improved genome release (version Mt4.0) for the model legume Medicago truncatula. BMC Genomics 15:312

Tejada-Jiménez M, Castro-Rodríguez R, Kryvoruchko I, Lucas MM, Udvardi M, Imperial J, González-Guerrero M (2015) Medicago truncatula Natural Resistance-Associated Macrophage Protein1 is required for iron uptake by rhizobia-infected nodule cells. Plant Physiol 168:258–272

Van Loon LC (2009) Role of iron in plant-microbe interactions. In: advances in botanical research, vol 51

Veerappan V, Jani M, Kadel K, Troiani T, Gale R, Mayes T, Shulaev E, Wen J, Mysore KS, Azad RK, Dickstein R (2016) Rapid identification of causative insertions underlying Medicago truncatula Tnt1 mutants defective in symbiotic nitrogen fixation from a forward genetic screen by whole genome sequencing. BMC Genomics 17:141. https://doi.org/110.1186/s12864-12016-12452-12865

Veerappan V, Kadel K, Alexis N, Scott A, Kryvoruchk I, Sinharoy S, Taylor M, Udvardi M, Dickstein R (2014) Keel petal incision: A simple and efficient technique for genetic crossing in Medicago truncatula. Plant Methods 10:11. https://doi.org/10.1186/1746-4811-1110-1111

Veereshlingam H, Haynes JG, Sherrier DJ, Penmetsa RV, Cook DR, Dickstein R (2004) nip, a symbiotic Medicago truncatula mutant that forms root nodules with aberrant infection threads and plant defense-like response. Plant Physiol 136:3692–3702

Walton JH, Kontra-Kováts G, Green RT, Domonkos Á, Horváth B, Brear EM, Franceschetti M, Kaló P, Balk J (2020) The Medicago truncatula Vacuolar iron Transporter-Like proteins VTL4 and VTL8 deliver iron to symbiotic bacteria at different stages of the infection process. New Phytol 228:651–666. https://doi.org/10.1111/nph.16735

Young ND, Debelle F, Oldroyd GED, Geurts R, Cannon SB, Udvardi MK, Benedito VA, Mayer KFX, Gouzy J, Schoof H, Van de Peer Y, Proost S, Cook DR, Meyers BC, Spannagl M, Cheung F, De Mita S, Krishnakumar V, Gundlach H, Zhou S, Mudge J, Bharti AK, Murray JD, Naoumkina MA, Rosen B, Silverstein KAT, Tang H, Rombauts S, Zhao PX, Zhou P, Barbe V, Bardou P, Bechner M, Bellec A, Berger A, Berges H, Bidwell S, Bisseling T, Choisne N, Couloux A, Denny R, Deshpande S, Dai X, Doyle JJ, Dudez A-M, Farmer AD, Fouteau S, Franken C, Gibelin C, Gish J, Goldstein S, Gonzalez AJ, Green PJ, Hallab A, Hartog M, Hua A, Humphray SJ, Jeong D-H, Jing Y, Jocker A, Kenton SM, Kim D-J, Klee K, Lai H, Lang C, Lin S, Macmil SL, Magdelenat G, Matthews L, McCorrison J, Monaghan EL, Mun J-H, Najar FZ, Nicholson C, Noirot C, O/'Bleness M, Paule CR, Poulain J, Prion F, Qin B, Qu C, Retzel EF, Riddle C, Sallet E, Samain S, Samson N, Sanders I, Saurat O, Scarpelli C, Schiex T, Segurens B, Severin AJ, Sherrier DJ, Shi R, Sims S, Singer SR, Sinharoy S, Sterck L, Viollet A, Wang B-B, Wang K, Wang M, Wang X, Warfsmann J, Weissenbach J, White DD, White JD, Wiley GB, Wincker P, Xing Y, Yang L, Yao Z, Ying F, Zhai J, Zhou L, Zuber A, Denarie J, Dixon RA, May GD, Schwartz DC, Rogers J, Quetier F, Town CD, Roe BA (2011) The Medicago genome provides insight into the evolution of rhizobial symbioses. Nature 480 (7378):520–524. http://www.nature.com/nature/journal/v480/n7378/abs/nature10625.html#supplementary-information

Regulation of Leaf Blade Development in *Medicago truncatula*

9

Hui Wang, Jianghua Chen, and Million Tadege

Abstract

Leaves are the main photosynthetic organs for most flowering plants serving as solar panels capturing solar energy and converting it into chemical energy on which most heterotrophic organisms including humans depend. Leaves are confronted with two conflicting choices; while the broad and flat surface improves photosynthetic efficiency via facilitating capture of solar energy, it inadvertently promotes water loss via transpiration. Thus, plants strictly control the shape and size of their leaves depending on the environments in which they grow. In both simple and compound leaf species, leaf development starts with recruitment of leaf primordium founder cells from the peripheral region of the shoot apical meristem (SAM). The founder cells then undergo a series of well-organized cell proliferation and cell differentiation programs to allow the formation of a characteristically expanded lamina along with three polarized planes: proximal–distal, medial–lateral, and adaxial–abaxial. The legume model species *Medicago truncatula* has a compound leaf with three leaflets. Thus, the leaf founder initials, in addition to orchestrating growth in the three axes of the blade, are required for the initiation of lateral leaflets. In this review, we will summarize the major molecular factors that regulate leaf blade expansion in *M. truncatula* and highlight some recent findings in how leaflet initiation may be controlled.

9.1 Introduction

Leaves are the main photosynthetic organs of flowering plants assimilating solar energy and CO_2 into sugars on which most heterotrophic organisms depend for their livelihood. Leaves can be simple with a single blade or compound with more than one blade called leaflets. In both simple and compound leaf species, leaf development starts with recruitment of leaf primordium founder cells from the peripheral region of the shoot apical meristem (SAM) and a group of undifferentiated pluripotent stem cells at the shoot tips of plants. These small population of

H. Wang (✉)
College of Grassland Science and Technology, China Agricultural University, Beijing 100193, China
e-mail: huiwang211@cau.edu.cn

J. Chen
CAS Key Laboratory of Topical Plant Resources and Sustainable Use, CAS Center for Excellence in Molecular Plant Sciences, Xishuangbanna Tropical Botanical Garden, Chinese Academy of Sciences, Kunming 650223, Yunnan, China

M. Tadege (✉)
Department of Plant and Soil Sciences, Institute for Agricultural Biosciences, Oklahoma State University, Ardmore, OK 73401, USA
e-mail: million.tadege@okstate.edu

S. Sinharoy et al. (eds.), *The Medicago truncatula Genome*, Compendium of Plant Genomes,
https://doi.org/10.1007/978-3-030-90757-0_9

cells are strictly regulated by a complex and interconnected transcriptional network of genes and hormones. The Class 1 KNOTTED1-LIKE HOMEOBOX (KNOX1) family transcriptional regulators are expressed throughout the SAM and maintain the undifferentiated state in the SAM. Leaf primordium initiation requires localized accumulation of the phytohormone auxin as auxin maxima (Reinhardt et al. 2000, 2003; Braybrook and Kuhlemeier 2010) and repression of *KNOX1* gene expression by the ASSYMMETRIC LEAVES1 and 2 (AS1/AS2) complex (Long et al. 1996; Uchida et al. 2007; Guo et al. 2008; Jun et al. 2010) at the primordium initiation site. *KNOX1* genes modulate the cytokinin/gibberellin (GA) ratio in the SAM by activating cytokinin biosynthesis and repressing GA biosynthesis or activating GA catabolism (Jasinski et al. 2005; Bolduc and Hake 2009).

Once the leaf primordium initial cells are recruited from the SAM, the primordium organizes itself into defined cell layers through hierarchically programed cell division and cell expansion patterns forming a flattened blade along with three distinct growth axes: proximal–distal (length direction), medial–lateral (width direction) and adaxial–abaxial also called dorsal–ventral (thickness direction). Several elegant studies in Antirrhinum and Arabidopsis over the last 25 years have clearly demonstrated that polarity patterning along the adaxial–abaxial axis of the leaf blade is required for blade outgrowth (Waites and Hudson 1995; Waites et al. 1998; Byrne et al. 2000; Kerstetter et al. 2001; McConnellet al. 2001; Emery et al. 2003; Iwakawa et al. 2007; Husbands et al. 2009). The adaxial factors such as *PHABULOSA* (*PHB*) and *ASYMMETRIC LEAVES2* (*AS2*) are expressed on the adaxial/upper side of the leaf blade (McConnell et al. 2001; Iwakawa et al. 2002), whereas the *KANADI* (*KAN*) and *YABBY* (*YAB*) families of transcription factors are expressed on the abaxial side (Siegfried et al. 1999; Kerstetter et al. 2001). The adaxial and abaxial factors interact antagonistically to prevent expression in each other's domains, and this interaction is required for proper blade outgrowth (Husbands et al. 2009; Kidner and Timmermans 2010; Szakonyi et al. 2010). Polarity patterning is further fine-tuned by auxin response factors (ARFs), miRNAs, and tasi-ARFs, which also show polarized expression patterns (Efroni et al. 2010; Husbands et al. 2015; Merelo et al. 2016).

Nevertheless, the phenotypes of some of the loss-of-function mutants of polarity genes in different plant species are variable and sometimes double and triple mutants are required to observe obvious leaf phenotypes. Recently, a new regulatory domain at the adaxial and abaxial juxtaposition was discovered (Tadege et al. 2011; Nakata et al. 2012), which is required for outgrowth of the leaf blade in the lateral direction. In this chapter, we will briefly revise the regulatory factors and their mechanism of action in the middle domain, as well as their connection to polarity factors with reference to *Medicago truncatula*. Also, since *M. truncatula* has trifoliate leaves, we will highlight some of the recent progress in the understanding of leaflet initiation, which has been obscured in the simple leaf Arabidopsis.

9.2 WOX Genes in Medio-Lateral Outgrowth of *M. Truncatula* Leaf Blade

WUSCHEL-related homeobox (WOX) family genes are plant-specific homeodomain transcription factors involved in a variety of plant developmental programs including shoot and root meristem maintenance, embryo development, and lateral organs growth (Mukherjee et al. 2009; van der Graaff et al. 2009). In Arabidopsis, the WOX family consists of 15 members, including the founding members of the WOX family WUSCHEL (WUS) and WOX1-WOX14 (Haecker et al. 2004). Arabidopsis WUS and WOX5 function in shoot and root meristem maintenance, respectively (Mayer et al. 1998; Sarkar et al. 2007).

WUS is expressed in the organizing center of SAM in Arabidopsis and the protein diffuses to neighboring stem cells in the central zone forming a gradient to function in pluripotent stem cells maintenance by activating *CLAVATA3* (*CLV3*) (Mayer et al. 1998; Yadav et al.

2010, 2011; Daum et al. 2014). CLV3, a peptide signal that activates CLV receptor kinase signaling, on the other hand, represses *WUS* to restrict its expression (Brand et al. 2000; Kayes and Clark 1998). This negative feedback loop regulation is required for meristem maintenance (Somssich et al. 2016). The *wus* mutant terminates the meristem and starts to regrow showing a 'stop-and-go' type of growth habit resulting in aberrant vegetative and inflorescence meristems (Laux et al. 1996; Kieffer et al. 2006; Wang et al. 2017). But the leaves of *wus* mutants display no discernible phenotype. However, we observed that when the *pressed flower* (*prs*) mutant is combined with the *wox1* and *wus* mutants, the *wox1 prs wus* triple mutant displayed much smaller and narrower leaf blade than the *wox1 prs* double mutant (Zhang and Tadege 2015), suggesting that Arabidopsis *WUS* may function in leaf development redundantly with other WOX genes.

In *M. truncatula*, the *WUS* homolog, *HEADLESS* (*HDL*) appears to have non-redundant function in leaf development. The *hdl* mutant displays a shorter and heart-shaped blade phenotype compared to wild type R108 (Meng et al. 2019; Wang et al. 2019), suggesting a role in proximo-distal growth. Unlike *WUS*, the *HDL* transcript is detected in leaf primordia in addition to the SAM (Meng et al. 2019), consistent with its direct involvement in leaf development. Interestingly, the *hdl* mutant also has other additional phenotypes compared to the *wus* mutant. The *hdl* mutant never make stems and as a result it never flowers (Meng et al. 2019; Tadege et al. 2015; Wang et al. 2019) but leaf primordia initiation is not affected as such, if any, more leaves are formed compared to WT. This finding suggests that the regions of the meristem may have different requirements for HDL function, in which the central meristem could completely fail while the peripheral meristem is fully functional. This also suggests that the aberrant leaf shape in the *hdl* mutant is not caused by failure in the recruitment of leaf founder initials from the SAM but rather by failure in specific functions of HDL during leaf proliferation and expansion after the leaf primordium is initiated. Nevertheless, the main function of HDL does seem to be in leaf development, as leaves in the *hdl* mutant are pretty normal despite modest alterations in shape. Other *WOX* genes may play a more determinant role in leaf blade development and the first such evidence came from sources other than Medicago.

In the maize *narrow sheath1* (*ns1*) and (*ns2*) double mutant, the lateral domain of the leaf blade is ablated leading to severely narrow blades (Scanlon et al. 1996; Scanlon and Freeling 1997). Both NS1 and NS2 were found to be homologs of Arabidopsis WOX3/PRS (Nardmann et al. 2004), demonstrating a key role for WOX genes in regulating medio-lateral outgrowth of the leaf blade. In the *ns1 ns2* double mutant, *KNOX* genes are not down regulated in the founder cell domain and as a result *NS1* and *NS2* are supposed to function in direct recruitment of marginal founder cells from the SAM (Scanlon 2000). Similarly, in rice, the homologous *nal2 anl3* double mutant causes narrow leaf and other pleotropic phenotypes (Ishiwata et al. 2013; Cho et al. 2013), confirming the function of *NS/WOX3* in leaf development at least in monocots.

In *M. truncatula*, medio-lateral outgrowth of the leaf blade is regulated by another *WOX* gene named *STENOFOLIA (STF)*. Loss of function of *STF* and its homolog *LAM1* in *Nicotiana sylvestris* results in severely narrow leaf blades and flower petals, and aberrant blade vasculature due mainly to defects in cell proliferation (McHale 1992; Tadege et al. 2011; Lin et al. 2013a). STF is a homolog of Arabidopsis WOX1, which redundantly with WOX3/PRS controls lateral expansion of the leaf blade but the *wox1* mutant alone has no visible phenotype (Vandenbussche et al. 2009; Nakata et al. 2012). Histological analysis by in situ hybridization and GUS staining demonstrated that *STF* is specifically expressed at the adaxial–abaxial junction of leaf primordia and expanding leaves but excluded from the SAM (Tadege et al. 2011). The Medicago *stf* and *N. sylvestris lam1* mutants are drastically affected in blade width (medio-lateral growth), the *lam1* mutant showing the more severe phenotype with nearly naked midrib and vestigial blade strips, but proximo-distal growth (leaf length) is unaffected (Tadege et al. 2011; Tadege 2014). The phenotypes are primarily caused by reduced cell proliferation. This demonstrates that a factor that is neither adaxial nor abaxial in expression pattern critically regulates medio-lateral growth by activating cell proliferation. Phylogenetic analysis revealed that STF/WOX1 homologs are specific to eudicots and the ancestral *Amborella trichopoda* but absent from monocots (Vandenbussche et al. 2009; Tadege et al. 2011; Zhang et al. 2014). Despite the *WOX1* and *WOX3/PRS* redundancy in Arabidopsis leaf blade development, most STF/WOX1 homologs, in addition to *lam1*, display phenotypes as single mutants including *maw* in petunia (Vandenbussche et al. 2009), *lath* in pea and *naol* in Lotus (Zhuang et al. 2012), *mf* in cucumber (Wang et al. 2020), and *Sllam1* in tomato (Wang et al. 2021). Moreover, the WOX3 homolog in Medicago has a flower but no leaf blade phenotype in the *lfl* mutant (Niu et al. 2015). These observations suggest that *STF* and its homologs are key regulators of lateral blade expansion in eudicots, but this function is performed by *WOX3* and its homologs in monocots.

9.3 Mechanism of STF Function

9.3.1 STF Acts as a Transcriptional Repressor

The STF protein has two highly conserved motifs of about 10 aa each in length near the 3' end; WUS box and STF box (Tadege et al. 2011). These motifs additively confer repressive activity to STF, and mutational analysis revealed that this activity is required for STF function in promoting blade outgrowth (Zhang et al. 2014). Thus, STF is a transcriptional repressor required for lateral organ (including leaves and petals) expansion and vascular patterning (Lin et al. 2013a, b; Zhang and Tadege 2015). To perform its repressive function, STF physically interacts with the MtTPL family of transcriptional corepressors with its WUS box and STF box (Zhang et al. 2014). The STF-MtTPL interaction is required for STF-mediated cell proliferation at the adaxial–abaxial junction and medio-lateral growth of the leaf blade, suggesting that STF probably works in a protein complex to carry out its central function of transcriptional repression to allow cell proliferation in the specified region that is not yet differentiated into adaxial or abaxial tissue. STF is also demonstrated to physically and genetically interact with transcriptional co-activators MtAN3 and MtLUG in the regulation of leaf and inflorescence development (Zhang et al. 2019), although the mechanism of this interaction requires further investigation.

9.3.2 Relationship Between STF and Polarity Factors

Because leaf polarity patterning is a prerequisite for blade lateral outgrowth, we wondered if and how STF affects polarity patterning. In the *stf* and *lam1* mutants, differentiation of adaxial and abaxial cells is weakened but not abolished (Tadege et al. 2011). Neither adaxialization nor abaxilization was observed in these mutants, suggesting that STF-mediated blade outgrowth may be orchestrated downstream of polarity specification. Genetic studies in Arabidopsis suggest that the middle domain-specific WOX1 and PRS may repress both adaxial and abaxial factors to establish boundary and prevent the mixing of these two cell types (Nakata et al. 2012; Nakata and Okada 2012, 2013). In Medicago, this proposal holds true at least for the STF and *MtAS2* interaction. MtAS2 is a LOB domain adaxial polarity factor required for leaf adaxial identity in Medicago. In vitro and in vivo studies uncovered that STF directly binds to the *MtAS2* promoter and prevents its expression at the adaxial–abaxial boundary in the leaf margin (Zhang et al. 2014). In the *stf* mutant leaf primordium, the expression of *MtAS2* expands to the adaxial–abaxial junction at the leaf margin, consistent with the Arabidopsis model. Interestingly, repression of the *MtAS2* promoter by STF in transgenic Arabidopsis cannot occur in the absence of TPL, and it also requires histone deacetylation at the target promoter site (Zhang et al. 2014). This indicates that the STF-mediated repression of the adaxial polarity factor *MtAS2* at the adaxial–abaxial junction of the leaf margin involves a transcriptional repressor complex that includes corepressors and chromatin modifiers.

In this connection, *MtWOX9-1* is identified to be an abaxial factor required for proper blade outgrowth in Medicago, and STF represses *MtWOX9-1* expression by directly binding to its promoter at multiple sites (Wolabu et al. 2021). It is, therefore, likely that STF establishes and maintains a cell proliferation zone at the adaxial–abaxial junction in the middle mesophyll and leaf margin by keeping adaxial and abaxial polarity factors away from this region. Preventing polarity factors from expressing in the margin may help to keep cells in the undifferentiated state so that they can continue dividing and supply cells for differentiation into both adaxial and abaxial tissues. In this way, the STF transcriptional repressor complex may maintain a cell proliferation zone at the adaxial–abaxial boundary and perhaps also contribute to cell proliferation per se.

Overexpression of *STF* in transgenic switchgrass, rice, wheat, and *Brachypodium* leads to increased biomass yield attributed to increased leaf lateral expansion caused by enhanced cell proliferation (Wang et al. 2017; Liu et al. 2018). At least in transgenic rice, STF directly binds to the promoters of some cytokinin oxidase/dehydrogenase enzymes (*CKXs*) that degrade cytokinin and represses their activity leading to increased active cytokinin contents (Wang et al. 2017), which promotes cell proliferation. This suggests that the STF repressor complex at the adaxial–abaxial junction maintains a cell proliferation zone and promotes cell proliferation at the same time. In Arabidopsis, *WOX1* and *PRS* are known to be activated by auxin and the auxin response factor MONOPTEROS (MP) and dominant negative expression of *MP* leads to narrow leaves (Guan et al. 2017). Although this has not been shown directly in Medicago, the involvement of auxin in the STF pathway was predicted from microarray experiments and *lam1* mutant treatment with auxin and cytokinin (Tadege et al. 2011; Tadege and Mysore 2011). Thus, phytohormones, especially the auxin–cytokinin crosstalk may be important for the STF-mediated regulation of medio-lateral growth through activation of cell proliferation.

9.4 Leaf Blade Length Growth Regulators in *M. Truncatula*

The fact that the *stf* mutant and even the more severe *lam1* mutant are not affected in leaf length suggests that proximo-distal and medio-lateral growth are likely controlled by separate genetic factors. However, mutants specifically affected only in leaf length are not described as such in Medicago. Mutant leaves with proximo-distal defect are generally small indicating that medio-lateral growth is also restricted. This might be because of the nature of leaf growth. When the primordium first emerges from the meristem, it appears as a peg-like structure and assumes a rod shape (Ge et al. 2014). As cell division and differentiation continues, the blade starts to flatten because of lateral expansion. It is also clear that *STF* is never expressed in any region of the meristem, central or peripheral, but the first detectable expression is in visible primordium, about stage P1. STF is, thus, unlikely to be involved in the recruitment of primordial founder cells from the meristem and in the initiation of the primordium. These two observations, the rod shape of the primordium at the beginning and appearance of *STF* after primordium initiation, suggest that proximo-distal growth is primary and medio-lateral growth is secondary. This assumption explains why proximo-distal growth defect has impact on medio-lateral growth but not necessarily vice versa.

Irrespective of the reason, there are Medicago mutants with small leaves that may be either genuinely affected in both length and width growth or caused by consequences of defects in proximo-distal growth alone. The *hdl* mutant may be an exception in this regard because the defect appears at the distal tip of the blade where it fails to elongate without any impact on leaf width (Meng et al. 2019; Wang et al. 2019) altering the blade shape. The *agamous-like flower* (*aglf*) mutant though described for its primary defects in floral organ development also appears to have defect specifically in leaf proximo-distal growth that results in short blade, rachis, and petiole (Zhang et al. 2019).

In Arabidopsis, two BTB/POZ domain-ankyrin repeat proteins BLADE-ON-PETIOLE1 (BOP1) and BOP2 are reported to regulate proximo-distal patterning (Ha et al. 2003; Hepworth et al. 2005). The Medicago *BOP1* homolog, *NOOT*, loss-of-function phenotype is primarily in the nodule, stipules and flower petals without clear indication of proximo-distal defect in the petiole or blade (Couzigou et al. 2012). Homologs of the *ROTUNDIFOLIA3* (*ROT3*) and *ROT4* genes, another group involved in proximo-distal polarity patterning in Arabidopsis, have not been characterized in Medicago. However, *ROT3* encodes a member of the cytochrome P450 family involved in brassinosteroid (BR) synthesis (Kim et al. 1998, 2005), and brassinosteroid signaling has been shown to regulate overall plant growth including leaf expansion in Medicago. The brassinosteroid receptor (*mtbri1*) mutants of Medicago are extremely dwarf in all aspects of plant architecture with nearly rounded leaf blades (Cheng et al. 2017), suggesting that the primary defect in the leaf blade may be in cell proliferation and/or expansion in the proximo-distal axis.

Dwarf and Increased Branching 1 (*DIB1*) a GA3ox1 enzyme (Zhang et al. 2020) and *mini plant 1* (*mnp1*) a CPS enzyme (Guo et al. 2020) in the gibberellin biosynthetic pathway are extremely dwarf plants with defects in cell length expansion, and show slightly more oval-shaped leaves than WT. The auxin efflux carrier mutant *smooth leaf margin1* (*slm1*) also called *mtpin10* has slightly shorter leaf blades but also with defects in leaflet number and leaf blade margin serration (Peng and Chen 2011; Zhou et al. 2011). Another Medicago gene, *SMALL LEAF AND BUSHY1* (*SLB1*)/ *MINI ORGAN1* (*MIO1*), is recently described to affect organ size (Yin et al. 2020; Zhou et al. 2021). *slb1/mio1* mutant leaves are small but relatively shorter, suggesting a primary defect in proximo-distal growth. *SLB1/MIO1* encodes an F-box protein assumed to be part of the SKP1/Cullin/F-box (SCF) E3 ubiquitin ligase complex and promotes leaf size by activating cell proliferation through targeting the cell division repressor *BIG SEEDS1* (*BS1*) for degradation (Yin et al. 2020; Zhou et al. 2021). *BS1* encodes a TIFY family transcriptional regulator that controls organ size by negatively regulating primary cell division (Ge et al. 2016).These observations suggest that the phytohormones, auxin, GA, brassinosteriodes and their crosstalk, and the ubiquitin ligase proteasome pathway are involved in modulating proximo-distal growth in Medicago leaves, but due to their effects on growth in the other axes as well, and consequences of proximo-distal growth defects on medio-lateral expansion, these defects generally result in small leaves affected in both length and width directions.

9.5 Control of Compound Leaf Development in *M. Truncatula*

M. truncatula has a characteristic trifoliolate (trifoliate) compound leaf with three leaflets; a single terminal leaflet (TL) distal to the petiole and two lateral leaflets (LL) below it separated from the TL by a short petiole-like structure called rachis, with the exception of the first leaf, which is always simple. In the *single leaflet1* (*sgl1*) mutant, all adult leaves are converted into simple leaves with single terminal leaflets (Wang et al. 2008). SGL1 is the homolog of Arabidopsis LFY, snapdragon FLO and pea UNI. *SGL1* is expressed in the SAM and leaf primordia (Wang et al. 2008), likely providing morphogenetic competence to primordial cells to initiate lateral leaflets. Another transcriptional regulator, FUSED COMPOUND LEAF1(FCL1), is required for boundary separation between leaflets as the leaflets appear fused in the *fcl1* mutant (Peng et al. 2011). *FCL1* encodes a class M KNOX protein that lacks a homeodomain (Peng et al. 2011) and presumably promotes cell proliferation at the boundary to enable competence and leaflet separation. *PALMATE-LIKE*

PENTAFOLIATA1 (*PALM1*) encoding a C2H2 zinc finger transcription factor is required to maintain the trifoliate identity by negatively regulating the expression of *SGL1* (Chen et al. 2010). In *palm1* mutants, two additional leaflets are initiated converting the three leaflet into five, in which the *SGL1* transcript is highly upregulated. Thus, PALM1 provides determinacy to leaflet initiation by spatial and temporal control of *SGL1* expression (Chen et al. 2010).

Another determinacy factor, *PINNATE-LIKE PENTAFOLIATA1* (*PINNA1*), has been described recently to regulate morphogenetic activity and leaflet number by directly repressing *SGL1* during *M. truncatula* leaf morphogenesis (He et al. 2020). *PINNA1* encodes a BEL-like homeodomain protein and functions alone in the TL region but synergizes with *PALM1* at the LL region to define the spatiotemporal expression pattern of *SGL1* (He et al. 2020). In the *pinna1* single mutant, the two new leaflets appear at the TL region making five leaflets, indicating that PALM1 function is sufficient to suppress extra leaflet initiation at the LL region, but in the *pinna1 palm1* double mutant, supernumerary leaflets are formed at the LL region, increasing the total number of leaflets to 12 (He et al. 2020). This indicates that the action of PINNA1 and PALM1 at the LL region is additive in suppressing lateral leaflet development, and this has been attributed to combined control of the indeterminacy factor SGL1 activity (He et al. 2020).

These observations together suggest that lateral leaflet initiation is a default pathway in leaf primordia orchestrated by *SGL1*, but *PINNA1* and *PALM1* control the spatial and temporal distribution of *SGL1* individually and in combination to achieve the characteristic trifoliate identity with one terminal and two lateral leaflets. Although *SGL1* is expressed in the SAM, it is not required for the formation of the first unifoliate leaf and the terminal leaflet since these structures appear unaffected in the *sgl1* mutant. It would be interesting to know the identity of the genetic factor that recruits leaf founder cells from the peripheral SAM in the first place and imparts leaf primordial identity in *M. truncatula* and other plants.

References

Bolduc N, Hake S (2009) The maize transcription factor KNOTTED1 directly regulates the gibberellin catabolism gene ga2ox1. Plant Cell 21:1647–1658

Brand U, Fletcher JC, Hobe M, Meyerowitz EM, Simon R (2000) Dependence of stem cell fate in Arabidopsis on a feedback loop regulated by CLV3 activity. Science 289:617–619

Braybrook SA, Kuhlemeier C (2010) How a plant builds leaves. Plant Cell 22(4):1006–1018

Byrne ME, Barley R, Curtis M, Arroyo JM, Dunham M, Hudson A, Martienssen RA (2000) Asymmetric leaves1mediates leaf patterning and stem cell function in Arabidopsis. Nature 408:967–971

Chen J, Yu J, Ge L, Wang H, Berbel A, Liu Y, Chen Y, Li G, Tadege M, Wen J, Cosson V, Mysore KS, Ratet P, Madueño F, Bai G, Chen R (2010) Control of dissected leaf morphology by a Cys(2)His(2) zinc finger transcription factor in the model legume *Medicago truncatula*. Proc Natl Acad Sci U S A 107(23):10754–10759

Cheng X, Gou X, Yin H, Mysore K, Li J, Wen J (2017) Functional characterisation of brassinosteroid receptor MtBRI1 in *Medicago truncatula*. Sci Rep 24; 7(1): 9327

Cho SH, Yoo SC, Zhang H, Pandeya D, Koh HJ, Hwang JY, Kim GT, Paek NC (2013) The rice narrow leaf2 and narrow leaf3 loci encode WUSCHEL-related homeobox 3A (OsWOX3A) and function in leaf, spikelet, tiller and lateral root development. New Phytol 198(4):1071–1084

Couzigou JM, Zhukov V, Mondy S, Abu el Heba G, Cosson V, Ellis TH, Ambrose M, Wen J, Tadege M, Tikhonovich I, Mysore KS, Putterill J, Hofer J, Borisov AY, Ratet P (2012) NODULE ROOT and COCHLEATA maintain nodule development and are legume orthologs of Arabidopsis BLADE-ON-PETIOLE genes. Plant Cell 24(11):4498–510

Daum G, Medzihradszky A, Suzaki T, Lohmann JU (2014) A mechanistic framework for noncell autonomous stem cell induction in Arabidopsis. Proc Natl Acad Sci USA 111:14619–14624

Efroni I, Eshed Y, Lifschitz E (2010) Morphogenesis of simple and compound leaves: a critical review. Plant Cell 22:1019–1032

Emery JF, Floyd SK, Alvarez J, Eshed Y, Hawker NP, Izhaki A, Baum SF, Bowman JL (2003) Radial patterning of Arabidopsis shoots by class III HD-ZIP and KANADI genes. Curr Biol 13:1768–1774

Ge L, Peng J, Berbel A, Madueño F, Chen R (2014) Regulation of compound leaf development by PHANTASTICA in *Medicago truncatula*. Plant Physiol 164 (1):216–228

Ge L, Yu J, Wang H, Luth D, Bai G, Wang K, Chen R (2016) Increasing seed size and quality by manipulating BIG SEEDS1 in legume species. Proc Natl Acad Sci USA 113(44):12414–12419

Guan C, Wu B, Yu T, Wang Q, Krogan NT, Liu X, Jiao Y (2017) Spatial auxin signaling controls leaf flattening in Arabidopsis. Curr Biol 27(19):2940–2950

Guo S, Zhang X, Bai Q, Zhao W, Fang Y, Zhou S, Zhao B, He L, Chen J (2020) Cloning and functional analysis of dwarf gene Mini Plant 1 (MNP1) in *Medicago truncatula*. Int J Mol Sci 21(14):4968

Guo M, Thomas J, Collins G, Timmermans MCP (2008) Direct repression of KNOX loci by the ASYMMETRIC LEAVES1complex of Arabidopsis. Plant Cell 20:48–58

Ha CM, Kim GT, Kim BC, Jun JH, Soh MS, Ueno Y, Machida Y, Tsukaya H, Nam HG (2003) The BLADE-ON-PETIOLE 1 gene controls leaf pattern formation through the modulation of meristematic activity in Arabidopsis. Development 130(1):161–172

Haecker A, Gross-Hardt R, Geiges B, Sarkar A, Breuninger H, Herrmann M, Laux T (2004) Expression dynamics of WOX genes mark cell fate decisions during early embryonic patterning in *Arabidopsis thaliana*. Development 131(3):657–668

He L, Liu Y, He H, Liu Y, Qi J, Zhang X, Li Y, Mao Y, Zhou S, Zheng X, Bai Q, Zhao B, Wang D, Wen J, Mysore K, Tadege M, Xia Y, Chen J (2020) A molecular framework underlying the compound leaf pattern of *Medicago truncatula*. Nat Plants. 6(5):511–521

Hepworth SR, Zhang Y, McKim S, Li X, Haughn GW (2005) BLADE-ON-PETIOLE-dependent signaling controls leaf and floral patterning in Arabidopsis. Plant Cell 17(5):1434–48

Husbands AY, Benkovics AH, Nogueira FT, Lodha M, Timmermans MC (2015) The ASYMMETRIC LEAVES complex employs multiple modes of regulation to affect adaxial-abaxial patterning and leaf complexity. Plant Cell 27(12):3321–3335

Husbands AY, Chitwood DH, Plavskin Y, Timmermans MC (2009) Signals and prepatterns: new insights into organ polarity in plants. Genes Dev 23:1986–1997

Ishiwata A, Ozawa M, Nagasaki H, Kato M, Noda Y, Yamaguchi T, Nosaka M, Shimizu-Sato S, Nagasaki A, Maekawa M, Hirano HY, Sato Y (2013) Two WUSCHEL-related homeobox genes, narrowleaf2 and narrow leaf3, control leaf width in rice. Plant Cell Physiol 54(5):779–792

Iwakawa H, Iwasaki M, Kojima S, Ueno Y, Soma T, Tanaka H et al (2007) Expression of the ASYMMETRIC LEAVES2 gene in the adaxial domain of Arabidopsis leaves represses cell proliferation in this domain and is critical for the development of properly expanded leaves. Plant J 51:173–184

Iwakawa H, Ueno Y, Semiarti E, Onouchi H, Kojima S, Tsukaya H et al (2002) The ASYMMETRIC LEAVES2 gene of *Arabidopsis thaliana*, required for formation of a symmetric flat leaf lamina, encodes a member of a novel family of proteins characterized by cysteine repeats and a leucine zipper. Plant Cell Physiol 43:467–478

Jasinski S, Piazza P, Craft J, Hay A, Woolley L, Rieu I, Phillips A, Hedden P, Tsiantis M (2005) KNOX action in Arabidopsis is mediated by coordinate regulation of cytokinin and gibberellin activities. Curr Biol 15:1560–1565

Jun JH, Ha CM, Fletcher JC (2010) BLADE-ON-PETIOLE1coordinates organ determinacy and axial polarity in Arabidopsis by directly activating ASYMMETRIC LEAVES2. Plant Cell 22:62–76

Kayes JM, Clark SE (1998) CLAVATA2, a regulator of meristem and organ development in Arabidopsis. Development 125(19):3843–3851

Kerstetter RA, Bollman K, Taylor RA, Bomblies K, Poethig RS (2001) KANADI regulates organ polarity in Arabidopsis. Nature 411:706–709

Kidner CA, Timmermans MC (2010) Signaling sides adaxial-abaxial patterning in leaves. Curr Top Dev Biol 91:141–168

Kieffer M, Stern Y, Cook H, Clerici E, Maulbetsch C, Laux T, Davies B (2006) Analysis of the transcription factor WUSCHEL and its functional homologue in Antirrhinum reveals a potential mechanism for their roles in meristem maintenance. Plant Cell 18:560–573

Kim GT, Fujioka S, Kozuka T, Tax FE, Takatsuto S, Yoshida S, Tsukaya H (2005) CYP90C1 and CYP90D1 are involved in different steps in the brassinosteroid biosynthesis pathway in *Arabidopsis thaliana*. Plant J 41(5):710–721

Kim GT, Tsukaya H, Uchimiya H (1998) The ROTUNDIFOLIA3 gene of Arabidopsis thaliana encodes a new member of the cytochrome P-450 family that is required for the regulated polar elongation of leaf cells. Genes Dev 12:2381–2391

Laux T, Mayer KF, Berger J, Jürgens G (1996) The WUSCHEL gene is required for shoot and floral meristem integrity in Arabidopsis. Development 122:87–96

Lin H, Niu L, McHale NA, Ohme-Takagi M, Mysore KS, Tadege M (2013b) Evolutionarily conserved repressive activity of WOX proteins mediates leaf blade outgrowth and floral organ development in plants. Proc Natl Acad Sci USA 110(1):366–371

Lin H, Niu L, Tadege M (2013a) STENOFOLIA acts as a repressor in regulating leaf blade outgrowth. Plant Signal Behav 8(6):e24464

Liu M, Lei L, Miao F, Powers C, Zhang X, Deng J, Tadege M, Carver BF, Yan L (2018) The STENOFOLIA gene from Medicago alters leaf width, flowering time and chlorophyll co-ntent in transgenic wheat. Plant Biotechnol J 16(1):186–196

Long JA, Moan EI, Medford JI, Barton MK (1996) A member of the KNOTTED class of homeodomain proteins encoded by the STM gene of Arabidopsis. Nature 379:66–69

Mayer KF, Schoof H, Haecker A, Lenhard M, Jürgens G, Laux T (1998) Role of WUSCHEL in regulating stem cell fate in the Arabidopsis shoot meristem. Cell 95:805–815

McConnell JR, Emery J, Eshed Y, Bao N, Bowman J, Barton MK (2001) Role of PHABULOSA and PHAVOLUTA in determining radial patterning in shoots. Nature 411:709–713

McHale NA (1992) A nuclear mutation blocking initiation of the lamina in leaves of Nicotiana sylvestris. Planta 186:355–360

Meng Y, Liu H, Wang H, Liu Y, Zhu B, Wang Z, Hou Y, Zhang P, Wen J, Yang H et al (2019) HEADLESS, a WUSCHEL homolog, uncovers novel aspects of shoot meristem regulation and leaf blade development in *Medicago truncatula*. J Exp Bot 70:149–163

Merelo P, Ram H, Pia Caggiano M, Ohno C, Ott F, Straub D, Graeff M, Cho SK, Yang SW, Wenkel S, Heisler MG (2016) Regulation of MIR165/166 by class II and class III homeodomain leucine zipper proteins establishes leaf polarity. Proc Natl Acad Sci USA 113(42):11973–11978

Mukherjee K, Brocchieri L, Bürglin TR (2009) A comprehensive classification and evolutionary analysis of plant homeobox genes. Mol Biol Evol 26 (12):2775–2794

Nakata M, Matsumoto N, Tsugeki R, Rikirsch E, Laux T, Okada K (2012) Roles of the middle domain-specific WUSCHEL-RELATED HOMEOBOX genes in early development of leaves in Arabidopsis. Plant Cell 24:519–535

Nakata M, Okada K (2013) Plants The Leaf Adaxial-Abaxial Boundary and Lamina Growth. (Basel) 2 (2):174–202

Nakata M, Okada K (2012) The three-domain model: a new model for the early development of leaves in *Arabidopsis thaliana*. Plant Signal Behav 7(11):1423–1427

Nardmann J, Ji J, Werr W, Scanlon MJ (2004) The maize duplicate genes narrow sheath1 and narrow sheath2 encode a conserved homeobox gene function in a lateral domain of shoot apical meristems. Development 131:2827–2839

Niu LF, Lin H, Zhang F, Watira TW, Li GF, Tang YH et al (2015) LOOSE FLOWER, a WUSCHEL-like Homeobox gene, is required for lateral fusion of floral organs in *Medicago truncatula*. Plant J 81(3):480–492

Peng J, Chen R (2011) Auxin efflux transporter MtPIN10 regulates compound leaf and flower development in *Medicago truncatula*. Plant Signal Behav 6(10):1537–1544

Peng J, Yu J, Wang H, Guo Y, Li G, Bai G, Chen R (2011) Regulation of compound leaf development in *Medicago truncatula* by fused compound leaf1, a class M KNOX gene. Plant Cell 23(11):3929–3943

Reinhardt D, Mandel T, Kuhlemeier C (2000) Auxin regulates the initiation and radial position of plant lateral organs. Plant Cell 12:507–518

Reinhardt D, Pesce ER, Stieger P, Mandel T, Baltensperger K, Bennett M, Traas J, Friml J, Kuhlemeier C (2003) Regulation of phyllotaxis by polar auxin transport. Nature 426:255–260

Sarkar AK, Luijten M, Miyashima S, Lenhard M, Hashimoto T, Nakajima K, Scheres B, Heidstra R, Laux T (2007) Conserved factors regulate signaling in *Arabidopsis thaliana* shoot and root stem cell organizers. Nature 446:811–814

Scanlon MJ, Freeling M (1997) Clonal sectors reveal that a specific meristematic domain is not utilized in the maize mutant narrow sheath. Dev Biol 182(1):52–66

Scanlon MJ, Schneeberger RG, Freeling M (1996) The maize mutant narrow sheath fails to establish leaf margin identity in a meristematic domain. Development 122(6):1683–1691

Scanlon MJ (2000) NARROW SHEATH1 functions from two meristematic foci during founder-cell recruitment in maize leaf development. Development 127:4573–4585

Siegfried KR, Eshed Y, Baum SF, Otsuga D, Drews GN, Bowman JL (1999) Members of the YABBY gene family specify abaxial cell fate in Arabidopsis. Development 126:4117–4128

Somssich M, Je BI, Simon R, Jackson D (2016) CLAVATA-WUSCHEL signaling in the shoot meristem. Development 143(18): 3238–3248

Szakonyi D, Moschopoulos A, Byrne ME (2010) Perspectives on leaf dorsoventral polarity. J Plant Res 123 (3):281–290

Tadege M (2014) Molecular insight into polarity-mediated lamina outgrowth. Int J Plant Biology Res 1(1):1005

Tadege M, Chen F, Murray J, Wen J, Ratet P, Udvardi MK, Dixon RA, Mysore KS (2015) Control of vegetative to reproductive phase transition improves biomass yield and simultaneously reduces lignin content in *Medicago truncatula*. Bioenergy Res 8:857–867

Tadege M, Lin H, Bedair M, Berbel A, Wen J, Rojas CM et al (2011) STENOFOLIA regulates blade outgrowth and leaf vascular patterning in *Medicago truncatula* and *Nicotiana sylvestris*. Plant Cell 23:2125–2142

Tadege M, Mysore KS (2011) Tnt1 retrotransposon tagging of STF in *Medicago truncatula* reveals tight coordination of metabolic, hormonal and developmental signals during leaf morphogenesis. Mob Genet Elements 1(4):301–303

Uchida N, Townsley B, Chung KH, Sinha N (2007) Regulation of SHOOT MERISTEMLESS genes via an upstream-conserved noncoding sequence coordinates leaf development. Proc Natl Acad Sci USA 104:15953–15958

van der Graaff E, Laux T (2009) Rensing SAThe WUS homeobox-containing (WOX) protein family. Genome Biol 10(12):248

Vandenbussche M, Horstman A, Zethof J, Koes R, Rijpkema AS, Gerats T (2009) Differential recruitment of WOX transcription factors for lateral development and organ fusion in Petunia and Arabidopsis. Plant Cell 21:2269–2283

Waites R, Hudson A (1995) Phantastica: A gene required for dorsoventrality of leaves in *Antirrhinum majus*. Development 121:2143–2154

Waites R, Selvadurai HRN, Oliver IR, Hudson A (1998) The PHANTASTICA gene encodes a MYB transcription factor involved in growth and dorsoventrality of lateral organs in *Antirrhinum*. Cell 93:779–789

Wang C, Zhao B, He L, Zhou S, Liu Y, Zhao W, Guo S, Wang R, Bai Q, Li Y, Wang D, Wu Q, Yang Y, Liu Y, Tadege M, Chen J (2021) The WOX family transcriptional regulator SlLAM1 controls compound leaf and floral organ development in *Solanum lycopersicum*. J Exp Bot 72(5):1822–1835

Wang H, Chen J, Wen J, Tadege M, Li G, Liu Y, Mysore KS, Ratet P, Chen R (2008) Control of compound leaf development by FLORICAULA/LEAFY ortholog SINGLE LEAFLET1 in Medic*ago truncatula*. Plant Physiol 146 (4):1759–1772

Wang H, Niu H, Li C, Shen G, Liu X, Weng Y, Wu T, Li Z (2020) WUSCHEL-related homeobox1 (WOX1) regulates vein patterning and leaf size in *Cucumis sativus*. Hortic Res. 7(1):182

Wang H, Niu L, Fu C, Meng Y, Sang D, Yin P, Wu J, Tang Y, Lu T, Wang ZY, Tadege M, Lin H (2017) Overexpression of the WOX gene STENOFOLIA improves biomass yield and sugar release in transgenic grasses and display altered cytokinin homeostasis. PLoS Genetics 13:e1006649

Wang H, Xu Y, Hong L, Zhang X, Wang X, Zhang J, Ding Z, Meng Z, Wang ZY, Long R, Zhou C (2019) HEADLESS regulates auxin response and compound leaf morphogenesis in *Medicago truncatula*. Front Plant Sci 10:1024

Wang HF, Xu Y, Hong L, Zhang X, Wang X, Zhang J, Ding Z, Meng Z, Zeng-Yu Wang ZY, Long R, Yang Q, Kong FJ, Han L, Zhou C (2019) HEADLESS regulates auxin response and compound leaf morphogenesis in *Medicago truncatula*. Front Plant Sci 16 (10):1024

Wolabu TW, Wang H, Tadesse D, Zhang F, Behzadirad M, Tvorogova VE, Abdelmageed H, Liu Y, Chen N, Chen J, Allen RD, Tadege M (2021) WOX9 functions antagonistic to STF and LAM1 to regulate leaf blade expansion in *Medicago truncatula* and *Nicotiana sylvestris*. New Phytol 229(3):1582–1597

Yadav RK, Perales M, Gruel J, Girke T, Jonsson H, Reddy GV (2011) WUSCHEL protein movement mediates stem cell homeostasis in theArabidopsis shoot apex. Genes Dev 25:2025–2030

Yadav RK, Tavakkoli M, Reddy GV (2010) WUSCHEL mediates stem cell homeostasis by regulating stem cell number and patterns of cell division and differentiation of stem cell progenitors. Development 137:3581–3589

Yin P, Ma Q, Wang H, Feng D, Wang X, Pei Y, Wen J, Tadege M, Niu L, Lin H (2020) SMALL LEAF AND BUSHY1 controls organ size and lateral branching by modulating the stability of BIG SEEDS1 in Medicago truncatula. New Phytol 226(5):1399–1412

Zhang F, Tadege M (2015) Repression of AS2 by WOX family transcription factors is required for leaf development in Medicago and Arabidopsis. Plant Signal Behav 10:e993291

Zhang F, Wang H, Kalve S, Wolabu TW, Nakashima J, Golz JF, Tadege M (2019) Control of leaf blade outgrowth and floral organ development by LEUNIG, ANGUSTIFOLIA3 and WOX transcriptional regulators. New Phytol 223:2024–2038

Zhang F, Wang Y, Li G, Tang Y, Kramer EM, Tadege M (2014) STENOFOLIA recruits TOPLESS to repress ASYMMETRIC LEAVES2 at the leaf margin and promote leaf blade outgrowth in *Medicago truncatula*. Plant Cell 26(2):650–664

Zhang X, He L, Zhao B, Zhou S, Li Y, He H, Bai Q, Zhao W, Guo S, Liu Y, Chen J (2020) Dwarf and Increased Branching 1 controls plant height and axillary bud outgrowth in *Medicago truncatula*. J Exp Bot 71(20):6355–6365

Zhou C, Han L, Wang ZY (2011) Potential but limited redundant roles of MtPIN4, MtPIN5 and MtPIN10/SLM1 in the development of *Medicago truncatula*. Plant Signal Behav 6(11):1834–1836

Zhou S, Yang T, Mao Y, Liu Y, Guo S, Wang R, Fangyue G, He L, Zhao B, Bai Q, Li Y, Zhang X, Wang D, Wang C, Wu Q, Yang Y, Liu Y, Tadege M, Chen J (2021) The F-box protein MIO1/SLB1 regulates organ size and leaf movement in *Medicago truncatula*. J Exp Bot 72(8):2995–3011

Zhuang LL, Ambrose M, Rameau C, Weng L, Yang J, Hu XH, Luo D, Li X (2012) LATHYROIDES, encoding a WUSCHEL-related Homeobox1 transcription factor, controls organ lateral growth, and regulates tendril and dorsal petal identities in garden pea (*Pisum sativum L.*). Molecular Plant 5:1333–1345

Function of *Medicago* WOX Genes and their Diversity

10

Hao Lin, Yingying Meng, Million Tadege, and Lifang Niu

Abstract

The *WUSCHEL*-related homebox (WOX) family genes play key roles in plant stem cell maintenance, embryonic patterning, and lateral organ development. However, the biological functions of WOX genes are poorly understood in legumes. In this chapter, we briefly summarize recent progress about the biological functions of WOX family genes and their diversity in the model legume *Medicago truncatula*.

10.1 Introduction

The WUSCHEL-related homebox (WOX) family genes encode a plant-specific family of the eukaryotic homeobox (HB) transcription factor (van der Graaff et al. 2009). Its members contain the typical homeodomain (HD) with 60–66 amino acid residues, and the sequence specificity of the HD domain makes the WOX gene family distinguished from other HB families (Gehring et al. 1990, 1994). According to evolutionary relationships and their distribution in the plant kingdom, WOX family genes can be classified into three clades: the ancient clade (found in vascular and nonvascular plants, including mosses and green algae), the intermediate clade (found in vascular plants including lycophytes), and the modern clade (found in seed plants) (Nardmann et al. 2009; van der Graaff et al. 2009). Previous studies in the model dicot *Arabidopsis thaliana* have reported that WOX family genes, including the founding member *WUSCHEL* (*WUS*) and *WOX1–WOX14* (Haecker et al. 2004), are important regulators that involve in the controlling of several key developmental processes, including stem cell maintenance in shoot and root meristems, embryo apical-basal polarity patterning, lateral organ development, as well as regeneration of isolated tissues and organs (Schoof et al. 2000; Lenhard et al. 2001; Sarkar et al. 2007; van der Graaff et al. 2009; Ueda et al. 2011).

In Arabidopsis, the founding member, *WUS*, which is specifically expressed in the organizing center of the shoot apical meristem (SAM), plays a central role in shoot stem cell identity and is required for maintenance of stem cell fate (Laux et al. 1996; Mayer et al. 1998). Ectopic expression of *WOX1* leads to abnormal meristem in Arabidopsis (Zhang et al. 2011). *WOX3*, called *PRESSED FLOWER* (*PRS*), specifically regulates the lateral axis-dependent development of

H. Lin · Y. Meng · L. Niu (✉)
Biotechnology Research Institute, Chinese Academy of Agricultural Sciences, Beijing 100081, China
e-mail: niulifang@caas.cn

M. Tadege
Department of Plant and Soil Sciences, Institute for Agricultural Biosciences, Oklahoma State University, 3210 Sam Noble Parkway, Ardmore, OK 73401, USA

S. Sinharoy et al. (eds.), *The Medicago truncatula Genome*, Compendium of Plant Genomes,
https://doi.org/10.1007/978-3-030-90757-0_10

flowers (Matsumoto and Okada 2001), while combining *prs* with *wox1* in the *wox1 prs* double mutant leads to a narrow leaf blade defect indicating that *WOX1* and *PRS* play functionally redundant roles in regulating Arabidopsis leaf blade development (Nakata et al. 2012). Arabidopsis *WOX2* and *WOX8* play critical functions in the regulation of early embryo patterning (Haecker et al. 2004; Ueda et al. 2011). *WOX4* imparts auxin responsiveness to regulate lateral plant growth (Suer et al. 2011). *WOX5* acts as a key regulator in determining the correct root patterning (Gonzali et al. 2005), while *WOX7* inhibits the development of lateral roots in a sugar-dependent manner (Kong et al. 2016). *WOX6/PFS2* plays essential roles in regulating ovule development and affects ovule patterning (Park et al. 2005). *WOX9/STIMPY* integrates developmental signals and cell cycle regulation to maintain cell division and prevent inappropriate differentiation in roots (Wu et al. 2005). *WOX11* and its close homolog *WOX12* were found to participate in Arabidopsis root organogenesis (Liu et al. 2014). The *WOX13* promotes replum formation in the Arabidopsis fruit (Romera-Branchat et al. 2013), while *WOX14* affects vascular cell differentiation by promoting bioactive gibberellin synthesis (Denis et al. 2017).

Legumes comprised one of the largest monophyletic families, with approximately 700 genera and 18,000 species (Dong et al. 2005), which is second only to grasses in terms of economic and nutritional value (Graham and Vance 2003). In addition to serving as important sources of protein and oil for the human diet, legumes are used as livestock forage and silage and as soil-enhancing green manure through fixing atmospheric nitrogen with rhizobial bacteria (Graham and Vance 2003). *Medicago truncatula* Gaertner, the so-called barrel medic, is a diploid (2n = 2x = 16) legume and is closely related to the majority of crop legumes, including alfalfa (*Medicago sativa*), the most important forage legume in the world (Barker et al. 1990). With the accomplishment of the *Medicago* genome sequencing and the development of abundant mutant resources (Tadege et al. 2008; Young et al. 2011; Chen and Chen 2018; Pecrix et al. 2018), *M. truncatula* is becoming a primary model for genomic and functional genomic research of legume biology. In this chapter, we will introduce recent progresses about the biological functions of WOX family genes and their diversity in the model legume *M. truncatula.*

10.2 Identification and Phylogenetic Analysis of *M. Truncatula* WOX Family Genes

To identify CKX family proteins in *M. truncatula*, we used Arabidopsis CKX protein sequences (WUS and WOX1–WOX14) as queries to search the *M. truncatula* genome (v4.0) (Young et al. 2011). A total of eighteen putative MtWOX sequences were obtained, including *Medtr8g107210.1/STENOFOLIA* (*STF*) (Tadege et al. 2011a), *Medtr7g060630.1/LOOSE FLOWER* (*LFL*) (Niu et al. 2015), *Medtr5g021930.1/HEADLESS* (*HDL*) (Meng et al. 2019), and *Medtr2g015000.1/MtWOX9-1* (Wolabu et al. 2020). These eighteen *MtWOX* genes are distributed on eight *M. truncatula* chromosomes (Fig. 10.1).

To uncover the evolutionary relationships of WOX family proteins in *M. truncatula* and Arabidopsis, the full-length protein sequences of 18 MtWOXs and 15 AtWOXs (WUS, WOX1–WOX14) were retrieved to construct an unrooted phylogenetic tree using the MEGA4 by the NJ method (Fig. 10.1). Among them, HDL is closely related to WUS and STF showing high similarity with Arabidopsis WOX1. LFL shares the highest identity with WOX3, while MtWOX9-1 is more closely related to Arabidopsis WOX9, respectively, (Fig. 10.1). These results indicate the evolutionary conservation among these WOX proteins in Arabidopsis and *M. truncatula*.

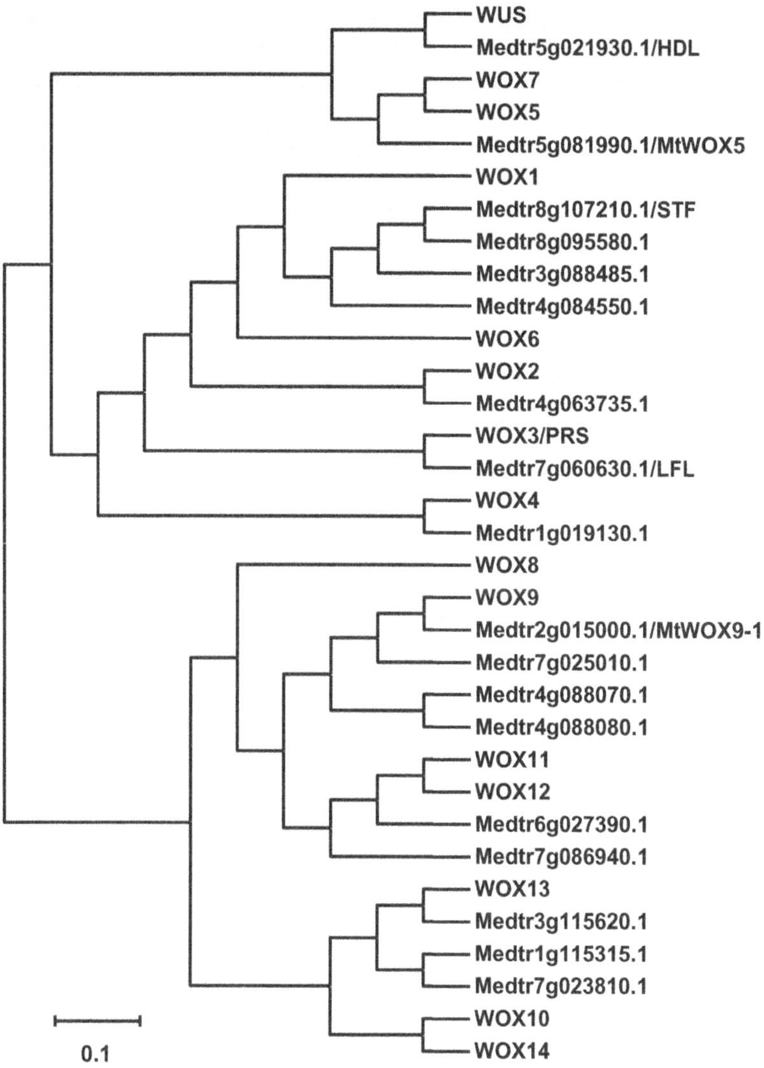

Fig. 10.1 Phylogenetic analysis of WOX family proteins in Arabidopsis and *M. truncatula*. Full-length amino acid sequences were aligned using ClustalW and the tree was constructed using MEGA4 with 1000 replicates to generate bootstrap values

10.3 *STENOFOLIA* Regulates Leaf Blade Outgrowth and Floral Organ Expansion in *M. Truncatula*

Identification and characterization of the narrow leaf mutant *stenofolia* (*stf*) revealed that *STENOFOLIA* (*STF*), a WUSCHEL-like homeobox transcriptional regulator, is required for *M. truncatula* leaf blade outgrowth and floral organ expansion (Tadege et al. 2011a, b). The *STF* gene is homologous to Arabidopsis *WOX1* (Zhang et al. 2011) and is expressed at the adaxial–abaxial boundary layer in leaf primordia, developing petals and carpels. Disruption of the single *STF* leads to severe defects in mediolateral (width) growth of the leaf blade and flower petal where both cell division and cell expansion are compromised. In addition, the disruption of vascular patterning is observed in *stf* leaves, indicating that *STF* is also required for leaf vasculature in *M. truncatula*. It is intriguing to note that the classical *laminaless1* (*lam1*) mutant of *Nicotiana sylvestris*, which is blocked in lamina expansion and stem elongation, is caused by deletion of the *STF* ortholog, suggesting *STF* plays a conserved role in regulating leaf blade outgrowth in dicots. The *lam1* mutant phenotype is much stronger than the *stf* mutant blade in severity, but the absence of visible stem in the former makes the *lam1* mutant appear different. In Arabidopsis, the corresponding leaf blade outgrowth function is carried out redundantly by *WOX1* and *WOX3/PRS,* as the *wox1* single mutant appears to be normal in blade expansion (Vandenbussche et al. 2009). However, despite the difference in severity, all these three mutants *stf, lam1*, and *wox1 prs* are similarly affected in mediolateral outgrowth, indicating a conserved function in these species. The *STF* gene also has an obvious function in floral organ development as the *stf* mutation results in narrow petals and female sterility (Tadege et al. 2011a). All the leaf blade, flower, and stem phenotypes of the *lam1* mutant can be fully complemented with the Medicago *STF* gene, suggesting that the absence of stem elongation in the *lam1* mutant is due to differences in species specific processes but not necessarily diversification of STF/LAM1 functions.

Biochemical assay and genetic complementation analyzes using the *lam1* mutant revealed that STF mainly acts as a transcriptional repressor in the regulation of leaf blade outgrowth and floral organ development (Lin et al. 2013a, b). Further comprehensive biochemical and genetic analyzes revealed that STF recruits TOPLESS corepressors (MtTPL), through its C-terminal WUS-box and STF-box motifs, to directly repress the expression of *ASYMMETRIC LEAVES2* (*MtAS2*) at the leaf margin to promote leaf blade expansion in *M truncatula* (Zhang et al. 2014; Zhang and Tadege 2015). Moreover, a recent study reported that the newly identified *M. truncatula* coactivator LEUNIG (MtLUG) interacts with another transcriptional coactivator ANGUSTIFOLIA3 (MtAN3), both enhances the *stf* and *lam1* mutant phenotypes when fused to the C-terminal end-deleted form of STF (STFdel), indicating complex regulatory mechanisms involving transcriptional repressors, corepressors and coactivators during leaf and flower morphogenesis (Zhang et al. 2019). Intriguingly, overexpression of the WOX gene *STF* can universally increase leaf blade width in *Brachypodium,* rice, wheat, and switchgrass, thus improves biomass yield and sugar release in these transgenic grasses probably through affecting cytokinin homeostasis (Wang et al. 2017; Liu et al. 2018), suggesting that the leaf regulator *STF* could serve as a powerful tool for genetic modification of biomass yield and sugar release in perennial and annual grasses.

10.4 *LOOSE FLOWER* is Required for Lateral Fusion of Floral Organs in *M. Truncatula*

The *M. truncatula* WOX gene *LOOSE FLOWER* (*LFL*) is homologous to Arabidopsis *PRESSED FLOWER* (*PRS*)/*WOX3* (Niu et al. 2015), which has been reported to be involved in the regulation of lateral axis-dependent development of Arabidopsis flowers (Matsumoto and Okada 2001). The loss-of-function mutation of *LFL* leads to a loose-flower phenotype, where severe defects are observed in the fusion of sepals, petals and the staminal tube, but leaf blades are apparently normal (Niu et al. 2015), indicating the functional conservation and diversification of WOX gene in the regulation of floral organ development in *M. truncatula.* Biochemical assay and genetic complementation analyzes showed that LFL mainly functions as a transcriptional repressor by recruiting MtTPL corepressors in the same manner as STF does, and can substitute for STF function in leaf and flower development when expressed under the promoter of *STF*, while the STF protein can also complement the floral phenotypes of the *lfl* mutant when expressed under the promoter of *LFL*, suggesting that the *M. truncatula STF* and *LFL* genes may employ a similar mechanism of action in organizing cell proliferation for lateral organ development. Further genetic analysis by generating the *lfl stf* double mutant revealed that *LFL* and *STF* act independently with minor redundant functions in flower development, but *LFL* has no obvious role in *M. truncatula* leaf development on its own or in combination with *STF,* suggesting that *LFL* and *STF* may have evolved different cis elements to acquire distinct functions.

10.5 *HEADLESS* Regulates Shoot Meristem and Leaf Shape in *M. Truncatula*

In *M. truncatula*, the WOX family gene *HEADLESS* (*HDL*) is a homolog of Arabidopsis *WUS*, which is required for the maintenance of Arabidopsis stem cell fate (Laux et al. 1996; Mayer et al. 1998). *HDL* plays an important role in the regulation of *M. truncatula* SAM maintenance (Meng et al. 2019). Disruption of *HDL* led to disorganized specification and arrest of the SAM and axillary meristems, resulting in the *hdl* mutant being locked in the vegetative phase without apparent stem elongation. Unlike the Arabidopsis *wus* mutant, the *M. truncatula hdl* never flowers and is stemless, but like WUS, HDL exhibits a repressive activity by recruiting MtTPL through its conserved WUS-box and EAR-like motifs and represses the expression of several type-A response regulators in the *M. truncatula* shoot apex (Kieffer et al. 2006; Ikeda et al. 2009; Causier et al. 2012; Dolzblasz et al. 2016; Meng et al. 2019). Notably, the leaves of *hdl* mutant are shorter in the proximal–distal axis due to reduced leaf tip elongation, which resulted in a higher blade width/length ratio and altered leaf shape, uncovering novel phenotypes not observed in Arabidopsis *wus* mutants. Further genetic analysis of the *hdl stf* double mutant suggested that *HDL* and *STF* act independently in leaf development, despite HDL could recruit MtTPL in the same manner as STF does. These observations indicate that *HDL* plays conserved and novel roles in the regulation of shoot meristem maintenance and leaf development in *M. truncatula*.

10.6 *MtWOX9* Functions Antagonistic to *STF* to Regulate Leaf Blade Expansion in *M. Truncatula*

The *M. truncatula* genome contains two Arabidopsis *WOX9*-like sequences named as *MtWOX9-1* and *MtWOX9-2*, which are classed in the intermediate clade. The *MtWOX9-1* gene is abaxially expressed in leaf primordia and strongly upregulated in the *stf* mutant leaves. Ectopic expression of *MtWOX9-1* driven by the promoter of *STF* can enhance the *stf* and l*am1* mutant phenotypes, while knockout of *NsWOX9* in *N. sylvestris* through genome editing technology led to the alteration of blade symmetry and expansion, suggesting that *MtWOX9* and *NsWOX9* function antagonistically to *STF* and *LAM1* to regulate leaf blade development in both *M. truncatula* and *N. sylvestris*. Moreover, the *MtWOX9* or *NsWOX9* expression is directly repressed by GR induction of *STF* or *LAM1* expression, and by *STF* in Dual Luciferase assay, indicating that direct repression of *MtWOX9*/*NsWOX9* by *STF/LAM1* is required for correct blade architecture and patterning in *M. truncatula* and *N. sylvestris*. Further EMSA and ChIP assays showed that STF directly binds to the proximal region of the *MtWOX9-1* promoter in *M. truncatula*, and represses its activity, suggesting that *STF* and *MtWOX9* function antagonistically to control leaf blade outgrowth (Wolabu et al. 2020).

10.7 *MtWOX5* Gene Expression is Associated with *M. Truncatula* Nodulation and Root Growth

In legumes, the symbiotic nodules are formed as a result of dedifferentiation and reactivation of cortical root cells. qRT-PCR analysis and promoter-reporter fusion experiment revealed that the *MtWOX5* gene was expressed during *M. truncatula* nodule organogenesis. Further expression analysis revealed that the transcription of *MtWOX5* was higher in nodules of supernodulating mutants, defective in autoregulation of nodulation (AON), than that in wild-type nodules, indicating that *WOX* genes are common regulators of cell proliferation in different systems (Osipova et al. 2012). Intriguingly, a recent study reported that the enhanced root system growth in *M. truncatula* after inoculation with selected bacteria includes an

Table 10.1 Identified *M. truncatula* WOX genes and their biological functions in plant growth regulation

Gene name	Accession number	Biological function	References
STENOFOLIA	JF276252	Leaf blade outgrowth, floral organ expansion	Tadege et al. (2011a, b); Lin et al. (2013a, b); Zhang et al. (2014); Zhang and Tadege (2015); Wang et al. (2017); Liu et al. (2018); Zhang et al. (2019)
LOOSE FLOWER	XM_003623010	Floral organ fusion	Niu et al. (2015)
HEADLESS	Medtr5g021930	Shoot meristem maintenance, leaf development	Meng et al. (2019)
MtWOX5	CU326389	Nodulation, root growth	Osipova et al. (2012); Kepczynska and Karczynski (2019)
MtWOX9-1	Medtr2g015000	Leaf development	Wolabu et al. (2020)

increase of nuclei in the cell cycle S phase and a reduction in phase G2 as well as an enhanced expression of the *WOX5* gene (Kepczynska and Karczynski 2019), suggesting that *MtWOX5* gene expression is highly associated with nodulation and root growth in *M. truncatula*.

10.8 Conclusions and Perspectives

In this chapter, we reviewed the recent discoveries on the functions of WOX genes in the model legume *M. truncatula*. The studies discussed above uncovered the functional diversity of *STENOFOLIA*, *MtWOX9, LOOSE FLOWER, HEADLESS,* and *MtWOX5* in the regulation of several aspects of plant growth and development including leaf development, floral organ fusion, meristem maintenance, and root nodulation (Table 10.1). Despite these enlightening advances, there are several interesting questions that still need to be clarified. For example, the WOX family comprises multiple members involved in diverse signaling pathways, but the molecular mechanism regarding how these pathways are regulated remains largely unclear. Individual WOX genes can regulate differentially to diverse signaling pathways. However, it is still unclear how the specificity of these different WOX genes is obtained. Moreover, apart from the known processes, what additional biological processes are regulated by WOX genes in legumes with distinct growth characteristics (e.g., nodulation) remains to be elucidated. With the completion of the *M. truncatula* genome sequence, the development of several genomics tools and the new technological advances (e.g., Genome editing technology), further identification and characterization of more WOX family genes and demonstrating their biological functions and genetic interactions with other signaling pathways in *M. truncatula* and other legume species will facilitate the comprehensive understanding of the biological function of WOX family genes in the regulatory network of various aspects of plant developmental dynamics.

Acknowledgements This work was supported by grants from the National Natural Science Foundation of China (32071864), Agricultural Science and Technology Innovation Program of CAAS (CAAS-ZDRW202009 and CAAS-ZDXT2019004), and Fundamental Research Funds for Central Non-profit Scientific Institution (Y2020YJ12 and No. 1610392020005).

References

Barker DG, Bianchi S, Blondon F, Dattee Y, Duc G, Essad S, Flament P, Gallusci P, Genier G, Guy P et al (1990) *Medicago truncatula*, a model plant for studying the molecular genetics of the *Rhizobium*-legume symbiosis. Plant Mol Biol Rep 8(1):40–49

Causier B, Ashworth M, Guo W, Davies B (2012) The TOPLESS interactome: a framework for gene repression in *Arabidopsis*. Plant Physiol 158(1):423–438

Chen Y, Chen R (2018) Physical mutagenesis in *Medicago truncatula* using fast neutron bombardment (FNB) for symbiosis and developmental biology studies. Methods Mol Biol 1822:61–69

Denis E, Kbiri N, Mary V, Claisse G, Conde ESN, Kreis M, Deveaux Y (2017) *WOX14* promotes bioactive gibberellin synthesis and vascular cell differentiation in *Arabidopsis*. Plant J 90(3):560–572

Dolzblasz A, Nardmann J, Clerici E, Causier B, van der Graaff E, Chen J, Davies B, Werr W, Laux T (2016) Stem cell regulation by *Arabidopsis WOX* genes. Mol Plant 9(7):1028–1039

Dong ZC, Zhao Z, Liu CW, Luo JH, Yang J, Huang WH, Hu XH, Wang TL, Luo D (2005) Floral patterning in *Lotus japonicus*. Plant Physiol 137(4):1272–1282

Gehring WJ, Muller M, Affolter M, Percival-Smith A, Billeter M, Qian YQ, Otting G, Wuthrich K (1990) The structure of the homeodomain and its functional implications. Trends Genet 6(10):323–329

Gehring WJ, Qian YQ, Billeter M, Furukubo-Tokunaga K, Schier AF, Resendez-Perez D, Affolter M, Otting G, Wuthrich K (1994) Homeodomain-DNA recognition. Cell 78(2):211 223

Gonzali S, Novi G, Loreti E, Paolicchi F, Poggi A, Alpi A, Perata P (2005) A turanose-insensitive mutant suggests a role for *WOX5* in auxin homeostasis in *Arabidopsis thaliana*. Plant J 44(4):633–645

Graham PH, Vance CP (2003) Legumes: importance and constraints to greater use. Plant Physiol 131(3):872–877

Haecker A, Gross-Hardt R, Geiges B, Sarkar A, Breuninger H, Herrmann M, Laux T (2004) Expression dynamics of WOX genes mark cell fate decisions during early embryonic patterning in *Arabidopsis thaliana*. Development 131(3):657–668

Ikeda M, Mitsuda N, Ohme-Takagi M (2009) Arabidopsis WUSCHEL is a bifunctional transcription factor that acts as a repressor in stem cell regulation and as an activator in floral patterning. Plant Cell 21 (11):3493–3505

Kepczynska E, Karczynski P (2019) *Medicago truncatula* root developmental changes by growth-promoting microbes isolated from Fabaceae, growing on organic farms, involve cell cycle changes and *WOX5* gene expression. Planta 251(1):25

Kieffer M, Stern Y, Cook H, Clerici E, Maulbetsch C, Laux T, Davies B (2006) Analysis of the transcription factor WUSCHEL and its functional homologue in *Antirrhinum* reveals a potential mechanism for their roles in meristem maintenance. Plant Cell 18(3):560–573

Kong D, Hao Y, Cui H (2016) The WUSCHEL related homeobox protein WOX7 regulates the sugar response of lateral root development in *Arabidopsis thaliana*. Mol Plant 9(2):261–270

Laux T, Mayer KF, Berger J, Jurgens G (1996) The *WUSCHEL* gene is required for shoot and floral meristem integrity in *Arabidopsis*. Development 122 (1):87–96

Lenhard M, Bohnert A, Jurgens G, Laux T (2001) Termination of stem cell maintenance in *Arabidopsis* floral meristems by interactions between *WUSCHEL* and *AGAMOUS*. Cell 105(6):805–814

Lin H, Niu L, Tadege M (2013) STENOFOLIA acts as a repressor in regulating leaf blade outgrowth. Plant Signal Behav 8(6):e24464

Lin H, Niu L, McHale NA, Ohme-Takagi M, Mysore KS, Tadege M (2013) Evolutionarily conserved repressive activity of WOX proteins mediates leaf blade outgrowth and floral organ development in plants. Proc Natl Acad Sci U S A 110(1):366–371

Liu J, Sheng L, Xu Y, Li J, Yang Z, Huang H, Xu L (2014) *WOX11* and *12* are involved in the first-step cell fate transition during de novo root organogenesis in *Arabidopsis*. Plant Cell 26(3):1081–1093

Liu M, Lei L, Miao F, Powers C, Zhang X, Deng J, Tadege M, Carver BF, Yan L (2018) The *STENOFOLIA* gene from *Medicago* alters leaf width, flowering time and chlorophyll content in transgenic wheat. Plant Biotechnol J 16(1):186–196

Matsumoto N, Okada K (2001) A homeobox gene, *PRESSED FLOWER*, regulates lateral axis-dependent development of *Arabidopsis* flowers. Genes Dev 15 (24):3355–3364

Mayer KF, Schoof H, Haecker A, Lenhard M, Jurgens G, Laux T (1998) Role of *WUSCHEL* in regulating stem cell fate in the *Arabidopsis* shoot meristem. Cell 95 (6):805–815

Meng Y, Liu H, Wang H, Liu Y, Zhu B, Wang Z, Hou Y, Zhang P, Wen J, Yang H et al (2019) *HEADLESS*, a *WUSCHEL* homolog, uncovers novel aspects of shoot meristem regulation and leaf blade development in *Medicago truncatula*. J Exp Bot 70(1):149–163

Nakata M, Matsumoto N, Tsugeki R, Rikirsch E, Laux T, Okada K (2012) Roles of the middle domain-specific *WUSCHEL-RELATED HOMEOBOX* genes in early development of leaves in Arabidopsis. Plant Cell 24 (2):519–535

Nardmann J, Reisewitz P, Werr W (2009) Discrete shoot and root stem cell-promoting WUS/WOX5 functions are an evolutionary innovation of angiosperms. Mol Biol Evol 26(8):1745–1755

Niu L, Lin H, Zhang F, Watira TW, Li G, Tang Y, Wen J, Ratet P, Mysore KS, Tadege M (2015) *LOOSE FLOWER*, a *WUSCHEL*-like homeobox gene, is required for lateral fusion of floral organs in *Medicago truncatula*. Plant J 81(3):480–492

Osipova MA, Mortier V, Demchenko KN, Tsyganov VE, Tikhonovich IA, Lutova LA, Dolgikh EA, Goormachtig S (2012) *WUSCHEL-RELATED HOMEOBOX5* gene expression and interaction of CLE peptides with components of the systemic control add two pieces to the puzzle of autoregulation of nodulation. Plant Physiol 158(3):1329–1341

Park SO, Zheng Z, Oppenheimer DG, Hauser BA (2005) The *PRETTY FEW SEEDS2* gene encodes an *Arabidopsis* homeodomain protein that regulates ovule development. Development 132(4):841–849

Pecrix Y, Staton SE, Sallet E, Lelandais-Briere C, Moreau S, Carrere S, Blein T, Jardinaud MF, Latrasse D, Zouine M et al (2018) Whole-genome landscape of *Medicago truncatula* symbiotic genes. Nat Plants 4(12):1017–1025

Romera-Branchat M, Ripoll JJ, Yanofsky MF, Pelaz S (2013) The *WOX13* homeobox gene promotes replum formation in the *Arabidopsis thaliana* fruit. Plant J 73 (1):37–49

Sarkar AK, Luijten M, Miyashima S, Lenhard M, Hashimoto T, Nakajima K, Scheres B, Heidstra R, Laux T (2007) Conserved factors regulate signalling in *Arabidopsis thaliana* shoot and root stem cell organizers. Nature 446(7137):811–814

Schoof H, Lenhard M, Haecker A, Mayer KF, Jurgens G, Laux T (2000) The stem cell population of *Arabidopsis* shoot meristems in maintained by a regulatory loop between the *CLAVATA* and *WUSCHEL* genes. Cell 100(6):635–644

Suer S, Agusti J, Sanchez P, Schwarz M, Greb T (2011) *WOX4* imparts auxin responsiveness to cambium cells in *Arabidopsis*. Plant Cell 23(9):3247–3259

Tadege M, Wen J, He J, Tu H, Kwak Y, Eschstruth A, Cayrel A, Endre G, Zhao PX, Chabaud M et al (2008) Large-scale insertional mutagenesis using the *Tnt1* retrotransposon in the model legume *Medicago truncatula*. Plant J 54(2):335–347

Tadege M, Lin H, Bedair M, Berbel A, Wen J, Rojas CM, Niu L, Tang Y, Sumner L, Ratet P et al (2011) *STENOFOLIA* regulates blade outgrowth and leaf vascular patterning in *Medicago truncatula* and *Nicotiana sylvestris*. Plant Cell 23(6):2125–2142

Tadege M, Lin H, Niu L, Mysore KS (2011) Control of dicot leaf blade expansion by a WOX gene. STF. Plant Signal Behav 6(11):1861–1864

Ueda M, Zhang Z, Laux T (2011) Transcriptional activation of Arabidopsis axis patterning genes *WOX8/9* links zygote polarity to embryo development. Dev Cell 20(2):264–270

van der Graaff E, Laux T, Rensing SA (2009) The WUS homeobox-containing (WOX) protein family. Genome Biol 10(12):248

Vandenbussche M, Horstman A, Zethof J, Koes R, Rijpkema AS, Gerats T (2009) Differential recruitment of WOX transcription factors for lateral development and organ fusion in Petunia and Arabidopsis. Plant Cell 21(8):2269–2283

Wang H, Niu L, Fu C, Meng Y, Sang D, Yin P, Wu J, Tang Y, Lu T, Wang ZY et al (2017) Overexpression of the WOX gene *STENOFOLIA* improves biomass yield and sugar release in transgenic grasses and display altered cytokinin homeostasis. PLoS Genet 13(3):e1006649

Wolabu TW, Wang H, Tadesse D, Zhang F, Behzadirad M, Tvorogova VE, Abdelmageed H, Liu Y, Chen N, Chen J, et al (2020) *WOX9* functions antagonistic to *STF* and *LAM1* to regulate leaf blade expansion in *Medicago truncatula* and *Nicotiana sylvestris. New Phytol*

Wu X, Dabi T, Weigel D (2005) Requirement of homeobox gene *STIMPY/WOX9* for *Arabidopsis* meristem growth and maintenance. Curr Biol 15(5):436–440

Young ND, Debelle F, Oldroyd GE, Geurts R, Cannon SB, Udvardi MK, Benedito VA, Mayer KF, Gouzy J, Schoof H et al (2011) The *Medicago* genome provides insight into the evolution of rhizobial symbioses. Nature 480(7378):520–524

Zhang F, Tadege M (2015) Repression of *AS2* by WOX family transcription factors is required for leaf development in *Medicago* and Arabidopsis. Plant Signal Behav 10(7):e993291

Zhang Y, Wu R, Qin G, Chen Z, Gu H, Qu LJ (2011) Over-expression of *WOX1* leads to defects in meristem development and polyamine homeostasis in *Arabidopsis*. J Integr Plant Biol 53(6):493–506

Zhang F, Wang Y, Li G, Tang Y, Kramer EM, Tadege M (2014) STENOFOLIA recruits TOPLESS to repress *ASYMMETRIC LEAVES2* at the leaf margin and promote leaf blade outgrowth in *Medicago truncatula*. Plant Cell 26(2):650–664

Zhang F, Wang H, Kalve S, Wolabu TW, Nakashima J, Golz JF, Tadege M (2019) Control of leaf blade outgrowth and floral organ development by LEUNIG, ANGUSTIFOLIA3 and WOX transcriptional regulators. New Phytol 223(4):2024–2038

Early Stages of Seed Development in *Medicago truncatula*: Lessons from Genomic Studies

11

Garima Chauhan, Prashant Yadav, Jaiana Malabarba, Jerome Verdier, and Kaustav Bandyopadhyay

Abstract

Seed quality and quantity are the two most economically important traits of legumes. As source of proteins, oils, and carbohydrates, legume seeds have received attention from researchers. Like other dicots, the embryo of *M. truncatula* develops from a globular stage, transitions through heart and torpedo stage, and finally turns into bent cotyledons. Availability of genome sequence, and other community resources like gene expression atlas, mutant collection, and HapMap collection has streamlined gene function discoveries for seed biologists. Early embryo patterning is regulated by *WUS/WOX*, Leucine Zippers, and Homeobox genes. LEAFY COTYLEDONs (*LECs*), *FUS3* (FUSCA 3), and *ABIs* (ABSCISIC ACID INSENSITIVE) are the main groups of genes which take over during seed filling. Seed filling is followed by acquisition of desiccation tolerance and longevity. Though this chapter limits itself within the discussion on early seed development, a considerable portion of the final seed size and quality traits is already determined in these stages.

11.1 Introduction:

When economic importance of plants is considered, the Leguminosae family stands second only to the Poaceae (or grasses) family (Graham and Vance 2003). Legumes provide food, edible oils, fodders, and medicines. Common bean (*Phaseolus vulgaris*), pea (*Pisum sativum*), soybean (*Glycine max*), chickpea (*Cicer arietinum*), pigeon pea (*Cajanus cajan*), broad bean (*Vicia faba*), cowpea (*Vigna unguiculata*), and lentil (*Lens esculenta*) are some of the globally important economic species. By symbiotically hosting nitrogen-fixing rhizobia into their root systems (root nodule symbiosis), legumes act as the flag bearers of sustainable and low-input agriculture. Many legumes also store secondary compounds with beneficial properties, including isoflavones (Dixon and Sumner 2003; Dixon and Pasinetti 2010). Moreover, many legumes store medicinally important compounds like L-DOPA in velvet bean (Lampariello et al. 2012). Finally, legumes can be used as raw material for biofuel production (Biswas et al. 2011). Amidst the plant tissues, seeds are often the most-utilized portion of a plant. In fact, most of the above-mentioned virtues of legumes come from their seeds (Fig. 11.1).

G. Chauhan
National Institute of Plant Genome Research, New Delhi, India

P. Yadav · K. Bandyopadhyay (✉)
Amity University Haryana, Gurgaon, India

J. Malabarba · J. Verdier
Institut Agro, INRAE, IRHS, Université d'Angers, SFR 4207 QuaSaV, 49071 Beaucouzé, France

S. Sinharoy et al. (eds.), *The Medicago truncatula Genome*, Compendium of Plant Genomes,
https://doi.org/10.1007/978-3-030-90757-0_11

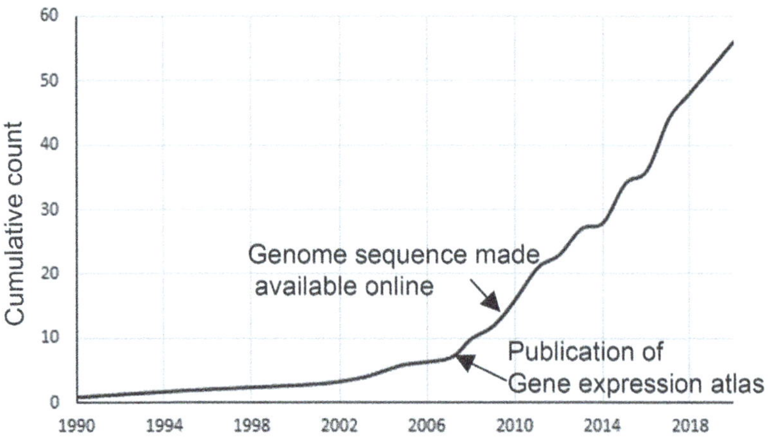

Fig. 11.1 Year-wise number of research articles on *M. truncatula* seed biology, as shown in PubMed

The annual, self-pollinating model legume *M. truncatula* is the closest relative of Queen of forage alfalfa (*Medicago sativa*). This model plant represents most of the economically important legumes of Papilionoideae sub-family. Developed inside a spiral pod, seeds of Medicago show typical morphological features of most dicot, and any legume in that sense. The seeds are composed of 35–45% proteins, 10% lipids, and only traces of starch. Globulins, and vicilins are the major storage proteins. The endosperm covers less than 10% of the fully developed seed (Mathesius et al. 2006). The stages of seed development in *M. truncatula* (Fig. 11.2) are similar to other legumes. Nonetheless, there is no clear boundary between the stages, i.e., the stages are overlapping. By 6 days after pollination (DAP), a globular stage embryo is visible. The embryo then transitions through the heart stage (9–10 DAP), and the torpedo stage (11–12 DAP). Cotyledons are formed at around 13–14 DAP. The cell division continues for another 4–6 days. Cells also start expanding and acquiring storage materials around 14 DAP (seed filling) which starts with massive RNA synthesis. Seed storage proteins (SSPs) start appearing from 16 DAP onward and their synthesis continues beyond 30 DAP. Surprisingly, seeds start losing water even before that (around 28 DAP), and by 32–34 DAP the

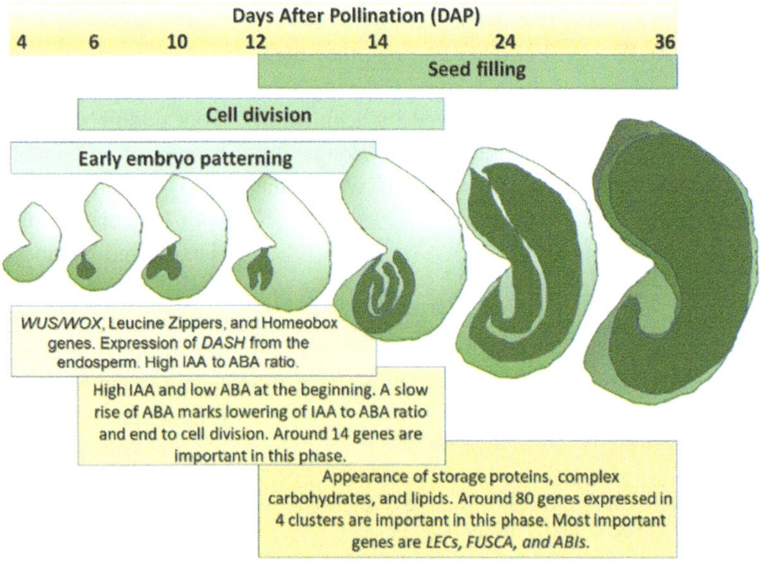

Fig. 11.2 Stages of early seed development of *M. truncatula*

water content becomes less than 50% of the early seed filling stage. 30–45 DAP marks the events of late maturation, i.e., acquisition of desiccation tolerance and longevity, and finally pod abscission (Gallardo et al. 2003, 2007; Verdier et al. 2008; Noguero et al. 2015). It should be remembered that different group of researchers have found variations in these time points, and these vary according to the growth and nutrient conditions. These time points also vary according to the genotype or ecotype of Medicago (Bandyopadhyay et al. 2016).

Being the model representing most of the crop legumes, Medicago has a number of useful tools and resources for genomic studies. Availability of detailed genetic and physical maps (Thoquet et al. 2002; Choi et al. 2004) is one of them. The Gene Expression Atlas (MtGEA) (Benedito et al. 2008) which is a spatio-temporal transcriptomic database helps in reverse genetic studies. Apart from covering various tissues and growth phases, the gene atlas also covers microarray experiments from plants growing under various environmental cues or pathogen challenge (He et al. 2009). For functional characteristics of genes, mutant analysis is indispensable. Different mutant populations including fast-neutron deletion, EMS, and transposon *Tnt1*-insertion mutants (Tadege et al. 2008) are available in Medicago. The genome sequence of Medicago was published in 2011 for the first time (Young et al. 2011), and as of today, the fifth version is much more detailed (Pecrix et al. 2018). Finally, availability of a large germplasm and HapMap collection has streamlined association genomic studies like GWAS (Stanton-Geddes et al. 2013). Like many other fields of study, research on seed development of Medicago has been tremendously benefited by genomic studies. These studies were only possible in the so-called genomic era, and as evident from Fig. 11.2, number of publications on seed development of *M. truncatula* as appeared in NCBI PubMed, escalated around 2010 (Fig. 11.1). In this chapter, we shall retrospect on the molecular details of seed development dissected by genomic studies. We shall also limit this chapter describing the events of early embryo development including seed filling. Late events of seed development, i.e., acquisition of desiccation tolerance or longevity will be discussed in detail in the next chapter.

11.2 Transcriptional Regulation of Early Embryo Patterning

Kurdyukov et al. (2014) have performed detailed transcriptional analysis during early embryo patterning in Medicago. This study has revealed how various TFs control the overall gene expression pattern during different stages of embryo development. They have identified 19 TF genes and tracked their expression pattern throughout this period. Some of these TF groups include six WUSCHEL/WUSCHEL-RELATED HOMEOBOX (WUS/WOX), three KNOTTED1-LIKE HOMEOBOX (KNOX) genes, five HOMEOBOX LEUCINE ZIPPER (HD-ZIP) genes, two AP2/ARF transcription factor called *SOMATIC EMBRYO-RELATED FACTOR 1 (MtSERF1)* and *BABY BOOM*, and one DOF domain TF.

MtWUS, MtHD-ZIP2,3,4 and 5, and MtKNOX7 control the ovule development and the early cell divisions up to the globular stage embryo. *MtWOX 1-like (STENFOLIA), MtWOX4-like, MtDOF1-like, and MtLMI1-like* control the vasculature development inside the early seed. The *WOX9-like gene and the MtSERF1* take control of the globular stage embryo, since their expression keep increasing since early globular embryo. This has been confirmed by situ hybridization resulting in transcripts localizing throughout globular-stage embryos and in the organizer region in the heart-stage embryo (Mantiri et al. 2008). *MtWOX5 and MtWUS* both start expressing at the heart stage (Chen et al. 2009), unlike their Arabidopsis counterparts that start expression at globular stage (Haecker et al. 2004). Most probably, these two genes are involved in the formation of apical meristems in both the poles of the heart stage embryo (Xu et al. 2018). The stem cell niche in the apical meristems of the later stage embryos (torpedo and early cotyledon) is maintained by *MtKNOX4 and MtKNOX6*. The torpedo stage embryo sees marked over-expression in the AP2 transcription factor *BABYBOOM*. Later stages of embryo development are controlled by *FUSCA3 (FUS3) and ABA INSENSITIVE3 (ABI3)*. As stated before, embryo patterning, and seed filling are overlapped in Medicago. Hence, it is difficult to dissect the exact role of these two in later stage cell division and are mainly attributed to seed filling.

Though the endosperm in most of the mature legume seeds is rudimentary, yet it plays a very important role in early embryo patterning. Verdier et al. (2008) have shown that a DOF type transcription factor which is expressed in endosperm, controls early embryo patterning. Mutant analysis shows that the absence of the gene *DOF Acting in Seed embryogenesis and Hormone accumulation (DASH)* results in the abortion of early embryo, specifically around the globular stage. Detailed transcriptomic analysis proved that a perturbation in Auxin transport/ signaling occurs in the early embryo in absence of *DASH*. This study proved beyond doubt that embryo patterning is also regulated by the endosperm (Noguero et al. 2015).

11.3 Regulation of Seed Filling

With a rapidly increasing population, it is crucial to increase the quality and quantity of energy-rich crops to achieve food security. Researchers across the globe are working to decipher the mechanism with which nutrient-rich biomolecules are accumulated in the economically important crops. After embryogenesis, the subsequent formation of seed involves embryo enlargement through cell expansion accompanied by rapid accumulation of proteins, lipids and carbohydrates as a reservoir of nitrogen and carbon. Furthermore, removal of water through desiccation will ensure continuous supply of energy for the germinating seed. Transcriptomic and proteomic studies carried out at in different developmental stages of seed development revealed that an array of transcription factors coordinate and regulate the overall seed filling processes (Gallardo et al. 2007; Verdier et al. 2008). Differential expression of various transcription factors in *M. truncatula* revealed that out of 169 transcription factor encoding genes, 41 are involved in coordinating the events of late embryogenesis, whereas 80 others are associated with storage compound accumulation and the remaining 48 were involved in providing desiccation tolerance to the maturing seed (Verdier et al. 2008; Verdier and Thompson 2008). Among them, the key regulators were *LEC1 (LEAFY COTYLEDON 1), LEC2 (LEAFY COTYLEDON 2), FUS3 (FUSCA 3) and ABI3 (ABSCISIC ACID INSENSITIVE 3) and WRI1 (WRINKLED1)*. These transcription factors bind to the cis-regulatory element present on the seed storage protein genes and regulate their expression. Furthermore, influence of cis-regulatory elements was also studied extensively. DNA–protein interaction studies done on various promoters revealed conserved sequences namely RY/G motif and B-box (Ezcurra et al. 2000). The RY element bind with the B3 domain carrying transcription factors, whereas the G element interacts with transcription factors carrying bZIP and bHLH domain (Kawagoe and Murai 1996). In the B-box, the presence of ABA response element (ABRE) domain allows the interaction with other classes of transcription factors such as MYB (Ezcurra et al. 2000).

Seed filling can be considered as the most important phase for seed quality and the economic value of a seed. Surprisingly, the final quality traits of a seed depend on early events of seed development. The auxin to ABA ratio during late embryogenesis/ early seed filling stage controls the cotyledonary cell division. The cell division continues as long as the seed maintains a high auxin to ABA ratio. With the appearance of ABA, cell division slowly diminishes (i.e., when IAA: ABA ratio decreases). More cotyledonary cells in this stage will attain a bigger size at the end. Thus, an increase in this cell division window can contribute to final seed size (Bandyopadhyay et al. 2016). This model is also supported by the fact that *SBT1.1*, a subtilase which controls cotyledon cell numbers, also contributes to final seed size and weight (D'Erfurth et al. 2012). *BIG SEED 1(BS1)* is a group II TIFY type of transcription regulator that has been shown to control primary cell division of similar stages. *Medicago BS1* is shown to interact with *Medicago NINJA*. The *Medicago BS1* mutant presents a bigger seed size (Ge et al. 2016). Therefore, *BS1* is a negative regulator of seed size. This is achieved by repressing *GIF and GRF* genes.

11.3.1 Transcriptional Regulation of Carbon Partitioning During Seed Development

The distribution of the carbon reserves to different parts of the plants determines the overall biomass. Leguminous seeds are rich in carbohydrates, proteins and oils. However, their proportion varies with species to species. In *M. truncatula*, protein and oil bodies are present in majority in the seeds. Apart from starch, complex carbohydrate in the form mucilage is also found predominantly in *Medicago orbiscularis*. In turn, the proportion of oil and protein decreases in the same variety. The abundance of mucilage in a seed provides additional tolerance to desiccation. *MtGLABRA2-like (MtGL2)* transcription factor works as a master regulator of carbon partitioning (Song et al. 2017). *MtGLABRA2-like (MtGL2)* is associated with an inhibition of transcriptional regulators of oil biosynthesis, *MtLEAFY COTYLEDON1-LIKE (MtL1L), MtABSCISIC ACID-INSENSITIVE3 (MtABI3), and MtWRINKLED-like (MtWRI)*. As a result, less oil is found in genotypes where there is an abundance of *MtGL2* expression.

11.3.2 Regulation of Storage Protein (SSPs) Synthesis

In dicotyledon seeds, majority of the proteins belong to 7S and 11S globulin classes (Abirached-Darmency et al. 2012). The legumes on the other hand, are known for having enhanced levels of legumin, vicilin, and convicilin proteins among the other classes of proteins (Gallardo et al. 2003). Storage proteins in *M. truncatula* appear in different clusters, as shown by the expression pattern of the SSP-coding genes. Vicilins appear at around 16–20 DAP, while legumin K (20–24 DAP) and legumin A (24–36 DAP) appear much later. Proteins first appear as small protein granules/bodies which would later fuse with each other to form larger protein bodies (Abirached-Darmency et al. 2012). The RY/G motif and the B-box motifs are extremely important in case of transcriptional regulation of storage protein synthesis, since these are found in the upstream regions of SSP-coding genes (Verdier and Thompson 2008). These elements (and hence the expression of SSP-coding genes) are regulated by four master regulatory genes. These are *LEC1, LEC2, ABI3 and FUS3*. This regulation of expression is continued until the master regulators are itself downregulated by late-appearing factors like *PICKLE (PKL) and ROM2*. Metabolomic and proteomic studies in Medicago have revealed compartmentalization of building blocks (mainly sulphur-containing amino acids and the enzymes needed for their biosynthesis) during seed filling. This is a remarkable example of tissue-specific regulation of nutrient availability between seed coat, the endosperm, and the cotyledons (Gallardo et al. 2007). Finally, ABI5 is also considered as a major hub for seed protein storage, as shown by GWAS studies in Medicago (Le Signor et al. 2017).

11.3.3 Oil and Lipid Storage

In the leguminous seeds, oil is often found as droplets or oil bodies embedded in the cytoplasm (Song et al. 2017). Triacylglycerols are the major components followed by layers of phospholipid enclosing the oil droplet (Tzen et al. 1993). Oil bodies are often found in vicinity of protein bodies. Oleosins are the integral proteins present along with other caleosin, and steroleosin (Chen et al. 2012). Production of seed oil/lipid requires utilization of the photosynthetic assimilates to biosynthesize the building blocks, fatty acids in the seed plastids. The acetyl Co-A produced as a result of Kreb's cycle is directly utilized for *de novo* fatty acid synthesis. Further processing takes place at the level of endoplasmic reticulum where the free triacylglycerols are assembled to form a central TAG matrix enclosed with a monolayer of phospholipid embedded with integral oleosin proteins (Chapman and Ohlrogge 2012). It has been demonstrated in other systems that the same master transcription regulators which control SSP biosynthesis (specifically the *LECs*) can also regulate lipid biogenesis (Song et al. 2017). Overexpression of *LEC1 and LEC2*-controlled genes *DGAT1 and WRI1* have shown significant promise in increasing oil content in other legumes (Roesler et al. 2016).

11.3.3.1 Regulation of Proanthrocyanidin Biosynthesis

Lethal pasture bloat is a life-threatening disease that occurs when cattle graze on pastures with young leguminous plants. The ingestion of excess protein rapidly increases the fermentation process which eventually leads to the formation of methane gas. Apart from proteins, legume crops also carry various flavonoids such as anthocyanins, proanthrocyanidins, flavonols, and isoflavanones. Among them, proanthrocyanidins (PAs) enhance medicinal and commercial value of legumes. PAs are also known to bind to proteins to reduce the fermentation in the grazing animals. Therefore, the PAs content in forage crops can be enhanced genetically to prevent lethal pasture bloat in grazing cattle. Overexpression or downregulation of positive and negative regulators can enhance the cellular PAs content. Genome wide screening of various transcription factors was done to determine the regulators of PA biosynthesis. Among them, MtPAR was identified which was highly expressed at 24 days after pollination. Further expression studies by qRT-PCR revealed its specific localization in the seed coat. MtPAR not only regulates PA biosynthesis but also increases PA level when ectopically expressed in alfalfa leaves (Verdier et al. 2012). MtPAR upregulates the expression of the enzymes in flavonoid-PA pathway by activating the gene WD40-1.

11.4 Concluding Remarks

The most economically important part of a legume is the seed. Decades of genetic research had already accumulated a plethora of knowledge of seed development in legumes like soybean and pea. The genomic era has streamlined gene function identification in model legumes. Nonetheless, validation of these candidate genes in crop legumes remains a major bottleneck, mainly due to lack of community resources. For instance, protein content in chickpea is much lower than in *Medicago*. Yet, a high-protein chickpea remains elusive. Therefore, the target of the next decade should be to improve seed quality traits by translation of the pool of candidate genes identified in *M. truncatula* (and other models like *Lotus japonicus*) into crop legumes.

Acknowledgements KB received a grant from SERB, India (Grant No. SRG/2019/0596).

References

Abirached-Darmency M, Dessaint F, Benlicha E, Schneider C (2012) Biogenesis of protein bodies during vicilin accumulation in Medicago truncatula immature seeds. BMC Res Notes 5:1–8. https://doi.org/10.1186/1756-0500-5-409

Bandyopadhyay K, Uluçay O, Şakiroğlu M et al (2016) Analysis of large seeds from three different *Medicago truncatula* ecotypes reveals a potential role of hormonal balance in final size determination of legume grains. Int J Mol Sci 17:1472. https://doi.org/10.3390/ijms17091472

Benedito VA, Torres-Jerez I, Murray JD et al (2008) A gene expression atlas of the model legume *Medicago truncatula*. Plant J 55:504–513. https://doi.org/10.1111/j.1365-313X.2008.03519.x

Biswas B, Scott PT, Gresshoff PM (2011) Tree legumes as feedstock for sustainable biofuel production: opportunities and challenges. J Plant Physiol 168:1877–1884

Chapman KD, Ohlrogge JB (2012) Compartmentation of triacylglycerol accumulation in plants. J Biol Chem 287:2288–2294

Chen SK, Kurdyukov S, Kereszt A et al (2009) The association of homeobox gene expression with stem cell formation and morphogenesis in cultured *Medicago truncatula*. Planta 230:827–840. https://doi.org/10.1007/s00425-009-0988-1

Chen DH, Chyan CL, Jiang PL et al (2012) The same oleosin isoforms are present in oil bodies of rice embryo and aleurone layer while caleosin exists only in those of the embryo. Plant Physiol Biochem 60:18–24. https://doi.org/10.1016/j.plaphy.2012.07.022

Choi HK, Kim D, Uhm T et al (2004) A sequence-based genetic map of *Medicago truncatula* and comparison of marker colinearity with *M. sativa*. Genetics 166:1463–1502. https://doi.org/10.1534/genetics.166.3.1463

D'Erfurth I, Le Signor C, Aubert G et al (2012) A role for an endosperm-localized subtilase in the control of seed size in legumes. New Phytol 196:738–751. https://doi.org/10.1111/j.1469-8137.2012.04296.x

Dixon RA, Pasinetti GM (2010) Flavonoids and isoflavonoids: from plant biology to agriculture and neuroscience. Plant Physiol 154:453–457. https://doi.org/10.1104/pp.110.161430

Dixon RA, Sumner LW (2003) Legume natural products: understanding and manipulating complex pathways for human and animal health. Plant Physiol 131:878–885

Ezcurra I, Wycliffe P, Nehlin L et al (2000) Transactivation of the *Brassica napus* napin promoter by ABI3 requires interaction of the conserved B2 and B3 domains of ABI3 with different cis-elements: B2 mediates activation through an ABRE, whereas B3 interacts with an RY/G-box. Plant J 24:57–66. https://doi.org/10.1046/j.1365-313X.2000.00857.x

Gallardo K, Le Signor C, Vandekerckhove J et al (2003) Proteomics of *Medicago truncatula* seed development establishes the time frame of diverse metabolic processes related to reserve accumulation. Plant Physiol 133:664–682. https://doi.org/10.1104/pp.103.025254

Gallardo K, Firnhaber C, Zuber H et al (2007) A combined proteome and ranscriptome analysis of developing *Medicago truncatula* seeds: evidence for metabolic specialization of maternal and filial tissues. Mol Cell Proteomics 6:2165–2179. https://doi.org/10.1074/mcp.M700171-MCP200

Ge L, Yu J, Wang H et al (2016) Increasing seed size and quality by manipulating BIG SEEDS1 in legume species. Proc Natl Acad Sci USA 113:12414–12419. https://doi.org/10.1073/pnas.1611763113

Graham PH, Vance CP (2003) Legumes: Importance and constraints to greater use. Plant Physiol 131:872–877

Haecker A, Groß-Hardt R, Geiges B et al (2004) Expression dynamics of WOX genes mark cell fate decisions during early embryonic patterning in *Arabidopsis thaliana*. Development 131:657–668. https://doi.org/10.1242/dev.00963

He J, Benedito VA, Wang M et al (2009) (2009) The *Medicago truncatula* gene expression atlas web server. BMC Bioinform 101(10):1–9. https://doi.org/10.1186/1471-2105-10-441

Kawagoe Y, Murai N (1996) A novel basic region/helix-loop-helix protein binds to a G-box motif CACGTG of the bean seed storage protein β-phaseolin gene. Plant Sci 116:47–57. https://doi.org/10.1016/0168-9452(96)04366-X

Kurdyukov S, Song Y, Sheahan MB, Rose RJ (2014) Transcriptional regulation of early embryo development in the model legume *Medicago truncatula*. Plant Cell Rep 33:349–362. https://doi.org/10.1007/s00299-013-1535-x

Lampariello L, Cortelazzo A, Guerranti R et al (2012) The magic velvet bean of *Mucuna pruriens*. J Tradit Complement Med 2:331–339. https://doi.org/10.1016/S2225-4110(16)30119-5

Le Signor C, Aim D, Bordat A et al (2017) Genome-wide association studies with proteomics data reveal genes important for synthesis, transport and packaging of globulins in legume seeds. New Phytol 214:1597–1613. https://doi.org/10.1111/nph.14500

Mantiri FR, Kurdyukov S, Lohar DP et al (2008) The transcription factor MtSERF1 of the ERF subfamily identified by transcriptional profiling is required for somatic embryogenesis induced by auxin plus cytokinin in *Medicago truncatula*. Plant Physiol 146:1622–1636. https://doi.org/10.1104/pp.107.110379

Mathesius U, Journet E, Sumner L (eds) (2006) The Medicago truncatula handbook, 1st edn. Noble Research Institute, Ardmore, OK

Noguero M, Le Signor C, Vernoud V et al (2015) DASH transcription factor impacts *Medicago truncatula* seed size by its action on embryo morphogenesis and auxin homeostasis. Plant J 81:453–466. https://doi.org/10.1111/tpj.12742

Pecrix Y, Staton SE, Sallet E et al (2018) Whole-genome landscape of *Medicago truncatula* symbiotic genes. Nat. Plants 4:1017–1025

Roesler K, Shen B, Bermudez E et al (2016) An improved variant of soybean type 1 diacylglycerol acyltransferase increases the oil content and decreases the soluble carbohydrate content of soybeans. Plant Physiol 171:878–893. https://doi.org/10.1104/pp.16.00315

Song Y, Wang XD, Rose RJ (2017) Oil body biogenesis and biotechnology in legume seeds. Plant Cell Rep 36:1519–1532

Stanton-Geddes J, Paape T, Epstein B et al (2013) Candidate genes and genetic architecture of symbiotic and agronomic traits revealed by whole-genome, sequence-based association genetics in *Medicago truncatula*. PLoS ONE 8:e65688. https://doi.org/10.1371/JOURNAL.PONE.0065688

Tadege M, Wen J, He J et al (2008) Large-scale insertional mutagenesis using the Tnt1 retrotransposon in the model legume *Medicago truncatula*. Plant J 54:335–347. https://doi.org/10.1111/j.1365-313X.2008.03418.x

Thoquet P, Ghérardi M, Journet EP, et al (2002) The molecular genetic linkage map of the model legume

Medicago truncatula: an essential tool for comparative legume genomics and the isolation of agronomically important genes. BMC Plant Biol 2. https://doi.org/10.1186/1471-2229-2-1

Tzen JTC, Cao YZ, Laurent P et al (1993) Lipids, proteins, and structure of seed oil bodies from diverse species. Plant Physiol 101:267–276. https://doi.org/10.1104/pp.101.1.267

Verdier J, Thompson RD (2008) Transcriptional regulation of storage protein synthesis during dicotyledon seed filling. Plant Cell Physiol 49:1263–1271

Verdier J, Kakar K, Gallardo K et al (2008) Gene expression profiling of *M. truncatula* transcription factors identifies putative regulators of grain legume seed filling. Plant Mol Biol 67:567–580. https://doi.org/10.1007/s11103-008-9320-x

Verdier J, Zhao J, Torres-Jerez I et al (2012) MtPAR MYB transcription factor acts as an on switch for proanthocyanidin biosynthesis in *Medicago truncatula*. Proc Natl Acad Sci USA 109:1766–1771. https://doi.org/10.1073/pnas.1120916109

Xu S, Chen X, Peng G et al (2018) An electrochemical biosensor for the detection of Pb2+ based on G-quadruplex DNA and gold nanoparticles. Anal Bioanal Chem 410:5879–5887. https://doi.org/10.1007/s00216-018-1204-6

Young ND, Debellé F, Oldroyd GED et al (2011) The *Medicago* genome provides insight into the evolution of rhizobial symbioses. Nature 480:520–524. https://doi.org/10.1038/nature10625

Seed Maturation Events in *Medicago truncatula*: Focus on Desiccation Tolerance

12

Jaiana Malabarba, Kaustav Bandyopadhyay, and Jerome Verdier

Abstract

During seed maturation phase, a diverse and important set of compounds need to be accumulated to guarantee the correct acquisition of the physiological mechanisms that will promote seed survival and future plant development. One of these mechanisms is the acquisition of desiccation tolerance (DT), which is the capacity that a tissue has to survive extreme dehydration (called desiccation, with a drastic loss of water content to levels below 0.1 g H_2O per gram of dry weight) and to remain alive in this dry state for an extended period of time. The ability of seeds to tolerate such stress is tightly regulated during seed development, being acquired at the onset of seed maturation, during the seed filling phase in *Medicago truncatula*, and later on, lost at early germination (DT switch). To date, a vast body of knowledge has been built around the understanding of DT resulting in the identification of compounds and genes induced during DT acquisition and potentially involved in protective cell functions allowing cellular survival in the dry state. Major efforts have focused on unraveling the nature of these protective compounds conferring DT, and the *Medicago truncatula* species has played a key role in these studies. Therefore, this review will focus on recent advances in desiccation tolerance comprehension focused on the use of *M. truncatula* for elucidating DT molecular mechanisms.

12.1 Introduction

The Angiosperm taxon is distinct from other plant taxa because of important autapomorphic characteristics and, among them, the presence of flowers and fruits. The name "Angiosperm" derives from the Greek word *angeion*, which means "vase" or "container", and *sperma*, which means "seed" (Raven et al. 2007). In this taxon, the seeds are enclosed within fruits that come from the development of the ovary after fertilization. Seeds consist of integuments, endosperm and embryo, each tissue aiming for embryo survival, thus ensuring the propagation and maintenance of its species (Berger et al. 2006). In *Medicago truncatula*, seed development can be divided spatially by the development of its three main tissues and temporally by embryogenesis and maturation phases. Spatially, in *M. truncatula* seeds, the integuments develop in the seed coat that becomes responsible for protecting the endosperm and the embryo from adverse

J. Malabarba (✉) · J. Verdier
Institut Agro, INRAE, IRHS, Université d'Angers, SFR 4207 QuaSaV, 49071 Beaucouzé, France

K. Bandyopadhyay
Amity Institute of Biotechnology, Amity University Haryana, Gurgaon, India

S. Sinharoy et al. (eds.), *The Medicago truncatula Genome*, Compendium of Plant Genomes,
https://doi.org/10.1007/978-3-030-90757-0_12

situations until there are favorable conditions for proper germination. The endosperm accumulates the necessary nutrients serving as a temporary storage compartment for embryo development, and will be consumed by the growing embryo. The embryo contains most of the storage molecules and upon maturation is divided into two parts, the radicle and the cotyledon, which will germinate to form a new plant (Raven et al. 2007). For the seed to fully develop, coordination of temporal and spatial growth between different tissues is necessary, in addition to homogeneous nutrition of all its components (Doktorgrades and Ungru 2010; Jiang and Lin 2013; Malabarba et al. 2017).

12.1.1 Seed Maturation Processes

Seed development is divided into embryogenesis and maturation phases. In *M. truncatula,* the embryogenesis phase is characterized by many rows of well-organized cell multiplication and differentiation, for tissue specification and patterning. From 0 to 5 days after pollination, the first embryo cell divisions occur and embryo reaches globular stage around 5–6 DAP. Then, it becomes heart shaped around 6–10 DAP, and it achieves a torpedo shape around 10–12 DAP (Verdier et al. 2013; Noguero et al. 2015). Once embryogenesis is completed, at 12 DAP, the seed maturation phase initiates with the elongation of embryo cells at around 28 DAP until reaching maturity (Wang et al. 2011).

The different physiological processes activated during seed maturation are necessary for the preparation of the seed tissues, in which seeds will acquire protective and developmental mechanisms to ensure life span. To summarize, seed maturation is marked by a series of processes that will serve two overlapping and synergic goals, which are the survival in the dry state and storability (i.e., desiccation tolerance and longevity) and the germination qualities such as rate, vigor and homogeneity (i.e., germination capacity, seed filling, and dormancy) (Sano et al. 2016; Leprince et al. 2016) (Fig. 12.1). Germination capacity and desiccation tolerance are qualitative traits, whereas effects of dormancy and longevity are quantitative traits.

From 16 to 36 DAP, the seeds undergo a long seed filling phase which is characterized by seed weight increase due to the accumulation of seed storage proteins (i.e., mainly vicilins and legumins) and other storage compounds, including oils and carbohydrates (Verdier et al. 2013), which will serve as reserve nutrients for germination (Fig. 12.1). Globulin class proteins are the main storage proteins in *M. truncatula* seeds, and they present a pattern of accumulation divided between vicilins and legumins storage. Vicilins started to accumulate as soon as 12–14 DAP with an accumulation peak at 24 DAP. On the other hand, legumins started to accumulate at 16 DAP, first with legumin K, reaching a maximum at 24 DAP, and later with legumin A (20–36 DAP) (Gallardo et al. 2003, 2007; Verdier et al. 2013).

Concomitantly to early seed filling, germination capacity is acquired from 18 to 20 DAP. Germination capacity is the ability of seeds to germinate and the degree of dormancy release defines germination rate, vigor (i.e., speed of germination) and the ability to produce homogeneous germination. Freshly harvested seeds acquire the capacity to germinate early during maturation from 16 DAP. Then, from 26 DAP onward, the germination speed decreases gradually, indicating the installation of dormancy (Verdier et al. 2013; Wang et al. 2011). Seed dormancy controls the timing of germination that prevents seed germination during unfavorable conditions, e.g., dry period or inadequate germination season (Bewley et al. 2013; Finch-Savage and Leubner-Metzger 2006). Seed dormancy is also a key mechanism to increase seed dispersal time, giving that the same species can disperse seeds with different levels of dormancy, which decreases the risk of losing a complete generation due to unexpected events (Dekkers et al. 2015; Hilhorst 2007). At late maturation, germination speed increases again (from 44 DAP) concomitantly with seed in dry state that indicates a partial release of dormancy, although the speed of seed germination is still much slower than in fully after-ripened seeds (Verdier et al. 2013).

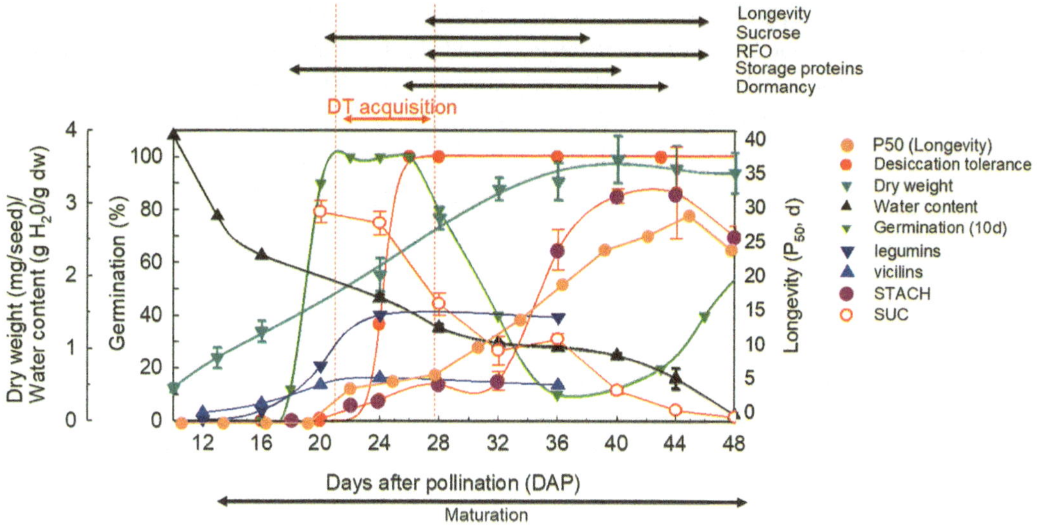

Fig. 12.1 ***Medicago truncatula*** **seed development.** From pollination, seed development is shown in a time scale of days (days after pollination, DAP) with seed maturation from 12 to 48 DAP, comprising seed filling and late maturation stages. Germination percentage, dry weigh and water content are shown on the left y-axis. Longevity is shown by the P50 value on the right y-axis. Compounds such as legumins, vicilins, stachyose (STACH) and sucrose (SUC) are indicated with their corresponding amounts at each time-points. DT acquisition is marked in red, and other metabolic and physiologic processes such as longevity, dormancy, sucrose, RFO, and storage proteins accumulation are indicated above the x-axis by black arrows. Data obtained in Vandecasteele et al. (2011), Verdier et al. (2013) and Righetti et al. (2015)

On the other hand, seed desiccation tolerance is acquired at early seed maturation from 22 to 28 DAP and represents the capacity of a seed to endure extreme dehydration and still be capable of germination and survival. This process is followed by the acquisition of seed longevity, which is the ability of a seed to be viable for a long period of time in the dry state. Therefore, desiccation tolerance is a prerequisite to longevity and long-term survival in the dry state (Oliver et al. 2000; Hoekstra et al. 2001). Both seed longevity and desiccation tolerance are acquired during seed maturation and specific protective mechanisms are necessary to the acquisition of these traits. Even if both mechanisms are slightly overlapping during the late maturation stage, seed desiccation tolerance is acquired early (from 22 to 28 DAP), then, followed by longevity which starts to be acquired at 28 DAP onwards. Within the acquisition of these processes, we observed an increase of the RFO metabolism with raffinose and stachyose increasing from 24 to 36 DAP at the expense of myo-inositol and sucrose (Fig. 12.1). The activation of the Raffinose Oligosaccharide Family (RFO) metabolism and the accumulation of the RFO molecules, such as stachyose in Medicago, are essential molecules for seed survival in the dry state to scavenge reactive oxygen species (ROS) and to replace water by the formation of a glassy state during desiccation tolerance to preserve cellular structures (i.e., allowing less molecular mobility inside the cells and by filling free volumes between molecules) (Verdier et al. 2013). RFOs act in complement of LEA (Late Embryogenesis Abundant) proteins, which also accumulate during water loss and act as a hydration buffer, therefore, protecting cellular membranes and other proteins (Tunnacliffe and Wise 2007). During seed maturation, sucrose, which is the transport form of carbohydrate and precursor of RFO, has a decreased accumulation from 24 DAP until 48 DAP and this decrease is correlated to the increase in longevity (Verdier et al. 2013). On the opposite, the increase in stachyose, a major soluble RFO in *M. truncatula* seed, is positively correlated with the increase in seed longevity. Stachyose increased slightly from 24 DAP and sharply from 32 and 36 DAP (Vandecasteele et al. 2011; Rosnoblet et al. 2007; Verdier et al. 2013). Lastly, at 44 DAP, pod abscission marks seed maturation stage, also called quiescent state, in which seeds display 5–10% of their water content (WC).

Taken together, seed maturation processes have the responsibility of ensuring the acquisition of mechanisms that will guarantee seed survival and germination for the future of the species and are key processes to ensure species propagation (Fenner and Thompson 2005; Gaff and Oliver 2013). Here, we will explore further the desiccation tolerance mechanism, from physiological and molecular perspectives.

12.2 Desiccation Tolerance

12.2.1 DT, Physiological Aspects

Desiccation tolerance (DT) is the capacity of an organism or tissue to endure extreme desiccation and yet survive and thrive. This means that desiccation tolerant tissues can undergo a dehydration less than 0.1 g of H_2O per gram of dry weight, or less than 10% of their dry weight, without accumulation of lethal damages (Ooms et al. 1993; Alpert 2005). This extreme survival mechanism was the keystone for plants to leave water-based environments and consequently colonize and conquer dry land (Linkies et al. 2010). The evolution of DT began possibly by the appearance of this mechanism in chlorophytic algae, and later with its complete development on Bryophytes, i.e., mosses and hornworts, to limit water loss in subaquatic habitats where dry events could occur (Leprince et al. 2016; Proctor et al. 2007; Costa et al. 2017). Whereas the majority of Bryophytes is desiccation tolerant, in vascular plants DT is nonexistent in gymnosperms and extremely rare on the vegetative tissues of angiosperms. Evolutionary studies showed that DT is absent in the vegetative tissues of basal angiosperms but is found in later angiosperms lineages, demonstrating that DT capacity was lost early in angiosperms evolution and regained later, at least in 10 different evolutive events, mainly on herbaceous lineages (Oliver et al. 2000, 2005). There are currently reported 330 species of angiosperms that present DT on their vegetative tissues, the so-called resurrection plants, which represents less than 0.2% of the total number of angiosperms (Proctor et al. 2007). On the contrary, the majority of angiosperms studied so far maintained DT in their vegetative propagules, producing seeds (95%) and also pollen (87%) that are desiccation tolerant (Gaff and Oliver 2013; Alpert 2006).

Indeed, seeds are mainly desiccation tolerant (DT) organisms (called orthodox seeds), but some are desiccation sensitive (DS) (called recalcitrant seeds) (Loi et al. 2013). Clearly, the maintenance of DT in the seeds is an evolutive advantage that permits plants to pass their genes in a safe manner by creating this shielding container, the seed, for protecting their heritage during extended periods (Dekkers et al. 2015). Therefore, orthodox seeds on dry state can survive a large range of environmental condition for many years or even centuries, ensuring their possible dispersion and germination on optimal conditions (Leprince et al. 2016; Dekkers et al. 2015). Fortunately, the main plant families of agronomical importance, Poaceae and Fabaceae, possess DT seeds. This ensures that staple crops seeds such as wheat, rice, corn, beans and soybean can be stored in dry state for long periods of time. Nevertheless, a large number of wet climate plant species have DS seeds, rendering impossible storability (Berjak and Pammenter 2008).

Desiccation tolerance is acquired during seed development, more specifically at the early stages of seed maturation, concomitantly with the filling phase (Figs. 12.1 and 12.2), from around 23 DAP in *M. truncatula*. Upon seed imbibition and germination, desiccation tolerance is lost along with water content increase and the continuity of metabolic processes that lead to the emergence of radicle and cotyledon development (Buitink et al. 2003, 2006; Maia et al. 2011).

Nonetheless, DT can also be re-induced after radicle protrusion. There is a short temporal window to re-induce desiccation tolerance (called DT window). In *M. truncatula*, 1–3 mm radicle length corresponded to the developmental stage, which DT can be re-induced by applying a mild osmotic stress using polyethylene glycol (PEG) or by using the phytohormone ABA (Buitink et al. 2003; Maia et al. 2011, 2014; Terrasson et al. 2013). This DT re-induction has been demonstrated for a number of species, including *Cucumis sativus, Impatiens walleriana, Medicago truncatula, Tabebuia impetiginosa* (Brazilian tree species) and also *Arabidopsis thaliana* (Bruggink and van der Toorn 1995; Buitink et al. 2003, 2006; Maia et al. 2011, 2014; Vieira et al. 2010).

The acquisition of DT is an active process, which takes developmental time to become effective, but it is a highly stable physiologic process. DT acquisition efficiency is not affected by abiotic stresses, such as heat and osmotic stresses (Righetti et al. 2015). When *M. truncatula* plants were grown at 26 °C (heat) or with a soil water potential at 20 MPa (osmotic stress), DT was acquired with the same timing and efficiency compared to 20 °C (control temperature) and standard watering conditions. Comparatively, the timing and the efficiency of longevity acquisition is influenced both by heat and by

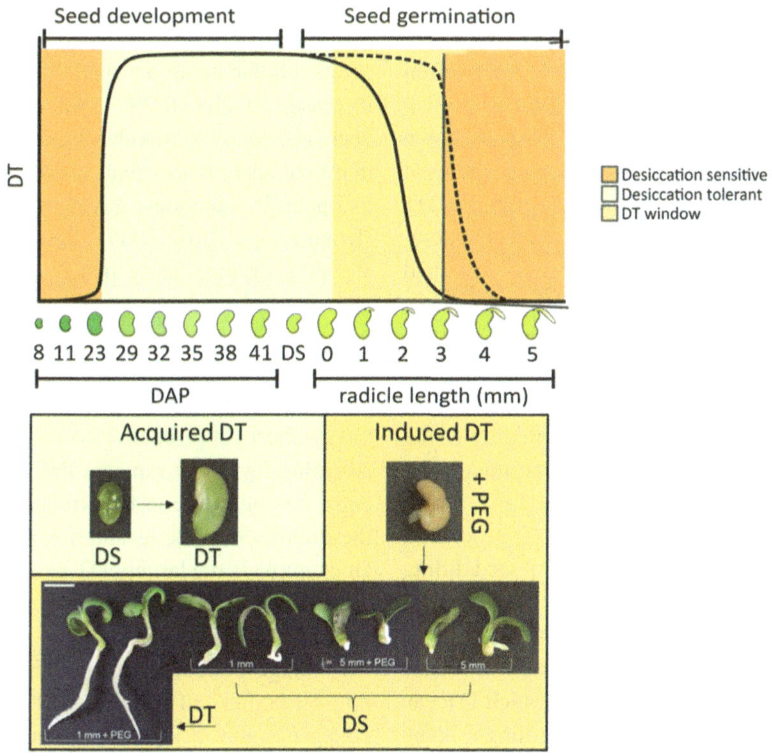

Fig. 12.2 Desiccation tolerance (DT) on *M. truncatula*. The upper panel shows a timeline of seed development and seed germination and its relation to desiccation. Desiccation sensitive stages are at the beginning of seed development and later upon germination, represented in dark-yellow. DT acquisition and loss are represented by a full line curve and desiccation tolerant tissues comprise the beige portion. The DT window is represented by the light-yellow portion and by the dashed curve. The panel below shows the division between acquired DT, during seed development, and induced DT, during seed germination with the application of a mild osmotic stress (PEG). The phenotype of germination seeds that were dried at 1 or 5 mm of radicle growth, and treated or not with PEG is shown below. Desiccation tolerant tissues (DT), desiccation sensitive tissues (DS)

osmotic stress which suggested that longevity and DT acquisition mechanisms have different regulatory processes (Righetti et al. 2015).

12.2.2 DT, Molecular Processes

Understanding the molecular processes of DT acquisition is a difficult task because DT overlaps with other seed maturation mechanisms (Fig. 12.1) (Sano et al. 2016; Leprince et al. 2016). To tackle this problem, *M. truncatula* is a great model to study how and why DT is acquired. The use of this legume model is justified by three points: i) Under standard temperature conditions, *M. truncatula* seed development lasts 48 DAP, including about 30 days related to seed maturation, in contrast to 5–6 days in *A. thaliana*, making it a better model to temporally dissociate between different seed maturation processes (Verdier et al. 2013; Leprince et al. 2016); (ii) environmental conditions was shown to strongly influence the developmental timing of *M. truncatula* seeds without affecting DT acquisition (Righetti et al. 2015). High temperature stress during seed development shortened seed maturation to 15 days after pollination, which impacts timings of accumulation of storage molecules, acquisition of longevity and dormancy, without impacting the timing or efficiency of DT acquisition. Therefore, by using sub-optimal conditions, DT could be uncoupled from other seed maturation processes, which provide a clearer dissection of DT processes, essentially by uncoupling DT from seed filling and acquisition of longevity molecular processes; (iii) Another major finding is the existence of the developmental window of desiccation tolerance (DT window) after germination in which DT can be re-induced in *M. truncatula* seedlings, being fairly easy to identify correct radicle size because of Medicago seed size compared to other model plants. All these advantages promoted the development of numerous studies of DT in *M. truncatula,* which investigated DT-associated genes and DT-required metabolic pathways.

Molecular processes during seed-acquired DT

Desiccation tolerance is one of the most important mechanisms that seeds need to put in place during seed maturation to be able to survive drying and for later survival at a dry state, and for this process to be well achieved, there is a need of metabolism coordination and regulation during desiccation. The coordination of preconditioning mechanisms, such as osmotic stress prior drying, a slow drying, and a control of development stages are necessary to acquire proper desiccation tolerance (Leprince et al. 2016). The avoidance of oxidative stress during drying and in the dry state is another important characteristic of DT capable plants (Leprince and Buitink 2015). As an example, the resurrection plant *Xerophyta viscosa* can diminish photosynthetic activity and decrease the generation of reactive oxygen species (Costa et al. 2017). To avoid cellular stress, antioxidant defenses need to increase, while ROS production needs to decrease. Many antioxidants are put in place for that goal such as ascorbate, glutathione, polyols, tocopherols, quinones, flavonoids and phenolics (Kranner and Birtić 2005). This happens clearly at seed drying, in which general metabolic activity and also photosynthesis is extremely reduced, along with ROS production (Bewley et al. 2013; Pammenter and Berjak 1999).

Along with stress avoidance, during dry state, the synthesis of protective compounds acts like a shielding layer that replaces the water molecules gaps, having a role in the structural state and on the stability of proteins and membranes function, an example is the largely DT-associated proteins, the LEAs (Costa et al. 2015; Leprince and Buitink 2015). Moreover, vacuoles will encapsulate the storage proteins, while lipids will be closed into oil bodies, all of those will prevent cellular collapse upon drying by filling up the cells (Buitink et al. 2000; Sano et al. 2016). As DT is related to protective compounds and constraints growth, it is hypothesized that the loss of DT on angiosperms vegetative part evolved probably due to the selection pressure during plant dry land conquest, which favored species with faster growth and larger height, therefore, displaying high dry-mass production (Alpert 2005, 2006; Illing et al. 2005; Oliver et al. 2000). Even with all the protective mechanisms in place, there will still be damaging occurring upon re-imbibition/rehydration. Therefore, molecules involved in repair mechanisms help to prevent damages. DT-associated repair molecules are synthesized before drying and/or during rehydration, as showed in bryophytes (Bewley et al. 2013; Oliver et al. 2005). Studies suggested that certain LEA proteins played a role in protective mechanisms during rehydration, with one seed-specific LEA protein from *M. truncatula*, PM25 that was shown to be capable of efficiently dissociate desiccation-induced protein aggregates (Boucher et al. 2010). Interestingly, during seed drying at late maturation stage, chromatin was shown to be more compact promoting nuclear size reduction, and upon germination, chromatin will be decompressed and nuclear size will return to normal (Van Zanten et al. 2011). DT loss has also been demonstrated to correlate with activation of cell cycle, cell division and DNA synthesis, however, difficult to dissociate between causes or consequences (Osborne and Boubriak 1994; Sargent et al. 1981).

Molecular processes during re-induced DT in germinating plantlets

As mentioned earlier, re-induction of DT is possible at early germination steps following a mild osmotic stress induced by PEG. It has been shown that PEG induces membrane modification, inhibits radicle growth, down-regulates genes related to energy metabolism and cell wall modification, up-regulates genes related to antioxidant activity, response to stress and seed storage, and induces synthesis of protective molecules, such as non-reducing sugars such RFOs and protective proteins, such as late embryogenesis abundant (LEA) and heat shock proteins (HSPs) (Buitink et al. 2003; Maia et al. 2011). The sucrose that accumulates in desiccation tolerant radicles is produced by mobilizing lipids and starch that were accumulated during seed filling (Buitink et al. 2006).

During the DT window, DNA synthesis and cell division were absent up to a radicle length of 2 mm. The restart of DNA synthesis and cell division were first detected in radicles with a length of 3 mm. The DT window closed entirely after 5 mm of radicle growth, in *M. truncatula,* therefore radicles completely lose DT, becoming desiccation sensitive (Fig. 12.2). Not even osmotic stress nor ABA application at this stage can reverse desiccation sensitivity (Buitink et al. 2003; Maia et al. 2011; Leprince et al. 2000; Vieira et al. 2010). Interestingly, during the DT window, it is possible to observe that different seed tissues possess different timing of DT window closing. Upon germination, *M. truncatula* radicles closed their DT window after 5 mm of growth, which is around 22 h from imbibition, but cotyledons closed their DT window 40 h after imbibition, being considerably more resistance to desiccation than radicles (Buitink et al. 2003). For seed establishment, a larger DT window can be a major ecological advantage for a proper plantlet emergence occurring during a hydric stress event.

Core DT-associated genes

The plant molecular mechanisms involved in desiccation tolerance have been extensively studied during the past two decades, in both vegetative parts and seeds, aiming to understand DT from physiological and molecular points of views (Ooms et al. 1993; Verdier et al. 2013; Righetti et al. 2015). These works allowed the development of a robust screening model and the identification of numerous DT-associated genes, mainly conferring protective cell functions (Terrasson et al. 2013; Costa et al. 2015; Verdier et al. 2013; Delahaie et al. 2013; González-Morales et al. 2016). Terrasson et al. (2013) identified about 200 genes associated with the acquisition of desiccation tolerance and thus potentially involved in the DT mechanisms based on microarray analysis in *M. truncatula*. A few years later, a similar experiment in Arabidopsis revealed a change of 1,177 gene expression during the acquisition of DT (Costa et al. 2015). These genes encoded mainly for proteins involved in the prevention of oxidative damage (e.g., peroxiredoxins and catalases), DNA repair, maintenance of protein (e.g., heat shock proteins and LEA proteins), membrane structure and synthesis of soluble sugars (e.g., sucrose and raffinose family oligosaccharides, RFOs). Furthermore, on the molecular level, DT induction displayed an early and a late gene response in Arabidopsis (Costa et al. 2015) and in *M. truncatula* (Buitink et al. 2006). In Medicago, the early response genes during induced DT belong to classes involved in early stress and adaptation responses (Buitink et al. 2006). The up-regulated DT-induced late response genes were comparable to late seed maturation genes; and downregulated genes were mainly from cell cycle, biogenesis, primary and energy metabolism classes (Buitink et al. 2006). However, the individual impacts of these cell protective genes on the acquisition of DT were unclear. These mechanisms appeared to act in synergy probably with a large redundancy between them, due to the necessity to protect all cellular compartments and biomolecules (Dekkers et al. 2015). The functional characterization of DT associated-genes have been initiated but, to date, their exact functions are still unknown for a large proportion of them or are under investigation.

Even if from a physiological perspective, drought and desiccation tolerances have different plant adaptation strategies, (i.e., drought tolerance will prevent water losses, whereas desiccation tolerance will prepare cells to remain alive with an extremely low amount of water), from a molecular point of view, they have many similarities, since both involved ABA signal transduction elements and induction of some similar transcription factors. Several studies revealed that there are ABA-dependent and ABA-independent regulatory pathways leading to DT and drought, and that both pathways might be partially overlapped between drought response and DT, since several regulatory genes typically expressed during abiotic/drought stresses were also upregulated during seed maturation and DT acquisition (Buitink et al. 2003, 2006; Maia et al. 2011). In DT, ABA treatment can alone re-induce DT on Arabidopsis germinated seeds which shows its effect during the DT window, and this induction is more dependent on ABA perception and signaling than on ABA content (Maia et al. 2011, 2014). Nevertheless, ABA appeared to be less important for DT acquisition during seed maturation since ABA sensitive or ABA synthesis defective mutants displayed DT seeds. However, these mutants were, often, impaired for DT re-induction at early germination such in *abi5* mutants in Arabidopsis and Medicago (Maia et al. 2014; Terrasson et al. 2013; Lopez-Molina et al. 2001). Another important point is that ABA synthesis and signaling genes are present in basal land plants, suggesting the role of this stress hormone in the evolution of drought and desiccation tolerance acquisition (Hauser et al. 2011; Ruszala et al. 2011).

An interesting work from Buitink et al. (2006) was able to characterize the regulatory processes and protective mechanisms that were induced or repressed both during the DT acquisition in seed maturation and during the DT re-induction at early germination, allowing to identify potential core DT-associated genes (i.e., necessary for DT). Microarrays data were used to identify common transcripts expressed during seed maturation tissues (before and after DT acquisition at 14 DAP versus 20 DAP) and upon germination with mild osmotic stress by PEG (72 h PEG-incubated 3-mm long germinated radicles versus non-treated radicles). The overlap of genes in these two DT states gave insight into the biological processes that are necessary to ensure DT in *M. truncatula*. A total of 267 genes were identified to be differentially expressed in both acquired and induced DT. From those, 169 genes were up-regulated and 98 genes were found downregulated (Fig. 12.3). Concerning the functional classes of the up-regulated genes, the LEA proteins appeared as one of the enriched functional classes, as well as hormones synthesis of cytokines and ethylene. In summary, enriched functional classes were from genes related to stress responses, response to abscisic acid stimulus, lipid localization, and seed development. The enriched downregulated processes were mainly related to response to heat stress, biogenesis, cell structure and cell transport. Furthermore, lipid metabolism was also impaired (Fig. 12.3). Outwardly, the ability to decrease metabolic activity is a prerequisite to desiccation

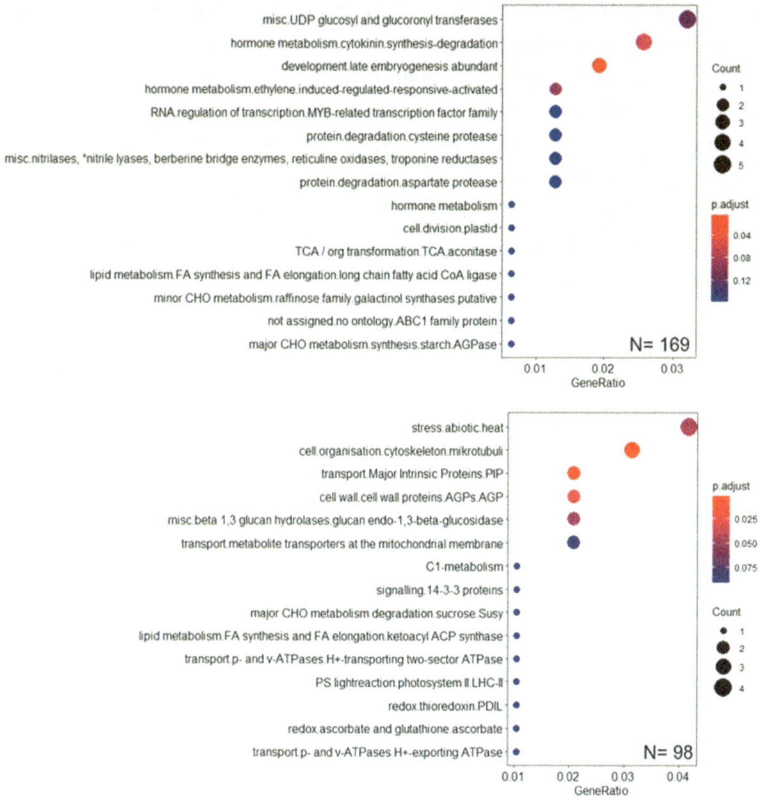

Fig. 12.3 ***Medicago truncatula*** **DT-associated genes functional classes, both in acquired and induced DT**. Differentially expressed upregulated (169) and downregulated (98) genes were used to perform enrichment analysis of Mapman functional categories using Cluster Profiler package, using data from Buitink et al. (2006). Dot color indicates p-values of enrichment analysis. Dot size indicates count numbers of genes belonging to different classes, with bigger dots having greater significance

tolerance acquisition as is the expression of DT-associated genes (Buitink et al. 2006). Similar transcriptomic approaches were also performed to identify set of upregulated genes during DT in Arabidopsis (Maia et al. 2011; Terrasson et al. 2013) and from a different study in *M. truncatula* (Terrasson et al. 2013). This latter work identified four times more downregulated (2829) than upregulated (740) genes. This study showed that upon induction of DT in germinated seeds, their tissues revert, at least in part, to an earlier developmental stage, (Buitink et al. 2006). Finally, another study constructed a co-expression network from *M. truncatula* seed data and identified different gene modules related to embryogenesis, seed filing, DT and seed drying (Verdier et al. 2013). The DT specific module comprised ABA-induced genes, LEAs and other stress related response corroborating with previous DT-associated genes identifications (Verdier et al. 2013).

12.2.3 Desiccation Tolerance Switch

As mentioned earlier, the ability to tolerate desiccation is acquired during the maturation stage of seed development, and gradually lost upon germination, until complete loss after 5 mm of radicle growth. Even if the physiological and molecular mechanisms involved in DT have been documented, the activation/repression of this major evolutive characteristic is still to be revealed.

The molecular mechanisms initiated during seed maturation are dependent of the LAFL development network, which is a transcriptional network controlled by the genes *LEC1* (*LEAFY COTYLEDON 1*), *ABI3* (*ABSCISIC ACID INSENSITIVE 3*), *FUS3* (*FUSCA 3*) and *LEC2* (*LEAFY COTYLEDON 2*) (Jia et al. 2014; Tian et al. 2020; Boulard et al. 2017). From these four transcription factors, three genes, *LEC2, ABI3* and *FUS3*, contain B3 domains, whereas

LEC1 is a CCAAT-box binding factor from the HAP3 family (Giraudat et al. 1992; Lotan et al. 1998; Stone et al. 2001). The abnormal expression of any of these genes has been shown to produce severe seed maturation phenotypes in Arabidopsis including loss of DT. Even if similar results could be observed in *M. truncatula*, to date, only *MtABI3* has been demonstrated to be essential in DT. Indeed, Medicago *abi3* mutants failed to acquire DT and did not survive seed drying, leading to loss of seed storage capability (Delahaie et al. 2013). Authors showed that for instance, they observed major defects in LEA protein accumulation in mutant lines (Delahaie et al. 2013). They also observed impairment in chlorophyll degradation, giving green color to mature seeds. Unfortunately, as described in Arabidopsis, *ABI3* is not specific to desiccation tolerance mechanism and represents a master regulator of seed maturation, broadly involved in many seed maturation processes, such as accumulation of different storage molecules, chlorophyll degradation, vascular tissue development, sensitivity to ABA (Santos-Mendoza et al. 2008; Costa et al. 2015; Nambara et al. 1992; Delmas et al. 2013).

Interestingly, the reprogramming of seed and pollen DT mechanisms was the genetic source for DT implementation to vegetative tissues (Gaff and Oliver 2013; Illing et al. 2005; Oliver et al. 2000). Indeed, DT is progressively lost early after germination, but can be induced up to a specific developmental stage at early germination via the application of a mild osmotic stress. During DT induction, similar mechanisms to those operating during seed maturation are activated (Buitink et al. 2003, 2006). These observations suggest that a molecular switch is able to de-repress the DT program upon an external environmental cue (i.e., mild osmotic stress) at early germination, then to definitely repress the DT program during seedling establishment. Recent studies confirmed this assumption as it has been recently shown that resurrection plants re-wired seed DT mechanisms in their vegetative tissues in response to water losses (VanBuren et al. 2017; Costa et al. 2017). Our unpublished preliminary studies revealed that the majority of putative DT-associated genes were mainly epigenetically repressed in desiccation sensitive tissues by a specific epigenetic mark. Indeed, out of 102 putative DT-associated genes, preliminary, identified using microarray studies from *A. thaliana* and *M. truncatula* (Buitink et al. 2006; Costa et al. 2015), 82% were specifically associated to the repressive histone mark, H3K27me3. These data suggest that the switch that activates and terminates the DT program might be firstly regulated at this epigenetic level. To complete these observations, a chromatin immunoprecipitation, followed by qPCR (ChIP-qPCR) was performed using H3K27me3 antibodies and confirmed that *ABI3* gene sequences were enriched in leaf and at early seed development (i.e., before DT acquisition) in comparison to late seed maturation (i.e., after DT acquisition), which confirmed repression of this regulatory gene by the H3K27me3 histone mark. Based on these results, we concluded that the repressive histone mark H3K27me3 was responsible to permanently repress the DT program inactivation in leaf and before the DT acquisition in seeds, therefore, in desiccation-sensitive tissues.

In this review, we propose a hypothetical model for DT-core genes epigenetic control in *M. truncatula*, based on the data gathered in the literature (Fig. 12.4). In this model, we can visualize the chromatin state depending on the related desiccation phenotype. On the core DT-associated genes, chromatin can be in a more condensed state, being facultative heterochromatin state, which represses genes expression or in a decondensed state, as euchromatin state, promoting the expression of these genes. During the vegetative growth, *M. truncatula* leaves are sensitive to desiccation and present a facultative heterochromatin on the DT-core genes due to the repressive histone marks, such as H3K27me3. On the other hand, during seed development, desiccation tolerance is acquired due to the expression of DT-core genes that are in euchromatin state due to the decrease and/or absence of repressive histone marks H3K27me3, and with an increased amount of active histone marks (i.e., DT switch ON). The existence of the DT window, which is capable of re-inducing the DT program at early germination following external environmental cue (i.e. mild osmotic stress) is a very interesting model regarding the mode of activation/repression of the DT switch and offers a wide variety of hypotheses regarding its regulation. Indeed, during this DT window, it is unlikely that strong repressive histone marks such as H3K27me3 repress DT-associated genes, as they are quickly re-inducible following mild osmotic stress, which suggests an intermediate repression of DT-associated genes to elucidate. Finally, after early germination phase (i.e., radicle length more than 5 mm in *M. truncatula*), the DT is switched OFF by the increasing the presence of repressive marks on core DT-associated genes, promoting sensitivity to desiccation (Fig. 12.4).

The confirmation or correction of these hypotheses on DT epigenetic control will certainly be brought to light in future works. Understanding how the DT switch is genetically turned ON will allow identification of potential pioneer. The concept of pioneer TFs was developed in the 1990s. Pioneer TFs are the first to engage in chromatin target sites (Mayran and Drouin 2018). Such initial binding can passively enhance transcription by reducing the number of additional factors that are needed to bind the DNA. In addition, pioneer TF binding can actively open the local chromatin and make it directly competent for binding of other TFs. Several animal pioneer TFs have been identified. However, the knowledge of plant pioneer TFs is much more limited and only four plant pioneer TFs have been yet identified to activate inflorescence meristem (Pajoro et al. 2014), flower organ specification (Sayou et al. 2016) and embryonic program (Tao et al. 2017) programs. The latter one is a seed-specific gene called *LEC1*, whose role as pioneer factor was recently revealed (Tao et al. 2017), which might

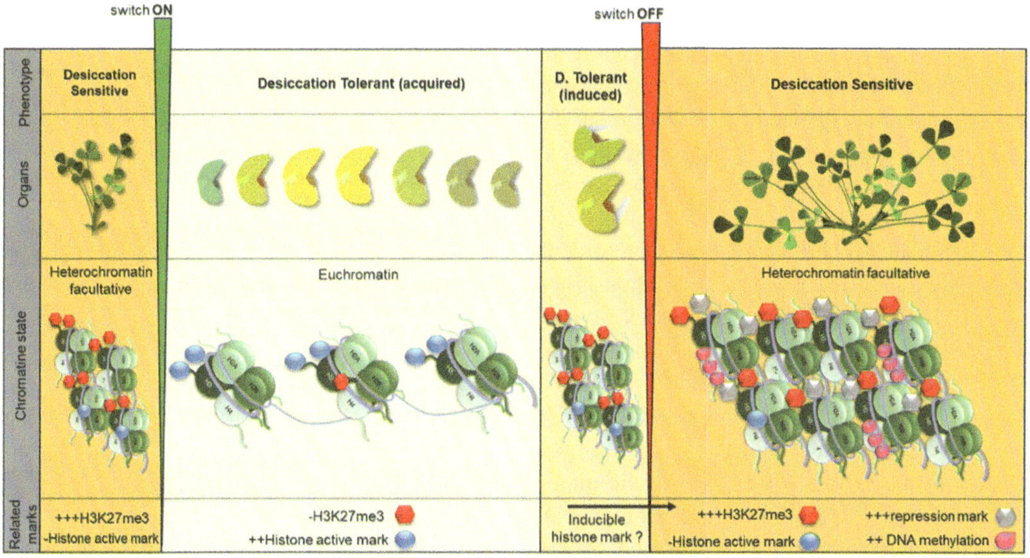

Fig. 12.4 Hypothetical model of epigenetic regulation of core DT-associated gene. On this panel, the DT-related phenotype, organs, chromatin state and epigenetic related marks are indicated on the left side. Desiccation sensitive stages are marked in dark-yellow, while desiccation tolerant stages are marked in beige. Concerning chromatin state, the nucleosomes are represented by histone octamers (green) and DNA is represented by gray lines. The histone marks are positioned on histone tails. The switch on and off are positioned at the beginning of DT acquisition in seeds and at the end of the DT window during germination, respectively

or might not play also a role in the activation of the DT switch and chromatin accessibility of DT-associated genes.

12.3 Conclusions

Drought and water scarcity are the greatest threats to agriculture worldwide. Projections indicate that in the next 30–90 years dry seasons will become more prevalent, more intense and longer-lasting, destabilizing food systems and threatening local to global food security. Staple crops do not possess vegetative parts that can survive extended periods of drought. However, their seeds are tolerant to desiccation. This means that theses seeds have the capacity to endure extreme dry and are capable of rehydration and resume of metabolic processes. For this to occur, a specify cellular machinery must be activated to limit cell damage during desiccation and to maintain its physiological integrity in dry state by enhancing molecular mechanisms for damage repair upon rehydration. In this context, scientists have been urged to understand desiccation tolerance mechanisms for selecting crop varieties that are more stress tolerant and able to survive under water-scarce conditions. The vast body of knowledge accumulated resulted in the identification of numerous DT-associated genes conferring protective cell functions. Given that staple crops possess all the genetic information necessary for desiccation tolerance, but only activate it in their seeds at a specific developmental stage, the comprehension of the epigenetic control of DT still needs efforts. In the last few years, new tools have become available enabling the investigation of the yet unexplored epigenetic molecular regulation of DT. By fully understanding DT regulation and function, several pivotal agricultural issues, such as food security and crop adaptation to climate change can be better inferred, creating a potential improvement of plant stress tolerance, and conservation of genetic resources, with easier management of desiccation sensitive seeds.

References

Alpert P (2006) Constraints of tolerance: why are desiccation-tolerant organisms so small or rare? J Exp Biol 209:1575–1584

Alpert P (2005) The limits and frontiers of desiccation-tolerant life. Integr Comp Biol 45:685–695

Berger F, Grini PE, Schnittger A (2006) Endosperm: an integrator of seed growth and development. Curr Opin Plant Biol 9:664–670

Berjak P, Pammenter N (2008) From Avicennia to Zizania: seed recalcitrance in perspective. Ann Bot 101:213–228

Bewley JD, Bradford KJ, Hilhorst HWM, Nonogaki H (2013) Seeds: physiology of development, germination and dormancy, 3rd edn

Boucher V, Buitink J, Lin X, Boudet J, Hoekstra FA, Hundertmark M, Renard D, Leprince O (2010) MtPM25 is an atypical hydrophobic late embryogenesis-abundant protein that dissociates cold and desiccation-aggregated proteins. Plant Cell Environ 33:418–430

Boulard C, Fatihi A, Lepiniec L, Dubreucq B (2017) Regulation and evolution of the interaction of the seed B3 transcription factors with NF-Y subunits. Biochim Biophys Acta—Gene Regul Mech 1860:1069–1078

Bruggink T, van der Toorn P (1995) Induction of desiccation tolerance in germinated seeds. Seed Sci Res 5:1–4

Buitink J, Leger JJ, Guisle I, Vu BL, Wuillème S, Lamirault G, Bars AL, Meur NL, Becker A, Küster H, Leprince O (2006) Transcriptome profiling uncovers metabolic and regulatory processes occurring during the transition from desiccation-sensitive to desiccation-tolerant stages in *Medicago truncatula* seeds. Plant J 47:735–750

Buitink J, Leprince O, Hoekstra FA (2000) Dehydration-induced redistribution of amphiphilic molecules between cytoplasm and lipids is associated with desiccation tolerance in seeds. Plant Physiol 124:1413–1425

Buitink J, Ly VB, Satour P, Leprince O (2003) The re-establishment of desiccation tolerance in germinated radicles of Medicago truncatula Gaertn seeds. Seed Sci Res 13:273–286

Costa M-CD et al (2017) A footprint of desiccation tolerance in the genome of Xerophyta viscosa. Nat. Plants 3:17038

Costa MCD, Righetti K, Nijveen H, Yazdanpanah F, Ligterink W, Buitink J, Hilhorst HWM (2015) A gene co-expression network predicts functional genes controlling the re-establishment of desiccation tolerance in germinated Arabidopsis thaliana seeds. Planta 242:435–449

Dekkers BJW, Costa MCD, Maia J, Bentsink L, Ligterink W, Hilhorst HWM (2015) Acquisition and loss of desiccation tolerance in seeds: from experimental model to biological relevance. Planta 241:563–577

Delahaie J, Hundertmark M, Bove J, Leprince O, Rogniaux H, Buitink J (2013) LEA polypeptide profiling of recalcitrant and orthodox legume seeds reveals ABI3-regulated LEA protein abundance linked to desiccation tolerance. J Exp Bot 64:4559–4573

Delmas F, Sankaranarayanan S, Deb S, Widdup E, Bournonville C, Bollier N, Northey JGB, McCourt P, Samuel MA (2013) ABI3 controls embryo degreening through Mendel's i locus. Proc Natl Acad Sci USA 110:1–7

Doktorgrades E, Ungru AH (2010) Analysis of communication pathways during seed development in Arabidopsis thaliana

Fenner M, Thompson K (2005) The ecology of seeds. Cambridge University Press

Finch-Savage WE, Leubner-Metzger G (2006) Seed dormancy and the control of germination. New Phytol 171:501–523

Gaff DF, Oliver M (2013) The evolution of desiccation tolerance in angiosperm plants: a rare yet common phenomenon. Funct Plant Biol 40:315–328

Gallardo K, Firnhaber C, Zuber H, Héricher D, Belghazi M, Henry C, Küster H, Thompson R (2007) A combined proteome and ranscriptome analysis of developing *Medicago truncatula* seeds: evidence for metabolic specialization of maternal and filial tissues. Mol Cell Proteomics 6:2165–2179

Gallardo K, Le Signor C, Vandekerckhove J, Thompson RD, Burstin J (2003) Proteomics of *Medicago truncatula* seed development establishes the time frame of diverse metabolic processes related to reserve accumulation. Plant Physiol 133:664–682

Giraudat J, Hauge BM, Valon C, Smalle J, Parcy F, Goodman HM (1992) Isolation of the Arabidopsis ABI3 gene by positional cloning. Plant Cell 4:1251–1261

González-Morales SI, Chávez-Montes RA, Hayano-Kanashiro C, Alejo-Jacuinde G, Rico-Cambron TY, de Folter S, Herrera-Estrella L (2016) Regulatory network analysis reveals novel regulators of seed desiccation tolerance in Arabidopsis thaliana. Proc Natl Acad Sci 113:E5232–E5241

Hauser F, Waadt R, Schroeder JI (2011) Evolution of abscisic acid synthesis and signaling mechanisms. Curr Biol 21:R346–R355

Hilhorst HWM (2007) Definitions and hypotheses of seed dormancy. Seed Dev Dormancy Germination 50–71

Hoekstra FA, Golovina EA, Buitink J (2001) Mechanism of plant desiccation tolerance. Trends Plant Sci 6:431–438

Illing N, Denby KJ, Collett H, Shen A, Farrant JM (2005) The signature of seeds in resurrection plants: a molecular and physiological comparison of desiccation tolerance in seeds and vegetative tissues. Integr Comp Biol 45:771–787

Jia H, Suzuki M, McCarty DR (2014) Regulation of the seed to seedling developmental phase transition by the LAFL and VAL transcription factor networks. Wiley Interdiscip Rev Dev Biol 3:135–145

Jiang WB, Lin WH (2013) Brassinosteroid functions in arabidopsis seed development. Plant Signal Behav 8

Kranner I, Birtić S (2005) A modulating role for antioxidants in desiccation tolerance. Integr Comp Biol 45:734–740

Leprince O, Buitink J (2015) Introduction to desiccation biology: from old borders to new frontiers. Planta 242:369–378

Leprince O, Harren FJM, Buitink J, Alberda M, Hoekstra FA (2000) Metabolic dysfunction and unabated respiration precede the loss of membrane integrity during dehydration of germinating radicles. Plant Physiol 122:597–608

Leprince O, Pellizzaro A, Berriri S, Buitink J (2016) Late seed maturation: drying without dying. J Exp Bot 68:827–841

Linkies A, Graeber K, Knight C, Leubner-Metzger G (2010) The evolution of seeds. New Phytol 186:817–831

Loi P, Iuso D, Czernik M, Zacchini F, Ptak G (2013) Towards storage of cells and gametes in dry form. Trends Biotechnol 31:688–695

Lopez-Molina L, Mongrand S, Chua NH (2001) A postgermination developmental arrest checkpoint is mediated by abscisic acid and requires the ABI5 transcription factor in Arabidopsis. Proc Natl Acad Sci USA 98:4782–4787

Lotan T, Ohto MA, Matsudaira Yee K, West MAL, Lo R, Kwong RW, Yamagishi K, Fischer RL, Goldberg RB, Harada JJ (1998) Arabidopsis LEAFY COTYLEDON1 is sufficient to induce embryo development in vegetative cells. Cell 93:1195–1205

Maia J, Dekkers BJW, Dolle MJ, Ligterink W, Hilhorst HWM (2014) Abscisic acid (ABA) sensitivity regulates desiccation tolerance in germinated Arabidopsis seeds. New Phytol 203:81–93

Maia J, Dekkers BJW, Provart NJ, Ligterink W, Hilhorst HWM (2011) The re-establishment of desiccation tolerance in germinated arabidopsis thaliana seeds and its associated transcriptome. PLoS One 6

Malabarba J, Buffon V, Mariath JEA, Gaeta ML, Dornelas MC, Margis-Pinheiro M, Pasquali G, Revers LF (2017) The MADS-box gene agamous-like 11 is essential for seed morphogenesis in grapevine. J Exp Bot 68:1493–1506

Mayran A, Drouin J (2018) Pioneer transcription factors shape the epigenetic landscape. J Biol Chem 293:13795–13804

Nambara E, Naito S, McCourt P (1992) A mutant of Arabidopsis which is defective in seed development and storage protein accumulation is a new abi3 allele. Plant J 2:435–441

Noguero M et al (2015) DASH transcription factor impacts *Medicago truncatula* seed size by its action on embryo morphogenesis and auxin homeostasis. Plant J 81:453–466

Oliver MJ, Tuba Z, Mishler BD (2000) The evolution of vegetative desiccation tolerance in land plants. Plant Ecol 151:85–100

Oliver MJ, Velten J, Mishler BD (2005) Desiccation tolerance in bryophytes: a reflection of the primitive strategy for plant survival in dehydrating habitats? Integr Comp Biol 45:788–799

Ooms J, Leon-Kloosterziel KM, Bartels D, Koornneef M, Karssen CM (1993) Acquisition of desiccation tolerance and longevity in seeds of Arabidopsis thaliana (A Comparative Study Using Abscisic Acid-Insensitive abi3 Mutants). Plant Physiol 102:1185–1191

Osborne DJ, Boubriak II (1994) Dna and desiccation tolerance. Seed Sci Res 4:175–185

Pajoro A et al (2014) Dynamics of chromatin accessibility and gene regulation by MADS-domain transcription factors in flower development. Genome Biol 15

Pammenter NW, Berjak P (1999) A review of recalcitrant seed physiology in relation to desiccation-tolerance mechanisms. Seed Sci Res 9:13–37

Proctor MCF, Oliver MJ, Wood AJ, Alpert P, Stark LR, Cleavitt NL, Mishler BD (2007) Desiccation-tolerance in bryophytes: a review. Bryologist 110:595–621

Raven, P.H., Evert, R.F., and Eichhorn, S.E. (2007). Biology of Plants.

Righetti K, Vu JL, Pelletier S, Vu BL, Glaab E, Lalanne D, Pasha A, Patel RV, Provart NJ, Verdier J, Leprince O, Buitink J (2015) Inference of longevity-related genes from a robust coexpression network of seed maturation identifies regulators linking seed storability to biotic defense-related pathways. Plant Cell 27:tpc.15.00632

Rosnoblet C, Aubry C, Leprince O, Vu BL, Rogniaux H, Buitink J (2007) The regulatory gamma subunit SNF4b of the sucrose non-fermenting-related kinase complex is involved in longevity and stachyose accumulation during maturation of *Medicago truncatula* seeds. Plant J 51:47–59

Ruszala EM, Beerling DJ, Franks PJ, Chater C, Casson SA, Gray JE, Hetherington AM (2011) Land plants acquired active stomatal control early in their evolutionary history. Curr Biol 21:1030–1035

Sano N, Rajjou L, North HM, Debeaujon I, Marion-Poll A, Seo M (2016) Staying alive: molecular aspects of seed longevity. Plant Cell Physiol 57:660–674

Santos-Mendoza M, Dubreucq B, Baud S, Parcy F, Caboche M, Lepiniec L (2008) Deciphering gene regulatory networks that control seed development and maturation in Arabidopsis. Plant J 54:608–620

Sargent JA, Mandi SS, Osborne DJ (1981) The loss of desiccation tolerance during germination: an ultrastructural and biochemical approach. Protoplasma 105:225–239

Sayou C et al (2016) A SAM oligomerization domain shapes the genomic binding landscape of the LEAFY transcription factor. Nat Commun 7:11222

Stone SL, Kwong LW, Yee KM, Pelletier J, Lepiniec L, Fischer RL, Goldberg RB, Harada JJ (2001) LEAFY COTYLEDON2 encodes a B3 domain transcription factor that induces embryo development. Proc Natl Acad Sci USA 98:11806–11811

Tao Z, Shen L, Gu X, Wang Y, Yu H, He Y (2017) Embryonic epigenetic reprogramming by a pioneer transcription factor in plants. Nature 551:124–128

Terrasson E, Buitink J, Righetti K, Ly VB, Pelletier S, Zinsmeister J, Lalanne D, Leprince O (2013) An emerging picture of the seed desiccome: confirmed regulators and newcomers identified using transcriptome comparison. Front Plant Sci 4:1–16

Tian R, Wang F, Zheng Q, Niza VMAGE, Downie AB, Perry SE (2020) Direct and indirect targets of the arabidopsis seed transcription factor ABSCISIC ACID INSENSITIVE3. Plant J 103:1679–1694

Tunnacliffe A, Wise MJ (2007) The continuing conundrum of the LEA proteins. Naturwissenschaften 94:791–812

VanBuren R, Wai CM, Zhang Q, Song X, Edger PP, Bryant D, Michael TP, Mockler TC, Bartels D (2017) Seed desiccation mechanisms co-opted for vegetative desiccation in the resurrection grass oropetium thomaeum. Plant Cell Environ 40:2292–2306

Vandecasteele C et al (2011) Quantitative trait loci analysis reveals a correlation between the ratio of sucrose/raffinose family oligosaccharides and seed vigour in *Medicago truncatula*. Plant Cell Environ 34:1473–1487

Verdier J et al (2013) A regulatory network-based approach dissects late maturation processes related to the acquisition of desiccation tolerance and longevity of *Medicago truncatula* seeds. PLANT Physiol 163:757–774

Vieira CV, da Silva EAA, de Alvarenga AA, de Castro EM, Toorop PE (2010) Stress-associated factors increase after desiccation of germinated seeds of Tabebuia impetiginosa Mart. Plant Growth Regul 62:257–263

Wang XD, Song Y, Sheahan MB, Garg ML, Rose RJ (2011) From embryo sac to oil and protein bodies: embryo development in the model legume *Medicago truncatula*. New Phytol 193:327–338

Van Zanten M, Koini MA, Geyer R, Liu Y, Brambilla V, Bartels D, Koornneef M, Fransz P, Soppe WJJ (2011) Seed maturation in Arabidopsis thaliana is characterized by nuclear size reduction and increased chromatin condensation. Proc Natl Acad Sci USA 108:20219–20224

Functional Genomic and Genetic Studies of Organ Size Control in *Medicago truncatula*: An Overview

13

Zeyong Zhang and Rujin Chen

Abstract

Diverse sizes of leaves and seeds are present in plants. Even though the molecular mechanisms that underlie organ size control remain largely to be investigated, especially in crop species including legumes, in the past years, the developmental processes of vegetative and reproductive organs and several central regulatory genes have been studied in a number of plant species. These genes are involved in regulating cell proliferation and/or cell expansion. According to their functions and interactions, the encoded proteins are grouped into regulatory modules. In *Arabidopsis thaliana*, at least six regulatory modules or key genes have been identified, including DA1–ENHANCER OF DA1 (EOD1), GROWTH REGULATING FACTORs (GRFs)–GRF-INTERACTING FACTORs (GIFs), SWITCH3 (SWI3)-SUCROSE NON-FERMENTING (SNF), GIBBERELLIN INSENSITIVE DWARF2 (GID2)–DELLA, KLUH (KLU), and KINASE-INDUCIBLE INTERACTING8/9 (KIX8/9)-PEAPOD (PPD). Here, we present a brief overview of the current understanding of the role of the KIX-PEAPOD (PPD)-STERILE APETALA (SAP) module in regulating cell proliferation and expansion during organ growth in plants with an emphasis on the model legume *Medicago truncatula*.

13.1 Introduction

In nature, plant diversity is reflected to some extent by the diverse forms, sizes and arrangements of their leaves and fruits on plants. The shape and size of plant leaves and seeds are causally related to the size, number, and arrangement of individual cells in the organ (Gonzalez et al. 2012; Czesnick and Lenhard 2015; Li et al. 2019a). In addition, the development of vegetative and reproductive organs is different from each other and may also differs dependent on plant species. Leaf primordia are initiated from the periphery zone (PZ) of the shoot apical meristem (SAM) (Du et al. 2018). In most eudicot species, leaf formation involves the recruitment of founder cells to PZ of SAM, initiation of outgrowth of leaf primordia, formation of the adaxial–abaxial and proximal–distal axes, and initiation of lamina and intercalary growth of leaf blades (Du et al. 2018; Poethig and Sussex 1985; Donnelly et al. 1999; Nakata et al. 2012; Andriankaja et al. 2012; Nakata and Okada

Z. Zhang · R. Chen (✉)
School of Life Sciences, Lanzhou University, Lanzhou 730000, China
e-mail: rjchen@lzu.edu.cn

R. Chen
MOE Key Laboratory of Cell Activities and Stress Adaptations, Lanzhou University, Lanzhou 730000, China

S. Sinharoy et al. (eds.), *The Medicago truncatula Genome*, Compendium of Plant Genomes,
https://doi.org/10.1007/978-3-030-90757-0_13

2013). It has been shown that *microRNA 396* (*miR396*)-*GRFs* and *miR319-The Class II TEOSINTE BRANCHED1/CYCLOIDEA/PROLIFERATING CELL FACTORs* (*TCPs*) modules play antagonistic roles in cell proliferation and expansion during intercalary growth (Tsukaya 2018). NGATHA (NGA) and TCP transcription factors are involved in regulating leaf margin expansion and differentiation by inhibiting *WUSCHEL-RELATED HOMEOBOX* (*WOX*) gene expression (Alvarez et al. 2016). The transcription factors GRFs delay transition from cell proliferation to cell differentiation during intercalary growth (Kim et al. 2003; Rodriguez et al. 2010). The interacting GRFs-ANGUSTIFOLIA3 (AN3; also called GIF1) module modulates the expression levels of core cell cycle genes including *CYCLIN*s and promotes cell proliferation (Debernardi et al. 2014). TCP4 directly activates the expression of *miR396*, which negatively regulates the expression level of *GRF*s (Rodriguez et al. 2010; Schommer et al. 2014).

Development of seeds differs from that of leaves. A mature seed consists of three parts: embryo, endosperm, and seed coat, the latter of which is derived from maternal tissues. The growth and final size of seeds are regulated by both maternal and zygotic tissues (Li et al. 2019a). Recent studies have shown that HAIKU (IKU) and several hormonal signaling pathways control the seed size by regulating the growth of endosperms (Garcia et al. 2003; Luo et al. 2005; Wang et al. 2010). The IKU pathway is regulated by abscisic acid (ABA) and brassinosteroids (BR) signaling pathways (Jiang et al. 2013; Cheng et al. 2014). In Arabidopsis, the IKU pathway includes the following key regulatory genes, *HAIKU1* (*IKU1*), *IKU2*, *MINISEED3* (*MINI3*) and *SHORT HYPOCOTYL UNDER BLUE1* (*SHB1*) (Garcia et al. 2003; Luo et al. 2005; Wang et al. 2010). Loss-of-function *iku1*, *iku2*, and *mini3* mutants produced smaller seeds, whereas gain-of-function mutants of *SHB1* produced larger seeds than control plants; SHB1 is recruited by MINI3 to promote the expression of *IKU2* and *MINI3*, which encode a Leucine-rich repeat (LRR) kinase and a WRKY transcription factor, respectively (Garcia et al. 2003; Luo et al. 2005; Wang et al. 2010). The bZIP transcription factor ABA-INSENSITIVE5 (ABI5) binds to the ABA response element (ABRE) in the promotor sequence of *SHB1* and represses its expression (Cheng et al. 2014). BR signaling also regulates endosperm development. The BR-inducible transcription factor, BRASSINAZOLE-RESISTANT1 (BZR1) binds to the promoter sequences of *SHB1*, *MINI3*, and *IKU2* and regulates their expression (Jiang et al. 2013). In addition, the cytokinin and auxin signaling pathways are also involved in regulating endosperm development. In Arabidopsis, *CYTOKININ OXIDASE/DEHYDROGENASE (CKX)* genes, which play a regulatory role in the maintenance of cytokinin homeostasis, are direct targets of MINI3 (Werner et al. 2003; Li et al. 2013). In rice, *THOUSAND-GRAIN WEIGHT6* (*TGW6*), encoding an IAA-glucose hydrolase, regulates endosperm development and controls the size of grains but not husks (Ishimaru et al. 2013).

Besides zygotic tissues, seed size is also regulated by maternal genotypes. In eudicot species, integuments can affect the final size of seeds, because integuments form the seed coat and set the volume of the cavity for development of the embryo and endosperm (Li et al. 2019a). In monocot species, seed size is controlled by growth of the spikelet hull. A number of genes that encode components of the ubiquitin-26S proteasomal pathway, including *DA1*, *SAMBA, ROOT PHOTOTROPISM 2* (*RPT2*) and *RICE OVARIAN TUMOR DOMAIN-CONTAINING UBIQUITIN ALDEHYDE-BINDING PROTEIN 1* (*OsOTUB1*) has been identified to play a key role in seed size control through maternal effects (Kurepa et al. 2009; Eloy et al. 2012; Zhang et al. 2015; Wang et al. 2017). In addition, G-proteins, Mitogen-activated protein kinases, and several transcriptional regulators are also involved in regulating seed size (Ashikari et al. 1999; Fujisawa et al. 1999). Interestingly, some regulatory modules such as the OsmiR396-OsGRF4-OsGIFs regulatory module (Duan et al. 2015; He et al. 2017) and the KIX-PEAPOD regulatory module (Ge et al. 2016; Naito et al. 2017; Li et al. 2019b), regulate not only seed development but also leaf size.

13.2 *M. truncatula BIG SEEDS 1* (*BS1*) is Required for the Development of Leaves and Seeds

Recent studies have shown that functions of the PEAPOD (PPD) module in regulating the growth of both vegetative and reproductive organs (leaves and seeds) is conserved in a number of eudicot species. *BIG SEEDS1* (*BS1*), the *M. truncatula PPD* ortholog, has been identified to play a key role in regulating growth of seeds, seed pods, leaves and other organs, through a regulatory module that targets the primary cell proliferation (Ge et al. 2016). Loss-of-function *bs1* mutants exhibited larger seeds, seed pods and leaves than that of WT plants. Interestingly, the developmental defects occur only at a later stage of leaf and seed development in the *bs1* mutants. For example, only after 9 days post-anthesis (9 DPA), seeds became significantly larger in *mtbs1-1* than WT. Furthermore, measurements of epidermal cells of mature seed coats, stipules and expanded leaves show that there were no significant differences in final cell size between *mtbs1-1* and WT plants, indicating that the larger leaf and seed phenotype was due to increased cell proliferation rather than cell expansion (Ge et al. 2016). MtBS1 interacts with the transcription co-suppressor, NINJA to suppress the expression of downstream genes (Ge et al. 2016). Down-regulation of two *GmBS* orthologs results in larger leaves and seeds in transgenic soybean lines than control plants. In addition, the expression of *GRF5*, *GIF1*, *CYCLIN D3;3* (*CYCD3;3*), and *HISTONE4* genes were significantly up-regulated in the knock-down transgenic soybean plants, similarly as that in the *mtbs1-1* mutant. The *BS1* orthologs in *P. sativum* and *Vigna mungo* have been shown to be involved in regulating organ size (Naito et al. 2017; Li et al. 2019b), supporting a conserved function of BS1 in regulating organ size in legume species.

In Arabidopsis, PPD1 and PPD2 (PPD1/2) control leaf size by targeting secondary morphogenesis (meristemoid proliferation) (White 2006; Wang et al. 2016). The phenotype of enlarged and dome-shaped leaves is observed in the *mtbs1*mutants, *GmBS1/2* knock-down plants and Arabidopsis *ppdl/2* mutant (Ge et al. 2016; White 2006; Wang et al. 2016). In Arabidopsis, both *ninja* mutants and overexpression of *CYCD3;2* transgenic plants exhibited dome-shaped leaves but no increases of the meristemoid cells (Wang et al. 2016; Karidas et al. 2015; Baekelandt et al. 2018). These results suggest that *CYCD3;2* is likely a PPD2 target that regulates leaf shape (Baekelandt et al. 2018). In addition, the Arabidopsis *ppd* mutants showed an increased number of epidermal and palisade cells (White 2006). This phenotype may be related to cell-to-cell movement of the transcription regulator AN3 (Andriankaja et al. 2012; Kawade et al. 2013). The expression levels of *GRF5* and *GIF1* were up-regulated in *mtbs1* seeds and AtPPD2 interacted with AtMYC3 and AtMYC4 to regulate the expression of *AN3/GIF1* and seed size (Liu et al. 2020). Whether this pathway is also involved in regulating leaf development remains to be investigated.

13.3 KIX8/9, SAP and Other PPD Interacting Proteins Are Involved in Regulating Leaf and Seed Development

SMALL LEAF AND BUSHY1 (*SLB1*), encoding an F-box protein and orthologous to *A. thaliana STERILE APETALA* (*AtSAP*), has been recently shown to regulate lateral branching and organ size in *Medicago truncatula* (Yin et al. 2020). Loss-of-function *slb1* mutant exhibited smaller leaves, flowers and seeds compared with those of WT plants (Yin et al. 2020). In contrast, *SLB1* overexpression plants exhibited larger organs including leaves, flowers, and seeds than that of WT. SLB1 interacts with BS1, ASK1/2 and CUL1 and MtKIX and regulates the stability of BS1 through the ubiquitin-26S proteasomal pathway (Yin et al. 2020; Zhou et al. 2021).

These results indicate that the KIX-PPD/BS1-SAP pathway involved in regulating leaf, and seed size is conserved in *Medicago truncatula.*

In addition, the expression level of *MtCYCD3;2*, *MtCDKB1;1,* and *MtE2Fb* is decreased, whereas the expression level of *MtKRP3*, encoding a repressor of cell proliferation, is increased in the *slb1* mutant (Zhou et al. 2021). Consistent with their role in cell cycle regulation, *MtCYCD3;2* and *MtH4* are expressed in SAM (Zhou et al. 2021). These results support the notion that SLB1 regulates leaf and seed size by regulating the stability of BS1, and therefore promoting the expression of core cell cycle genes (Fig. 13.1). In several other eudicot species, the *SLB1* orthologs such as *LITTLELEAF* (*LL*) in cucumber (*Cucumis sativus*), *BIG LEAF* (*BL*) in popular (*Populus tremula*), *CrSAP* in Red Shepherd's Purse (*Capsella rubella*) have also been shown to play a similar role in controlling leaf, seed or fruit size (Sicard et al. 2016). In *C. sativus*, LL has also been shown to regulate lateral branching (Yang et al. 2018). Gain-of-function of *PtBL* exhibited an increased number of adventitious roots but reduced development of xylem tissues (Yordanov et al. 2017).

AtKIX8/9 and their orthologs in a number of other plant species such as *LjBIO* in *L. japonicas*, *PsBIO* in *P. sativum*, *SlKIX8/9* in tomato (*Solanum lycopersicum*) have been shown to function similarly in regulating leaf and seed size by targeting cell proliferation (Li et al. 2019b; Swinnen et al. 2020). Loss-of-function *Mtbio* mutants exhibited larger leaves, seeds and seed pods than that of WT, whereas overexpression of *MtBIO* plants exhibited smaller leaves and seeds compared to that of WT, indicating that *MtBIO* plays a similar role in regulating leaf, seed and pod size as *BIO* orthologs in other species (our unpublished results). Interestingly, it has been shown that *BIO* plays a role in regulating floral symmetry in both *L. japonicas* and *P. sativum* (Li et al. 2019b).

LATHYROIDES (LATH), a WUSCHEL-RELATED HOMEOBOX (WOX1) transcription factor has been shown to interact with PPD and KIX and plays a role in PPD-mediated leaf development in *L. japonicus* and *P. sativum*. In *P. sativum*, the LATH-BIO-ELE1 module regulates the expression of the downstream genes, *GRF5* and *AN3*, and plays a key role in organ development (Li et al. 2019b; Zhuang et al. 2012).

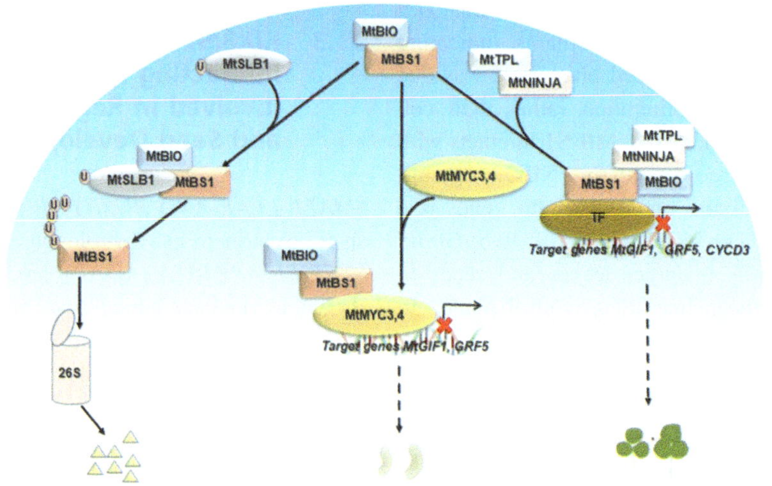

Fig. 13.1 The MtBIO-BS1-SLB1 module regulates organ development in *Medicago truncatula*. The BS1-BIO complex interacts with NINJA and TPL to control leaf growth by repressing the expression of target genes such as *GIF1*, *GRF5*, and *CYCD3*, and BS1-BIO also can interact with MYC3 andMYC4 to control seed development by repressing the expression of *GIF1* and *GRF5*. SLB1 promotes organ development by targeting BS1 and BIO proteins for degradation

13.4 Downstream Targets of the KIX-PPD-SAP Module in Leaf and Seed Development

In Arabidopsis, loss-of-function mutants of *PPD2* exhibited propeller-like rosettes and dome-shaped leaves (White 2006; Wang et al. 2016). Similar phenotypes were also exhibited in *ninjia* mutants and *CYCD3;2* overexpression plants. Thus, *CYCD3;2* may be one of the downstream targets of the KIX-PPD-SAP module (Wang et al. 2016; Karidas et al. 2015; Baekelandt et al. 2018). Recently, it has been shown that *AS1* (*ASYMMETRIC LEAF1*) is also likely a direct target of PPD2 based on tandem chromatin affinity purification-sequencing (TChAP-seq) results and gene co-expression analyses. In addition, loss-of-function *as2* mutants exhibited propeller-like rosettes and dome-shaped leaves (Gonzalez et al. 2015). PPD2, through interactions with the transcription factors, MYC3 and MYC4, directly regulate the expression of *AN3/GIF1* in Arabidopsis (Liu et al. 2020). In *M. truncatula*, the expression level of *GIF1*, *GRF5* and some core cell cycle related genes was significantly upregulated in the *mtbs1* mutants (Ge et al. 2016). Whether transcription factors such as GIFs, GRFs, MYC3/4 play a role in regulating the size of vegetative and reproductive organs remains to be investigated.

13.5 Concluding Remarks

Here, we presented a brief overview of the role of the KIX-PPD-SAP module in the development of vegetative and reproductive organs in eudicot species. Experimental results suggest that the KIX-PPD-SAP pathway is conserved in regulating vegetative and reproductive organ development mainly by targeting cell proliferation. However, the developmental processes that are regulated by the PPD pathway differ depending on plant species. In *A. thaliana*, the PPD pathway primarily regulates meristemoid proliferation and also cell expansion during leaf development; whereas in *M. truncatula*, the PPD pathway regulates only primary cell proliferation during leaf and seed development. The PPD pathway is also involved in regulating other developmental processes such as lateral branching and floral symmetry. In future, research aimed at identifying additional PPD regulatory network components and associated molecular processes will help us to further understand the role of the KIX-PPD-SAP module in regulating diverse developmental processes and to develop effective strategies to improve crop yield by targeting PPD-regulated cell proliferation and expansion.

Acknowledgements This work was supported by the Fundamental Research Funds for the Central Universities (lzujbky-2019-72 and lzujbky-2020-sp04) and Guo Chang Education Fund Project of Lanzhou University.

References

Alvarez JP, Furumizu C, Efroni I, Eshed Y, Bowman JL (2016) Active suppression of a leaf meristem orchestrates determinate leaf growth. eLife 5:e15023

Andriankaja M, Dhondt S, De Bodt S, Vanhaeren H, Coppens F, De Milde L, Mühlenbock P, Skirycz A, Gonzalez N, Beemster GT, Inzé D (2012) Exit from proliferation during leaf development in *Arabidopsis thaliana*: a not-so-gradual process. Dev Cell 22(1):64–78

Ashikari M, Wu J, Yano M, Sasaki T, Yoshimura A (1999) Rice gibberellin-insensitive dwarf mutant gene *Dwarf 1* encodes the α-subunit of GTP-binding protein. Proc Natl Acad Sci USA 96(18):10284–10289

Baekelandt A, Pauwels L, Wang ZB, Li N, De Milde L, Natran A, Vermeersch M, Li YH, Goossens A, Inzé D, Gonzalez N (2018) Arabidopsis leaf flatness is regulated by PPD2 and NINJA through repression of *CYCLIN D3* genes. Plant Physiol 178(1):217–232

Cheng ZJ, Zhao XY, Shao XX, Wang F, Zhou C, Liu YG, Zhang Y, Zhang XS (2014) Abscisic acid regulates early seed development in *Arabidopsis* by ABI5-mediated transcription of *SHORT HYPOCOTYL UNDER BLUE1*. Plant Cell 26(3):1053–1068

Czesnick H, Lenhard M (2015) Size control in plants–lessons from leaves and flowers. Cold Spring Harbor Perspect Biol 7(8):a019190

Debernardi JM, Mecchia MA, Vercruyssen L, Smaczniak C, Kaufmann K, Inzé D, Rodriguez RE, Palatnik JF (2014) Post-transcriptional control of GRF transcription factors by microRNA *miR396* and GIF co-activator affects leaf size and longevity. Plant J 79(3):413–426

Donnelly PM, Bonetta D, Tsukaya H, Dengler RE, Dengler NG (1999) Cell cycling and cell enlargement in developing leaves of *Arabidopsis*. Dev Biol 215 (2):407–419

Du F, Guan C, Jiao Y (2018) Molecular mechanisms of leaf morphogenesis. Mol Plant 11(9):1117–1134

Duan P, Ni S, Wang J, Zhang B, Xu R, Wang Y, Chen H, Zhu X, Li Y (2015) Regulation of OsGRF4 by OsmiR396 controls grain size and yield in rice. Nat Plants 2:15203

Eloy NB, Gonzalez N, Van Leene J, Maleux K, Vanhaeren H, De Milde L, Dhondt S, Vercruysse L, Witters E, Mercier R, Cromer L, Beemster GT, Remaut H, Van Montagu MC, De Jaeger G, Ferreira PC, Inzé D (2012) SAMBA, a plant-specific anaphase-promoting complex/cyclosome regulator is involved in early development and A-type cyclin stabilization. Proc Natl Acad Sci USA 109 (34):13853–13858

Fujisawa Y, Kato T, Ohki S, Ishikawa A, Kitano H, Sasaki T, Asahi T, Iwasaki Y (1999) Suppression of the heterotrimeric G protein causes abnormal morphology, including dwarfism, in rice. Proc Natl Acad Sci USA 96(13):7575–7580

Garcia D, Saingery V, Chambrier P, Mayer U, Jurgens G, Berger F (2003) Arabidopsis *haiku* mutants reveal new controls of seed size by endosperm. Plant Physiol 131(4):1661–1670

Ge LF, Yu JB, Wang HL, Luth D, Bai GH, Wang K, Chen RJ (2016) Increasing seed size and quality by manipulating *BIG SEEDS1* in legume species. Proc Natl Acad Sci USA 113(44):12414–12419

Gonzalez N, Pauwels L, Baekelandt A, De Milde L, Van Leene J, Besbrugge N, Heyndrickx KS, Cuellar Perez A, Durand AN, De Clercq R, De Slijke EV, Bossche RV, Eeckhout D, Gevaert K, Vandepoele K, De Jaeger G, Goossens A, Inzé D (2015) A repressor protein complex regulates leaf growth in *Arabidopsis*. Plant Cell 27(8):2273–2287

Gonzalez N, Vanhaeren H, Inzé D (2012) Leaf size control: complex coordination of cell division and expansion. Trends Plant Sci 17(6):332–340

He ZS, Zeng J, Ren Y, Chen D, Li WJ, Gao FY, Cao Y, Luo T, Yuan GQ, Wu XH, Liang YY, Deng QM, Wang SQ, Zheng AP, Zhu J, Liu HN, Wang LX, Li P, Li SC (2017) OsGIF1 positively regulates the sizes of stems, leaves, and grains in rice. Front Plant Sci 8:1730

Ishimaru K, Hirotsu N, Madoka Y, Murakami N, Hara N, Onodera H, Kashiwagi T, Ujiie K, Shimizu B, Onishi A, Miyagawa H, Katoh E (2013) Loss of function of the IAA-glucose hydrolase gene *TGW6* enhances rice grain weight and increases yield. Nat Genet 45(6):707–711

Jiang WB, Huang HY, Hu YW, Zhu SW, Wang ZY, Lin WH (2013) Brassinosteroid regulates seed size and shape in *Arabidopsis*. Plant Physiol 162(4):1965–1977

Karidas P, Challa KR, Nath U (2015) The tarani mutation alters surface curvature in *Arabidopsis* leaves by perturbing the patterns of surface expansion and cell division. J Exp Bot 66(7):2107–2122

Kawade K, Horiguchi G, Usami T, Hirai MY, Tsukaya H (2013) ANGUSTIFOLIA3 signaling coordinates proliferation between clonally distinct cells in leaves. Curr Biol 23(9):788–792

Kim JH, Choi D, Kende H (2003) The AtGRF family of putative transcription factors is involved in leaf and cotyledon growth in *Arabidopsis*. Plant J 36(1):94–104

Kurepa J, Wang S, Li Y, Zaitlin D, Pierce AJ, Smalle JA (2009) Loss of 26S proteasome function leads to increased cell size and decreased cell number in Arabidopsis shoot organs. Plant Physiol 150(1):178–189

Li N, Xu R, Li Y (2019a) Molecular networks of seed size control in plants. Annu Rev Plant Biol 70:435–463

Li X, Liu W, Zhuang LL, Zhu Y, Wang F, Chen T, Yang J, Ambrose M, Hu ZB, Weller JL, Luo D (2019b) BIGGER ORGANS and ELEPHANT EARLIKE LEAF1 control organ size and floral organ internal asymmetry in pea. J Exp Bot 70(1):179–191

Li J, Nie X, Tan JL, Berger F (2013) Integration of epigenetic and genetic controls of seed size by cytokinin in *Arabidopsis*. Proc Natl Acad Sci USA 110(38):15479–15484

Liu ZP, Li N, Zhang YY, Li YH (2020) Transcriptional repression of *GIF1* by the *KIX–PPD–MYC* repressor complex controls seed size in *Arabidopsis*. Nat Commun 11(1):1846

Luo M, Dennis ES, Berger F, Peacock WJ, Chaudhury A (2005) *MINISEED3* (*MINI3*), a *WRKY* family gene, and *HAIKU2* (*IKU2*), a *LEUCINE-RICH REPEAT* (*LRR*) *KINASE* gene, are regulators of seed size in *Arabidopsis*. Proc Natl Acad Sci USA 102 (48):17531–17536

Naito K, Takahashi Y, Chaitieng B, Hirano K, Kaga A, Takagi K, Ogiso-Tanaka E, Thavarasook C, Ishimoto M, Tomooka N (2017) Multiple organ gigantism caused by mutation in *VmPPD* gene in blackgram (*Vigna mungo*). Breed Sci 67(2):151–158

Nakata M, Nakata M, Matsumoto N, Tsugeki R, Rikirsch E, Laux T, Okada K (2012) Roles of the middle domain-specific WUSCHEL-RELATED HOMEOBOX genes in early development of leaves in *Arabidopsis*. Plant Cell 24(2):519–535

Nakata M, Okada K (2013) The leaf adaxial-abaxial boundary and lamina growth. Plants (basel) 2(2):174–202

Poethig RS, Sussex IM (1985) The cellular parameters of leaf development in tobacco: a clonal analysis. Planta 165(2):170–184

Rodriguez RE, Mecchia MA, Debernardi JM, Schommer C, Weigel D, Palatnik JF (2010) Control of cell proliferation in *Arabidopsis thaliana* by microRNA *miR396*. Development 137(1):103–112

Schommer C, Debernardi JM, Bresso EG, Rodriguez RE, Palatnik JF (2014) Repression of cell proliferation by *miR319*-regulated TCP4. Mol Plant 7(10):1533–1544

Sicard A, Kappel C, Lee YW, Woźniak NJ, Marona C, Stinchcombe JR, Wright SI, Lenhard M (2016) Standing genetic variation in a tissue specific enhancer underlies selfing-syndrome evolution in *Capsella*. Proc Natl Acad Sci USA 113(48):13911–13916

Swinnen G, Baekelandt A, Clercq RD, Doorsselaere JV, Pauwels L (2020) KIX8 and KIX9 are conserved repressors of organ size in the asterid species tomato. BioRxiv. https://doi.org/10.1101/2020.02.07.938977

Tsukaya H (2018) Leaf shape diversity with an emphasis on leaf contour variation, developmental background, and adaptation. Semin Cell Dev Biol 79:48–57

Wang A, Garcia D, Zhang HY, Feng K, Chaudhury A, Berger F, Peacock WJ, Dennis ES, Luo M (2010) The VQ motif protein IKU1 regulates endosperm growth and seed size in *Arabidopsis*. Plant J 63 (4):670–679

Wang ZB, Li N, Jiang S, Gonzalez N, Huang XH, Wang YC, Inzé D, Li YY (2016) SCF^{SAP} controls organ size by targeting PPD proteins for degradation in *Arabidopsis thaliana*. Nat Commun 7:11192

Wang SK, Wu K, Yuan QB, Liu XY, Liu ZB, Lin XY, Zeng RZ, Zhu HT, Dong GJ, Qian Q, Zhang GQ, Fu XD (2017) Non-canonical regulation of SPL transcription factors by a human OTUB1-like deubiquitinase defines a new plant type rice associated with higher grain yield. Cell Res 27(9):1142–1156

Werner T, Motyka V, Laucou V, Smets R, Van Onckelen H, Schmülling T (2003) Cytokinin-deficient transgenic Arabidopsis plants show multiple developmental alterations indicating opposite functions of cytokinins in the regulation of shoot and root meristem activity. Plant Cell 15(11):2532–2350

White DW (2006) PEAPOD regulates lamina size and curvature in *Arabidopsis*. Proc Natl Acad Sci U S A 103(35):13238–13243

Yang LM, Liu HQ, Zhao JY, Pan YP, Cheng SY, Lietzow CD, Wen CL, Zhang XL, Weng YQ (2018) LITTLELEAF (LL) encodes a WD40 repeat domain-containing protein associated with organ size variation in cucumber. Plant J 95:834–847

Yin PC, Ma QX, Wang H, Feng D, Wang XB, Pei YX, Wen JQ, Tadege M, Niu LF, Lin H (2020) *SMALL LEAF AND BUSHY1* controls organ size and lateral branching by modulating the stability of BIG SEEDS1 in *Medicago truncatula*. New Phytol 226(5):1399–1412

Yordanov YS, Ma C, Yordanova E, Meilan R, Strauss SH, Busov VB (2017) BIG LEAF is a regulator of organ size and adventitious root formation in poplar. PLoS One 12(7):e0180527

Zhang Y, Du L, Xu R, Cui RF, Hao JJ, Sun CX, Li YH (2015) Transcription factors SOD7/NGAL2 and DPA4/ NGAL3 act redundantly to regulate seed size by directly repressing KLU expression in *Arabidopsis thaliana*. Plant Cell 27(3):620–632

Zhou SL, Yang TQ, Mao YW, Liu Y, Guo SQ, Wang RR, Fangyue GW, He LL, Zhao BL, Bai QZ, Li YH, Zhang XJ, Wang DF, Wang CQ, Wu Q, Yang YF, Liu Y, Tadege M, Chen JH (2021) The F-box protein MIO1/SLB1 regulates organ size and leaf movement in *Medicago truncatula*. J Exp Bot 72(8):2995–3011

Zhuang LL, Ambrose M, Rameau C, Weng L, Yang J, Hu XH, Luo D, Li X (2012) *LATHYROIDES*, encoding a WUSCHEL-related Homeobox1 transcription factor, controls organ lateral growth, and regulates tendril and dorsal petal identities in garden pea (*Pisum sativum* L.). Mol Plant 5(6):1333–1345

The manufacturer's authorised representative in the EU is Springer Nature Customer Service Centre GmbH, Europaplatz 3, 69115 Heidelberg, Germany. If you have any concerns regarding our products, please contact ProductSafety@springernature.com

Printed and bound by CPI Group (UK) Ltd, Croydon, CR0 4YY
30/09/2026
02225710-0013